The Global Climate System
Patterns, Processes, and Teleconnections

Over the last 20 years, developments in climatology have provided an amazing array of explanations for the pattern of world climates. This textbook examines the Earth's climate systems in light of this incredible growth in data availability, data retrieval systems, and satellite and computer applications. It considers regional climate anomalies, developments in teleconnections, unusual sequences of recent climate change, and human impacts on the climate system. The physical climate forms the main part of the book, but social and economic aspects of the global climate system are also considered. This textbook has been derived from the authors' extensive experience of teaching climatology and atmospheric science. Each chapter contains an essay by a specialist in the field to enhance the understanding of selected topics. An extensive bibliography and lists of websites are included for further study. This textbook will be invaluable to advanced students of climatology and atmospheric science.

HOWARD A. BRIDGMAN is currently a Conjoint Professor at the University of Newcastle in Australia, having retired at the Associate Professor level in February 2005. He has held visiting scientist positions at Indiana University, USA, the University of East Anglia, UK, the National Oceanographic and Atmospheric Administration, Boulder, Colorado, USA, the Atmospheric Environment Service in Canada, and the Illinois State Water Survey, USA.

He has written, edited or contributed to eleven other books on subjects including air pollution, applied climatology and climates of the Southern Hemisphere. He has published many articles in the field's leading journals.

JOHN E. OLIVER was educated in England and the United States, obtaining his Ph.D. at Columbia University, where he served on the faculty, before joining Indiana State University. Prior to his appointment as Emeritus Professor, he was Professor of Physical Geography and Director of the University Climate Laboratory at Indiana State. He also served as Department Chairperson and Associate Dean of Arts and Sciences.

He has published twelve books and his work on applied climatology and historic climates has appeared in a wide range of journals. He was founding editor, with Antony Orme, of the journal *Physical Geography*, for which until recently he served as editor for climatology. In 1998 he was awarded the first Lifetime Achievement Award from the Climatology Group of the Association of American Geographers.

The Global Climate System

Patterns, Processes, and Teleconnections

Howard A. Bridgman

School of Environmental and Life Sciences
University of Newcastle, Australia

John E. Oliver

Department of Geography, Geology and Anthropology
Indiana State University, USA

With contributions from
Michael Glantz, National Center for Atmospheric
Research, USA
Randall Cerveny, Arizona State University, USA
Robert Allan, Hadley Centre, UK
Paul Mausel, Indiana State University, USA
Dengsheng Lu, Indiana University, USA
Nelson Dias, Universidade de Taubaté, Brazil
Brian Giles, University of Birmingham, UK
Gerd Wendler, University of Alaska, USA
Gregory Zielinski, University of Maine, USA
Sue Grimmond, Indiana University, USA
and King's College London, UK
Stanley Changnon, University of Illinois, USA
William Lau, NASA Goddard Space Flight Center, USA

CAMBRIDGE
UNIVERSITY PRESS

CAMBRIDGE UNIVERSITY PRESS

Cambridge, New York, Melbourne, Madrid, Cape Town, Singapore, São Paulo

Cambridge University Press
The Edinburgh Building, Cambridge CB2 2RU, UK

Published in the United States of America by Cambridge University Press, New York

www.cambridge.org
Information on this title: www.cambridge.org/9780521826426

First published 2006

Printed in the United Kingdom at the University Press, Cambridge

A catalog record for this publication is available from the British Library

ISBN-13 978-0-521-82642-6 hardback
ISBN-10 0-521-82642-X hardback

Contents

Contributors

Michael Glantz is a senior scientist at the National Center for Atmospheric Research, Boulder, Colorado, USA, and is an expert on climate change impacts on society and lifestyle.

Robert Allan is a senior scientist at the Hadley Centre, Met Office, United Kingdom, and is an expert on El Niño–Southern Oscillation, its teleconnections and its climate impacts.

Randall Cerveny is a Professor in Geography at Arizona State University, Phoenix, Arizona, USA, and is an expert on tropical circulations and climates of South America.

Paul Mausel is a Professor at Indiana State University, Terre Haute, Indiana, USA, and is an expert on remote sensing, and interpretations of biospheric and atmospheric changes from satellite data.

Dengsheng Lu is a research scientist in the Center for the Study of Institutions, Population, and Environmental Change at Indiana University and is an expert in remote sensing.

Nelson Dias is a research associate at the Universidade de Taubaté in Brazil, and researches changes to the Amazon rainforest using remote sensing techniques.

Brian Giles is a retired Professor from the School of Geography, Geology and Environmental Sciences at the University of Birmingham, UK, and is an expert on synoptic meteorology and NCEP/NCAR reanalysis. He currently lives in Takapuna, New Zealand.

Gerd Wendler is a Professor and Director of the Arctic Research Institute at the University of Alaska, Fairbanks, Alaska, USA, and is an expert on synoptic climatology of the Arctic and Antarctic regions.

Gregory Zielinski is a scientist at the Institute for Quaternary and Climate Studies at the University of Maine, Orono, Maine, USA, and is an expert on Holocene paleoclimates and proxy interpretations of climate change.

Sue Grimmond is a Professor in the Environmental Monitoring and Modelling Group, Department of Geography, King's College London, UK, and is an expert on urban climate and urban impacts on energy and water balances.

Stanley Changnon is retired as Director of the Illinois State Water Survey, Champaign-Urbana, Illinois, USA, and is currently Emeritus Professor of Geography at the University of Illinois. His expertise is in water and climate change, and the impacts of weather hazards on economics and society.

William Lau is Head of the Climate and Radiation Branch, NASA Goddard Space Flight Center, Greenbelt, Maryland, USA, and is an expert on climate modeling.

Preface

As graduate students in the 1960s and 1970s, the authors became attracted to the exciting world of the atmosphere and climatology through both lectures and textbooks. The approach to climatology at that time is best described as "global descriptive," where we were introduced to climate patterns and regimes across the Earth, and what then were known as the explanations behind them. One of the best books for studying advanced climatology was *The Earth's Problem Climates* (University of Wisconsin Press, 1966), by Glenn Trewartha, a well-known and respected climatologist from the University of Wisconsin. In this book we explored, both geographically and systematically, the climate patterns and anomalies across the continents. We were introduced to the nature of the Atacama Desert, the climatic anomalies of northeast Brazil, the temperature extremes of central Siberia, and the monsoon variations in India and China, among other aspects. Trewartha's book was reprinted in 1981, but sadly the new version did not properly include new research and findings on global climate patterns. For example, despite recognition by the mid 1970s of its essential importance to global climatic variability, there was no discussion of the El Niño–Southern Oscillation!

During the decades of the 1970s, 1980s, and 1990s, there has been an explosion in climatic research and a new breadth and depth of understanding about climatology and the atmosphere. There have also been a number of excellent books published in the area of climatology. Almost all of these can be grouped into one of two categories: (a) introductory to intermediate textbooks, to support teaching, which basically assume little or no background knowledge in climate or atmospheric studies; and (b) detailed books on either a climatic topic or a geographical area, based on extensive summaries of research publications. Examples of the latter include Elsevier's *World Survey of Climatology* series; *El Niño: Historical and Palaeoclimatic Aspects of the Southern Oscillation* (editors Diaz and Markgraf); *Antarctic Meteorology and Climatology* (King and Turner); *El Niño Southern Oscillation and Climate Variability* (Allen, Lindesay, and Parker); and *Climates of the Southern Continents* (editors Hobbs, Lindesay, and Bridgman). There is currently no book that provides a synthesis and overview of this information, filling the gap left by *The Earth's Problem Climates*.

It is our purpose in *The Global Climate System* to fill this gap, providing a book that can be used as background to climate research, as well as a text for advanced climatology studies at senior undergraduate and graduate levels. We have, combined, over 50 years teaching experience in climate, atmospheric sciences and weather, and written or co-authored 12 books on climate, climatology, and the atmosphere.

Global climates mostly follow a semi-predictable pattern based upon the receipts of energy and moisture distribution, with modifications based upon the non-homogeneity of the Earth's surface. But within these arrangements of climate are areas that are atypical of the expected pattern. In the preface to the second edition of *The Earth's Problem Climates*, Glenn Trewartha wrote, "In the nearly two decades that have elapsed since the initial publication of this book, new information as well as new climatic data have become available concerning some of the earth's unusual climates." As noted, in the more than two decades since Trewartha wrote these words there has been an incredible growth in information, information technology, data availability, and rapid data retrieval systems. Satellite and computer applications have led to a modern climatology whose methods were not available when Trewartha penned his first edition. Given such developments, it is appropriate that a timely reexamination of the Earth's climate system should be undertaken. Some examples include:

1. Regional climates that cannot be well explained in the context of their surrounding climates. Such anomalies are dealt with by considering continental areas within the division of tropical, middle-latitude and polar climates.
2. The recent developments in teleconnections open an array of climatic observations that are not readily explained. Thus, new understandings of climate interactions, such as those arising for example from possible impacts of ENSO events, are explored.
3. Intense inquiry into processes and nature of climate change has opened new vistas for its study. However, within the sequence of change there are times and events that do not appear to follow an expected pattern.
4. Both the human inputs into climate and the impacts of climate upon humans provide an extensive area of study. In the urban environment, massive interruptions of the natural systems provide an arena in which many seemingly anomalous conditions occur. At the same time, problem climates also influence the social and economic well-being of many people.

We cannot cover the full details of the entire climate system in this book. The range of knowledge about the climate system is increasing too rapidly. Instead, we explore a range of aspects and topics, to show current understanding, but also to encourage interest and further research, from both the scientist and the student. To help achieve this aim, we have enlisted the input of respected scholars who contribute essays dealing with their areas of expertise. These essays are merged into each chapter in the hope that the text is a continuum of information. Each author was given some very general instructions about the aim of the book, the expected size of the essay, and the number of supporting figures and tables. Further specifics were intentionally left out, to allow the authors freedom to develop their essays in their own style. Initially we had hoped to have essayists from a range of different geographical locations around the world. The final list, nine from the USA, two from the UK, and one from Brazil, does not quite meet that aim, but we are very pleased with the outcome. The essays are shaded, to distinguish them from the material written by us.

We would like to thank the University of Newcastle and Indiana State University for their support, especially for study leave trips for both authors. We thank our support cartographers, Olivier Rey-Lescure at Newcastle and Lu Tao at Indiana State. Last, but not least, we thank our wives, who had a wonderful time socializing in the second half of 2004, allowing us to work uninterrupted on the manuscript.

Abbreviations

AAO	Antarctic Oscillation
ABRACOS	Anglo-Brazilian Amazonian Climate Observation Study
ACSYS	Arctic Climate System Study
ACW	Antarctic Circumpolar Wave
AGB	Above Ground Biomass
AGCM	Atmospheric General Circulation Model
ALPEX	Alpine Experiment of 1982
AM	Asian Monsoon
AMIP	Atmospheric Model Intercomparison Project (NCEP/DOE)
AMO	Atlantic Multidecadal Oscillation
AO	Arctic Oscillation
AUHI	Atmospheric Urban Heat Island
AVHRR	Advanced Very High Resolution Radiometer (satellite)
AWS	Automatic Weather Station
BUFR	Binary Universal Format Representation of the WMO
CACGP	Commission on Atmospheric Chemistry and Global Pollution
CCN	Cloud Condensation Nuclei
CCSP	Climate Change Science Program
CET	Central England Temperature Series
CliC	Climate and Cryosphere
CLIVAR	Climate Variability and Predictability
CMAP	CPC Merged Analysis of Precipitation
CMIP	Coupled Model Intercomparison Project
COADS	Comprehensive Ocean-Atmosphere Data Set
CPC	Climate Prediction Center
CPT	Circumpolar Trough
CPV	Circumpolar Vortex
CRU	Climatic Research Unit, University of East Anglia
DOE	Department of Energy
ECA	European Climate Assessment

ECMWF	European Centre for Medium-Range Weather Forecasts
ENSO	El Niño–Southern Oscillation
EOF	Empirical Orthogonal Function
FGGE	First GARP Global Experiment
GAIM	Global Analysis, Integration, and Modelling Program
GARP	Global Atmospheric Research Program
GATE	GARP Global Atlantic Experiment
GCM	General Circulation Model
	Global Climate Model
GCTE	Global Chemistry Tropospheric Experiment
GDP	Gross Domestic Product
GEOS	Goddard Earth Observing System
GEWEX	Global Energy and Water Cycle Experiment
GIS	Geographic Information System(s)
GISP2	Greenland Ice Sheet Project 2
GNP	Gross National Product
GRIB	Grided Binary representation (WMO)
GRIP	Greenland Ice Core Project
GURME	Global Urban Research Meteorology and Environmental Project
HadCRUT	Climatic Research Unit's land surface air temperatures
HadSST	Hadley Centre monthly gridded Sea Surface Temperatures
HRC	Highly Reflective Clouds
H/W	Height to Width ratio
IAMAS	International Association of Meteorology and Atmospheric Science
ICSU	International Council for Science
IGAC	International Global Atmospheric Chemistry Program
IGBP	International Geosphere/Biosphere Program
IGY	International Geophysical Year
IHDP	International Hydrological Development Program
ILEAPS	Integrated Land Ecosystem–Atmospheric Processes Study
INPE	Instituto Nacional de Pesquisas Espaciais (National Institute for Space Research, the Brazilian government)
IPCC	Intergovernmental Panel on Climate Change
IPCC DDC	Intergovernmental Panel on Climate Change Data Distribution Centre
IPO	Interdecadal Pacific Oscillation

IRD	Ice-Rafted Debris
ISL	Inertial Sub-Layer (urban)
ITC or ITCZ	Intertropical Convergence Zone
IUGG	International Union of Geodesy and Geophysics
JMA	Japanese Meteorological Agency
JRA-25	Japanese Re-Analysis 25 years
LBA	Large-scale Biosphere–Atmosphere Experiment in Amazonia
LF ENSO	Low-Frequency ENSO, 2.5 to 7 years
LFV	Local Fractional Variance
LIA	Little Ice Age
LULC	Land Use/Land Cover
MAP	Merged Analysis of Precipitation
MC	Maritime Continent
METROMEX	METROpolitan Meteorological EXperiment
MIP	Model Intercomparison Projects
MJO	Madden–Julian Oscillation
MMIP	Monsoon Model Intercomparison Project
MSLP	Mean Sea Level Pressure
MTM-SVD	Multi-Taper Method Singular Value Decomposition
MWP	Medieval Warm Period
NAO	North Atlantic Oscillation
NASA/DAO	National Aeronautics and Space Administration/Data Assimilation Office of the Goddard Laboratory for Atmospheres
NCAR	National Center for Atmospheric Research
NCEP/DOE AMIP-II	Reanalysis or Reanalysis 2
NCEP/NCAR	National Centers for Environmental Prediction/National Center for Atmospheric Research
NCEP/NCAR-40	Reanalysis project 1957–1996
NEE	Net Ecosystem Exchange (of CO_2)
NGDC	National Geophysical Data Center
NH	Northern Hemisphere
NMC	National Meteorological Center, USA
NOAA	National Oceanographic and Atmospheric Administration, USA
NPO	North Pacific Oscillation
NWS	National Weather Service, USA
OLR	Outgoing Longwave Radiation
PAGES	Past Global Changes
PDO	Pacific Decadal Oscillation
PDV	Pacific Decadal Variation

PILPS	Project of Intercomparison of Land Parameterization Schemes
PMIP	Paleoclimate Model Intercomparison Project
PNA	Pacific North American Oscillation
PNJ	Polar Night Jet
PSCs	Polar Stratospheric Clouds
QB ENSO	Quasi-Biennial ENSO, 2 to 2.5 years
QBO	Quasi-Biennial Oscillation
RSL	Roughness Sub-Layer (urban)
SAM	South Asian Monsoon
SAO	Semi-Annual Oscillation
SAR	Synthetic Aperture Radar
SCORE	Scientific Committee on Ocean Research
SEAM	South East Asian Monsoon
SEB	Surface Energy Balance
SH	Southern Hemisphere
SMIP	Seasonal Model Intercomparison Project
SO	Southern Oscillation
SOI	Southern Oscillation Index
SOLAS	Surface Ocean–Lower Atmosphere Study
SPARC	Stratospheric Processes and their Role in Climate
SPCZ	South Pacific Convergence Zone
SS1	Initial secondary succession
SS2	Secondary succession forest
SS3	Succession to mature forest
SST	Sea Surface Temperature
STHP	Subtropical High Pressure
SUHI	Surface Urban Heat Island
SVF	Sky View Factor (urban)
THC	Global Thermohaline Circulation
TM	Thematic Mapper, Landsat satellite sensor, resolution 30 m
TOGA	Tropical Ocean Global Atmosphere
TOMS	Total Ozone Monitoring Spectrometer
TOVS/SSU	TIROS Operational Vertical Sounder/Stratospheric Sounding Unit
TPI	Trans-Polar Index (Southern Hemisphere)
TRMM	Tropical Rainfall Measuring Mission
TRUCE	Tropical Urban Climate Experiment
UBL	Urban Boundary Layer
UCI	Urban Cool Island
UCL	Urban Canopy Layer

UHI	Urban Heat Island
UHIC	Urban Heat Island Circulation
UME	Urban Moisture Excess
UNCCD	United Nations Convention to Combat Desertification
UNCED	United Nations Conference on Environment and Development
UNEP	United Nations Environment Programme
VOC	Volatile Organic Compounds
WCRP	World Climate Research Programme
WETAMC	Wet season Atmospheric Mesoscale Campaign (Amazon Basin)
WMO	World Meteorological Organization

See also Table 10.1.

Chapter 1
Introduction

1.1 The climate system

Climate is a function not only of the atmosphere but is rather the response to linkages and couplings between the atmosphere, the hydrosphere, the biosphere, and the geosphere. Each of these realms influences any prevailing climate and changes in any one can lead to changes in another. Figure 1.1 provides in schematic form the major couplings between the various components of the climate system. A climate-systems approach avoids the isolation of considering only individual climatic or atmospheric components. This approach recognizes the importance of forcing factors, which create changes on scales from long-term transitional to short-term sudden, and that the climate system is highly non-linear. According to Steffen (2001), a systems approach also recognizes the complex interaction between components, and links between the other great systems of the Earth, and the ways in which humans affect climate through the socioeconomic system. Ignoring such interactions may create inaccuracies and misinterpretations of climate system impacts at different spatial scales.

In examining any component of the Earth's atmosphere, its systems and its couplings, basic knowledge of the energy and mass budgets is critical. Information concerning these is given in most introductory texts (Oliver and Hidore 2002; Barry and Chorley 1998) and they are not reiterated in detail here. Rather, the following provides a brief summary of major concepts.

1.1.1 Energy and mass exchanges

Energy

Every object above the temperature of absolute zero $-273\,^{\circ}\mathrm{C}$ radiates energy to its environment. It radiates energy in the form of electromagnetic waves that travel at the speed of light. Energy transferred in the form of waves has characteristics that depend upon wavelength, amplitude, and frequency.

The characteristics of the radiation emitted by an object vary as the fourth power of the absolute temperature (degrees Kelvin). The hotter an object, the greater the flow of energy from it. The Stefan–Boltzmann Law expresses this

Figure 1.1   A simplified
and schematic
representation of the Earth's
climate system.

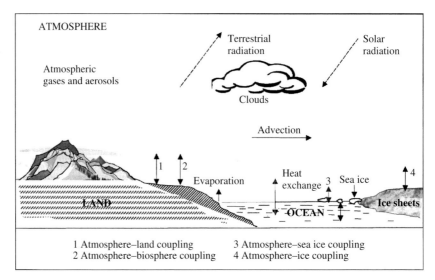

Figure 1.1   A simplified and schematic representation of the Earth's climate system.

relationship by the equation $F = \sigma T^4$ where F is the flux of radiation emitted per square meter, σ is a constant $(5.67 \times 10^{-8}$ W m^{-2}k^{-4} in SI units), and T is an object's surface temperature in degrees Kelvin.

Applying this law, the average temperature at the surface of the Sun is 6000 K. The average temperature of Earth is 288 K. The temperature at the surface of the Sun is more than 20 times as high as that of Earth. Twenty raised to the fourth power is 160 000. Therefore, the Sun emits 160 000 times as much radiation per unit area as the Earth. The Sun emits radiation in a continuous range of electromagnetic waves ranging from long radio waves with wavelengths of 10^5 meters down to very short waves such as gamma rays, which are less than 10^{-4} micrometers in length.

Another law of radiant energy (Wien's Law) states that the wavelength of maximum intensity of radiation is inversely proportional to the absolute temperature. Thus the higher the temperature, the shorter the wavelength at which maximum radiation intensity occurs. This is given by $\lambda_{max} = 2897/T$ where $T =$ temperature in degrees Kelvin, and wavelength is in micrometers.

For the Sun, λ_{max} is 2897/6000 which equals 0.48 μm. For the Earth λ_{max} is given by 2897/288, a wavelength of 10 μm. Thus the Sun radiates mostly in the visible portion of the electromagnetic spectrum and the Earth in the infrared (Figure 1.2). There is a thus a fundamental difference between solar and terrestrial radiation and the ways in which each interacts with the atmosphere and Earth's surface.

Utilization of these laws, and knowledge of Earth–Sun relations, enables the computation of the amount of energy arriving, the solar constant, and the nature of solar and terrestrial radiation. These are used to derive budgets of energy exchanges over the Earth's surface. Box 1.1 provides basic information on this using the customary symbols.

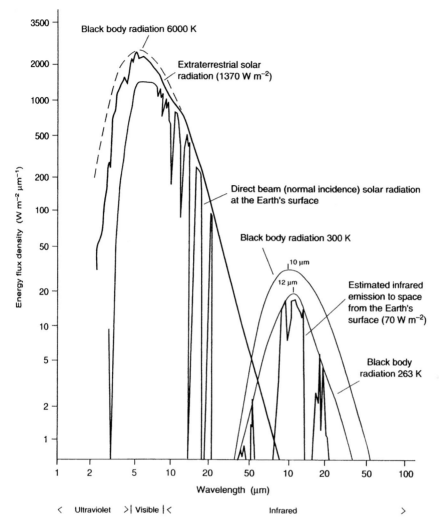

Figure 1.2 Wavelength characteristics of solar and terrestrial radiation. Note the difference between extraterrestrial solar radiation and that incident at the Earth's surface indicating atmospheric absorption of both short-wave ultraviolet and infrared radiant energy. Earth emits energy largely in the infrared portion of the spectrum. (After Sellers 1965)

The climate at any location is ultimately related to net radiation (Q^*) and is a function of a number of interacting variables. First, incoming solar radiation varies with latitude, being greatest at the equator and least at the poles. Hence, climate varies with latitude. Second, energy transformations at the surface are completely different over ice, water, and land, while also varying with topography, land use, and land cover. Climates will thus vary between such surfaces. The variation associated with such surfaces is seen in the heat budget equation.

The heat budget explains the relative partitioning between sensible heat and latent heat transfers in a given environment. In a moist environment a large part of available energy is used for evaporation with less available for sensible heat.

Background Box 1.1

Energy flow representation

The exchanges and flows associated with energy inputs into the Earth-atmosphere system is represented by a series of symbolic equations. Use of the equations permits easy calculation once values are input.

Shortwave solar radiation ($K\downarrow$) reaching the surface is made up of the vertical radiation (S) and diffuse radiation (D):

$$K\downarrow = S + D$$

Some of the energy is reflected back to space ($K\uparrow$) so that net shortwave radiation (K^*) is the difference between the two:

$$K^* = K\downarrow - K\uparrow$$

Net longwave, terrestrial radiation (L^*) comprises downward atmospheric radiation ($L\downarrow$) less upward terrestrial radiation ($L\uparrow$):

$$L^* = L\downarrow - L\uparrow$$

The amount of energy available at any surface is thus the sum of K^* and L^*. This is net all-wave radiation (Q^*):

$$Q^* = K^* + L^*$$

which may also be given as

$$Q^* = (K\downarrow - K\uparrow) + (L\downarrow - L\uparrow)$$

Q^* may be positive or negative.

High positive values will occur during high sun periods when $K\downarrow$ is at its maximum and atmospheric radiation, $L\downarrow$, exceeds outgoing radiation, $L\uparrow$.

Negative values require outgoing values to be greater than incoming. This happens, for example, on clear nights when $L\uparrow$ is larger than other values.

On a long-term basis, Q^* will vary with latitude and surface type.

The heat budget

Consider a column of the Earth's surface extending down to where vertical heat exchange no longer occurs (Figure 1.3). The net rate (G) at which heat in this column changes depends upon the following:

Net radiation $(K\uparrow - K\downarrow) + (L\uparrow - L\downarrow)$
Latent heat transfer (LE)
Sensible heat transfer (H)
Horizontal heat transfer (S)

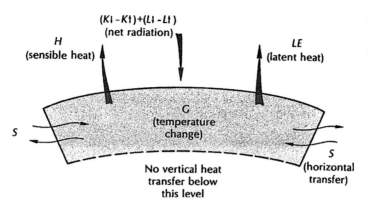

Figure 1.3 Model of energy transfer in the atmospheric system.

In symbolic form:

$$G = (K\uparrow - K\downarrow) + (L\uparrow - L\downarrow) - LE - H \pm S$$

Since

$$(K\uparrow - K\downarrow) + (L\uparrow - L\downarrow) = Q*$$

then

$$G = Q* - LE - H \pm S$$

in terms of $Q*$

$$Q* = G + LE + H \pm S$$

The column will not experience a net change in temperature over an annual period; that is, it is neither gaining nor losing heat over that time, so $G = 0$ and can be dropped from the equation.

$$Q* = LE + H \pm S$$

This equation will apply to a mobile column, such as the oceans. On land, where subsurface flow of heat is negligible, S will be unimportant. The land heat budget becomes

$$Q* = LE + H$$

The ratio between LE and H is given as the *Bowen Ratio*.

(After Oliver and Hidore 2002)

The opposite is true in dry environments. The ratio of one to another is expressed by the Bowen Ratio; a high value would indicate that large amounts of energy are available for sensible heat, a low value indicates that much available energy is used for latent heat transfer. This partially explains why desert regions, which

Figure 1.4 Estimated amounts of water involved in the global hydrologic cycle.

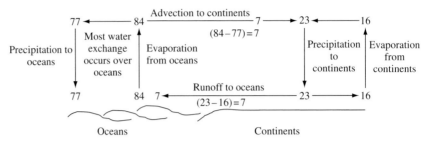

100 units = mean annual global precipitation = 85.7 cm (33.8 in)

have high Bowen Ratios, can attain much higher temperatures than those in a maritime environment.

Water and its changes of state

The significance of water as an atmospheric variable is a result of its unique physical properties. Water is the only substance that exists as a gas, liquid, and solid at temperatures found at the Earth's surface. This special property enables water to cycle over the Earth's surface. Figure 1.4 illustrates the relative partitioning of water in the hydrologic cycle. As can be seen, a large proportion of the exchanges occur over the world oceans. While changing from one form to another, water in its various forms acts as an important vehicle for the transfer of energy in the atmosphere.

The chemical symbol of water, H_2O, is probably the best known of all chemical symbols. Water in all of its states has the same atomic content, the only difference is the arrangement of the molecules. At low temperatures the bonds binding the water molecules are firm and pack tightly in a fixed geometric pattern in the solid phase. As temperature increases, the available energy causes bonds to form, break, and form again. This permits flow to occur and represents the liquid phase of water. At higher temperatures and with more energy, the bonding between the water molecules breaks down and the molecules move in a disorganized manner. This is the gas phase. If the temperature decreases, the molecules will revert to a less energetic phase and reverse the processes. Gas will change to liquid and liquid to solid.

The processes of melting, evaporation, and sublimation from solid to liquid to gas phase result in absorbed energy. This added energy causes the molecules to change their bonding pattern. The amount of energy incorporated is large for the changes to the water vapor stage, and much lower for the change from ice to water.

The energy absorbed is latent energy and goes back to the environment when the phase changes reverse. When water vapor changes to liquid, it releases the

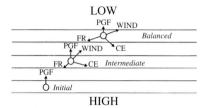 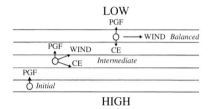

Figure 1.5 Idealized diagrams showing how interacting forces result in surface friction layer winds (left) and upper air, geostrophic winds (right). PGF, pressure gradient force; FR, friction; CE, coriolis effect.

energy originally absorbed and retained as latent heat. The same is true when water freezes and water vapor sublimates to ice.

The significance of the release of latent heat shows in many ways. For example, it plays a critical role in the redistribution of heat energy over the Earth's surface. Because of the high evaporation in low latitudes, air transported to higher latitudes carries latent heat with it. The vapor in this air condenses and releases energy to warm the atmosphere in higher latitudes.

Air in motion

Newton's first law of motion deals with inertia. It states that a body will change its velocity of motion only if acted upon by an unbalanced force. In effect, if something is in motion, it will keep going until a force modifies its motion. On Earth, a parcel of air seldom moves continuously and in a straight line. This is because, as Newton's second law states, the acceleration of any body, in this case the parcel of air, is directly proportional to the magnitude of the net forces acting upon it and inversely proportional to its mass. Note that these laws concern acceleration, which is change of velocity with time.

By identifying the forces that act upon a parcel of air, it becomes possible to understand more fully the processes that lead to the acceleration (or deceleration) of air. If we consider a unit parcel of air ($m = 1$), then Newton's second law becomes

$$\text{Acceleration} = \text{Sum of forces}$$
$$\text{or} \quad F_{a} = \sum F$$

The $\sum F$ is made up of the atmospheric forces so that:

$$\text{Acceleration} = \text{Pressure gradient force} + \text{Coriolis force}$$
$$+ \text{Frictional forces} + \text{Rotational forces}$$

The understanding and evaluation of each of these forces (or accelerations) provide the key to winds that blow, at both the surface and aloft, over the globe. This is schematically illustrated in Figure 1.5 where the interacting forces are shown to produce friction winds at lower levels of the atmosphere, and geostrophic winds aloft.

1.2 Patterns, processes, and teleconnections

Since early times humans must have been aware of their climatic environment. Agriculturalists were faced with the impact of changing seasons, hunters followed migrating herds and fishermen experienced the vagaries of stormy seasons at sea. From such a beginning the study of climate, climatology, has evolved through numerous stages to the rigorous science that it now is. To assess aspects of the current understandings, it is useful to consider the global climate system in terms of its patterns, processes, and teleconnections.

1.2.1 Patterns

In reviewing the history of climatology Oliver (1991) notes that the development of any discipline is closely associated with the logical organization, the classification, of the elements that are studied as part of that discipline. Such is very true of climatology, for the classification process dominated the discipline from the late nineteenth to the middle of the twentieth centuries. The effort and thought that went into studies have provided the modern climatologists with the basic ideas of the various patterns of climate that exist over the Earth's surface. However, the zonal patterns that were originally postulated have been shown to be a major oversimplification, and have led to many misunderstandings about the nature of climates in various regions of the world. As an example, the climates that are grouped as the "tropical rainforest climate" are no longer perceived as monotonous, readily explained climates such as they were once described.

It follows that one emphasis of this work, the patterns of climate over the Earth, need be examined in the light of new ideas and findings. To this end, the first chapters of this text use the long-recognized patterns – the tropics, mid-latitudes and polar realms – to identify climate types, but look at them in a way that brings together the dynamic understanding that has been the focus of recent research.

But patterns are not just spatial; temporal patterns of climate must also be considered. Climate has varied in the past on many time scales. There have been long periods, more than 50 million years, of relatively undisturbed climates when conditions were warmer than the current climate (Ruddiman 2001). These time spans have been interrupted by shorter periods, a few million years or so, of quite variable climates. For about the past 2 million years climate has been in a disturbed period with ice ages alternating with milder interglacials. In this work, only a short temporal pattern is examined in detail. The last 1000 years or so is selected because of the impact of changing or variable climates upon people and their environments.

1.2.2 Process

The second emphasis of this work concerns the processes that produce a climate. A dictionary definition of process states that it is a natural or involuntary course of action or a series of changes. In the milieu of climate, process may be regarded as a continuum of energy flow wherein available energy is utilized to maintain the climate system. The resulting global and local energy and mass budgets eventually provide the key to ongoing processes. Background Box 1.1 provides an example of the standard symbols used to depict energy flows in the environment and the relationship to the heat budget. Changes in energy flows then lead to changes in the nature of a climate and its resulting impact upon the human environment. Such is considered in a number of ways in this work. As already noted, the human response to changes over the last 1000 years is considered a temporal pattern. It also represents a change in the processes resulting in that climate. Another area where change is seen is in the urban environment. The buildings that comprise a town or city create conditions that result in a totally modified energy budget. The construction of an environment of cement and macadam results in changing moisture flows and patterns (Bonan 2002 and Chapter 7). Urban climatology has become a major area of specialization.

One result of the intensive study of process is the development of the concept that any climate process that occurs at a given location does not vary or change independently of other, often far distant, processes. This has led to an area of research that deals collectively with teleconnections.

1.2.3 Teleconnections

Teleconnection is a term used to describe the tendency for atmospheric circulation patterns to be related, either directly or indirectly, over large and spatially non-contiguous areas. The AMS *Glossary of Weather and Climate* (Geer 1996) defined it as a linkage between weather changes occurring in widely separated regions of the globe. Both definitions emphasize a relationship of distant processes. However, the word "teleconnection" was not used in a climate context until it appeared in the mid 1930s (Ångström 1935), and even until the 1980s was not a commonly used term in the climatic literature.

As stressed throughout this book, teleconnections are often associated with atmospheric oscillations. Any phenomenon that tends to vary above or below a mean value in some sort of periodic way is properly designated as an oscillation. If the oscillation has a recognizable periodicity, then it may be called a cycle, but few atmospheric oscillations are considered true cycles. This is illustrated by the early problems in predicting the best-publicized oscillation, the Southern Oscillation and El Niño (Chapter 2). Were this totally predictable then many of its far-reaching impacts could be forecast.

1.2.4 People and climate

The most important practical reason to understand the climate system is the link with people, their activities, and their decision making. The relationships between people and climate may be approached in many ways, often under the heading of Applied Climatology. Art, architecture, comfort, health, religion, and warfare are but a few of the topics considered (Oliver 1991). Of particular interest in this work is the role of people in what may be termed "problem climates". The best known treatment of this topic is to be found in Trewartha's book *The Earth's Problem Climates* (1966). This work explored anomalies across the globe but, obviously, lacked explanations based upon the information that is available today. Nonetheless, the very title begs the question of what is a problem climate? This is thoughtfully considered in the following essay by Michael Glantz.

Given the growth of world population and the immense impact upon the ecological systems of the Earth, the study of some problem climates faces a challenge. Is a recognized problem, such as a long-term drought, the result of natural climatic variation or is it the result of human activity.

1.3 ESSAY: Problem climates or problem societies?

Michael Glantz, *National Center for Atmospheric Research*

It cannot be denied that climate issues have made it to the top of the list of things to talk about. Those things to talk about include climate change to be sure, but also every week there is likely to be a weather or climate extreme occurring somewhere on the globe. At different times of the year we hear about adverse impacts of climate on agriculture, e.g. droughts in the out-of-phase growing seasons in the Northern and Southern Hemispheres.

In addition to local, regional, and global concerns about specific climate and weather extremes in their own right, there is a deepening concern world-wide on the part of the public, political figures, and scientific researchers about the adverse effects of a variable and changing climate on human activities and the resources on which they depend: food production and security, water resources, energy production, public health and safety, early warning, economy and environment. Each of these concerns also raises serious ethical and equity concerns.

The idea for this essay (and its title) is the result of three merging climate-related concerns: (1) an apparent overemphasis on blaming climate for many of society's woes such as food and water shortages and surpluses, and public health and safety problems; (2) an apparent overemphasis on the speculation about global warming and its impacts on societies (usually adverse and way out into the future); and (3) an apparent underemphasis on society's ability to influence the behavior of the atmosphere on all time and geographic scales. As a result of this underemphasis, societies are not forceful or determined enough to pursue changes in societal behavior in an attempt to minimize human influences on the atmosphere and therefore on local to global climate.

1.3.1 Introduction to the notion of problem climates

- **Climate** encompasses *variability* from season to season and from year to year, *fluctuations* are on the order of decades, *change* is on the order of centuries, and *extreme meteorological events* are extreme weather events or climate anomalies. Each of these forms of climate is appearing at the top of governmental lists of concerns about global environmental, demographic, and technological change.
- **Problem** is defined as a question raised for inquiry, consideration or solution; an intricate unsettled question; a source of perplexity, distress, or vexation (in this sense, problematic); synonym – a mystery.

In a recent report on climate change in the United States (NRC 2002), a graphic was used to depict the climate system. The graphic included the ocean, the atmosphere, ice, cloud systems, incoming solar radiation, and outgoing longwave radiation. Clearly, this has been the traditional view of the climate system. These elements interacted in a variety of ways from local to global levels producing regional to global climate regimes. Today though, this traditional view is no longer correct. Human activities are now affecting the environment (land, ocean, atmosphere) in ways that affect the climate. Headlines on climate these days often focus on global warming-related issues: greenhouse gas emissions (especially carbon dioxide), sea level rise, tropical deforestation, and so forth. Humans have become a forcing factor with respect to climate. What that means is that climate science has the obligation to improve our understanding of the climate system and to understand the contribution to the climate of each of its components (snow and ice, vegetation and forecasts, clouds systems, the oceans, and also society).

While meteorologists and climatologists are primarily concerned with the science of atmospheric processes, individuals as well as policy makers tend to be more concerned about the interactions between climate and society in general, and more specifically between climate extremes and human activities. Societies have tended to look at climate in at least one of three ways. Climate is seen as a hazard, as a constraint, and as a resource. In each society climate is a mix of these three, but the proportions among them can vary from one country to the next. Societies see climate as a hazard; its anomalies can lead to death, destruction, and misery. Governments at least in theory have the responsibility to protect their citizenry from climate and climate-related disasters. Climate as a constraint refers to the limitations that the physical climate places on human activities specifically and on economic development in general. In attempts to overcome climate-related constraints, societies have resorted to various technologies to reduce those limits: heating, refrigeration, transportation, irrigation, genetic manipulation of agricultural products, aquaculture, containment of rivers, etc.

Climate as a resource is taken for granted. Countries with climates that have been favorable to agriculture and animal husbandry often take their good climate conditions as normal. But climate information is also a resource. Forecasts are a resource. International consultants are a resource, and so forth. Governments usually leave it to the private sector to enhance the value of climate to their specific activities.

"Problem climate": its first use

About forty years ago (1966) geographer Glenn Trewartha published his book, *The Earth's Problem Climates*. Trewartha's selection of what he considered to be the Earth's "problem climates" was based on information available before 1960. He described a problem climate as one that did not really conform to what might be expected for a given latitude: "Were the earth's surface homogeneous (either land or water) and lacking terrain irregularities, it may be presumed that atmospheric pressure, winds, temperature, and precipitation would be arranged in zonal or east–west belts" (p. 3). He focused on "regional climatic aberrations," explicitly noting that he was writing for physical scientists, not for the general public. "It is designed to meet the needs of those interested in the professional aspects of climate rather than of laymen. A methodical description of all the earth's climates is not attempted, *for many areas are climatically so normal or usual that they require little comment* in a book which professes to emphasize the exceptional" (p. 6) [italics added].

Is such a statement still valid, given what we have learned about climate since 1960? Are there really areas on the globe that could be viewed as "climatically so normal or usual that they require little comment?" Are there *exceptional* "problem climates?" Should we also be asking questions about societies' role, if any, in the existence of problem climates?

1.3.2 What is normal climate?

Maps such as those originally produced in 1914 by Köppen and later modified by Trewartha, among others, depict the wide variety of normal climate types on continents around the globe. Australia provides an example representation of its so-called normal climate regime (Figure 1.6).

Such maps, while useful for educational purposes, are highly generalized and do not capture the full range of climate behavior such as its anomalies. Over time, the borders between climate zones will most likely shift, making the value of these maps mainly as snapshots of climate regimes for given periods of time. Nevertheless, they do provide a starting place for discussion of climate regions on a worldwide basis.

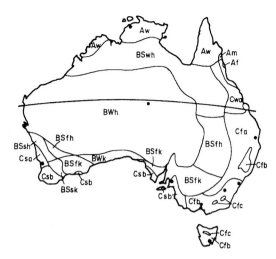

Figure 1.6 A climatic classification of Australia illustrates an example of so-called normal climatic regimes.

Normal climate is more than just average conditions. It includes extremes as well. There is a general view that Australia in the time of El Niño is under severe drought conditions. The 1997–8 El Niño was the most intense in the twentieth century. The map in Figure 1.7 shows the wide range of weather and climate conditions that can occur on the same continent, Australia, during an El Niño year. This is shown to reinforce the view that while climatological averages are useful for some purposes, by no means do they tell the whole climate story for a given country.

Climate can be defined either statistically or perceptually (Tribbia 2002).

Statistical definition

The International Research Institute for climate prediction defines normal rainfall for use in forecasting, for example, as follows:

"Normal" rainfall is defined as the average rainfall for 30 years for the period 1961 to 1990.

- "Above Normal" corresponds to one-third of the observations of which cumulative totals of rainfall were the highest (33%).
- "Below Normal" corresponds to one-third of the observations of which the cumulative rainfall totals were least (33%).
- "Near Normal" corresponds to the group of remaining years.

(from: iri.columbia.edu/climate/forecast/sup/May01_Afr/index_eng.html)

Officially, normal climate is designated by the UN WMO as the most recent three-decade averages of temperature, rainfall, etc. (1961–90; in May 2001 this normal was replaced by statistics for the 1971–2000 period). Others, however, tend to view normal as based on the statistics of the entire period (i.e. time-series) on record for a given location. Both approaches are

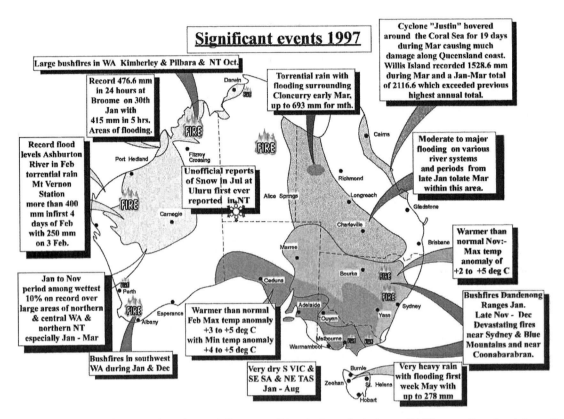

Figure 1.7 The wide range of weather and climate conditions in Australia during an El Niño year. (Based on information originally from www.bom.gov.au)

used. However, as Katz (personal communication) noted, "the entire record is more often relied on when estimating extremes (as opposed to averages)".

The better the instruments that are used to measure climate characteristics, the more reliable the information collected. The longer the record, the better the analysis is likely to be. Data for many places, however, are not very reliable, because data collection can be and has been disturbed by (a) moving the location of measurement, (b) urban modification of the local climate, (c) disruption of data collection as a result of conflict, or (d) failure of the measuring devices, etc. Nevertheless, the past hundred years or so of data is considered to be relatively robust and reliable for a statistical definition and assessment of normal local to national climates for several countries. Normal climate, however, can also be based on an individual's perceptions of actual climate conditions.

Perceptual definition

Most people weigh the climate anomalies that occurred earlier in history less heavily than recent events. They also weigh more heavily the ones they have witnessed than the ones they hear or read about. This tendency is reinforced by the local media, when they report for specific geographic locations, for example, that "this is the worst drought in 8 years" or "the heaviest flooding in 3 years" or "the hottest summer in 5 years." While these may be interesting facts, they are not very useful when it comes to understanding the behavior of the regional climate system. More serious are media reports of extremes that are unusual occurrences on the multi-decade scale.

Most likely, people do not recall the societal inconveniences of the intensity of the drought in the US northeast in the mid 1960s that prompted President Johnson to create a drought task force to identify ways to mitigate the severity of its impacts, if not to avoid future droughts in the urban centers. Yet, when drought struck New York City in 2001, policy makers, the public, and the media viewed the recent urban drought as an unprecedented event. As another example, people believe that the winters were snowier or that snowdrifts were higher in the past, when they were younger, than today. However, such perceptions of reality need to be compared to the actual climate record.

Perceptions about what constitutes normal climate conditions can be manipulated. For example, in the 1800s railroad companies sold land in the US West advertising the land as fertile for agriculture, as they expanded their rail lines westward into arid and semiarid areas. This could be called "greenwashing", where a government tried to convince people to settle in new areas where the climate-related conditions might not be conducive to sustained human activities such as agriculture.

Burroughs (2002) noted: "This personalized outlook on climate tends also to view any unpleasant event as being way outside past experience. Fanned in part by media hype, every storm, flood, heat wave or snowstorm is seen as having exceptional characteristics . . . In many instances however, unpleasant weather is nothing more than part of the normal fluctuations that make up climate . . . our memories are often of snowy winters, balmier springs, long hot summers or sunlit autumns. Unfortunately, these recollections have much more to do with how our memories embellish features of long ago and little to do with real climate change." Thus, what people believe to be normal climate in their area may not be normal at all. A sign captures what I think people need to keep in mind when it comes to their regional climate: "*Don't believe everything you think.*"

Normal climates' extremes
Anomalies denote the departure of an element (rainfall, temperature, etc.) from its long-period average value for the location concerned. For example, if the maximum temperature for June in Melbourne was 1°C higher than the long-term average for this month, the anomaly would be +1°C (http://www.bom.gov.au/climate/glossary/anomaly.shtml). Anomalies are of concern

because their impacts on societies, economies, and environments can be disruptive if not devastating, as the climate histories of most regions have shown. Anomalies are also influenced by regional and local factors as well as sea surface temperature changes in the Pacific and in other oceans.

In industrialized societies, meteorological extremes are also very disruptive. For example, a major storm system in the eastern half of the USA, called Superstorm93, encompassed 26 states and in a matter of a couple of days, left more than 280 people dead and caused an estimated $2 billion in damage. Its spatial extent affected Cuba as well as eastern Canada. As another example, the 1998 ice storm in Quebec caused relatively few deaths but generated considerable misery and suffering when electric power lines were toppled due to excessive ice accretion, causing loss of electricity for several weeks in the middle of winter. Also, the 1988 drought in the US Midwest, America's breadbasket, was estimated to have cost $40 billion, the costliest "natural" disaster in US history (see Section 8.4).

In developing countries, natural disasters can be very costly in terms of lives lost and in terms of loss of livelihood. Hurricane Mitch (1998) caused the death of more than 10 000 Hondurans; mudslides in Venezuela (1999) resulted in the death of more than 50 000; a tropical cyclone in 1970 was responsible for the deaths of more than 300 000 people in East Pakistan (now Bangladesh); drought-linked famine in Ethiopia in the early 1970s claimed more than one million victims. The list of such climate-system-related episodes is quite lengthy.

In addition to the immediate death and destruction, disruption of family and village life and widespread illness can plague the affected societies well into the future. For example, because of the predominant dependence of people in sub-Saharan Africa on rain-fed subsistence agricultural production, each year for many farm families there is what is called "a hunger season," a period where they must work the hardest during the pre-harvest time but their nutritional intake is poor. Thus, any disruption of the natural flow of the seasons can lead to a situation in which men are often forced to abandon their villages and families for varying lengths of time in search of food or funds. Some never return home from the urban slums or from refugee camps.

Policy makers at various levels of government can rise or fall, depending on whether or how they choose to deal with such extremes, such as droughts or floods. US city mayors have been voted out of office because of poor political responses to forecasts or impacts of blizzards or ice storms (e.g. as happened in Chicago and Denver). There are also examples from Africa of drought- and flood-related political changes of governments (e.g. either by coup, as in West Africa and Ethiopia in the mid 1970s, or by election).

1.3.3 What are problem climates?

There are at least two ways to look at the term problem climates: from a physical perspective and from an anthropogenic perspective. Climate processes are **physical** in that they center on the physical characteristics of the atmosphere. They are **anthropocentric** because climate processes intersect with human activities and the resources on which those activities depend.

Physical perspective

The physical climate can be viewed as a problem if the scientific basis for understanding it is highly uncertain. The climate is always changing on time scales that range from months to centuries and beyond. Knowledge about those changes is increasing through research and observations, as tools for researching and monitoring improve. Climate anomalies that might have surprised us decades ago no longer do, because we have now witnessed their occurrence. A good example is the 1982–3 El Niño that was called the El Niño of the century. The belief that such a label generated was that societies were safe from the return of an event of such magnitude for another hundred years. However, the 1997–8 El Niño was so surprisingly intense that scientists labeled it as the real El Niño of the twentieth century.

Climate changes, in the form of the atmosphere warming by a few degrees Celsius, generate a different set of ideas about what constitutes a problem climate (IPCC 2001). In a way, climate changes (at present global warming) force researchers and policy makers alike to enter into uncharted waters (i.e. an increased level of scientific uncertainty), because there is no precedent in recorded history for the current level of trace greenhouse gases in the atmosphere, especially carbon dioxide. Scientists expect that with global warming the nature of extreme climate and weather anomalies will change: extremes are likely to change in location, intensity, timing, and duration. Even in locations where people do not believe that they are living under a problem climate regime, that regime could change, and not necessarily for the better.

Australian meteorologist Neville Nicholls (2003, personal communication) noted the following: "The future climate is obviously the most important 'problem' climate, since we can't be sure how it will change. So we need to adapt as it is changing and that is proving to be very difficult. The recent fires in eastern Australia (2002–3) show how a changing climate is a problem climate. Last year's drought was much worse than previous droughts with similar low rainfall because it was much hotter than previous droughts (with consequently higher evaporation). This dried out the forests and made the year set for the enormous fires that took place (and the fires were of an immense size). We hadn't adapted our approach to fires to keep pace with the changing climate that is causing more ferocious fires."

The arid and semi-arid West African Sahel has a problem climate. It suffers not only from the extremes (droughts and floods) but, like other arid zones, also suffers when average conditions prevail. It is a characteristic of arid regions that rainfall is skewed toward dryness with a few high rainfall events being balanced out by a larger number of below average conditions. Therefore, average conditions could be harmful. Northeast Brazil is another area with a problem climate. Bangladesh is plagued with floods and droughts; Indonesia with floods, droughts, and fires; Papua New Guinea with drought and frost. In some cases an entire country can be said to have a problem with its climate regime, and in most cases there are smaller areas within a country that have problem climates.

Anthropocentric perspective

When researchers are asked what the phrase "problem climates" brings to mind, they most often respond by noting that climate is only a problem if it affects people in adverse ways. Several note the statement about a tree falling in the forest: when a tree falls in a forest and no one is there to hear it, does it make a sound? In other words, problem climates are only those climates that cause problems for activities that people and societies want to carry out. As a variation of this view, one can find examples of where there had been no human activities in a given area and, consequently, the climate was not viewed as a problem. Yet, as people move into areas that are marginal for human activities from a climate perspective, the interactions between society and the climate system become problematic: more crop failures, for example, because the soils or rainfall conditions were not suited to the selected crops or land-use practices. This particular process has been referred to as "drought follows the plow" (Glantz 1994).

Problem climates, then, are generated not only by changes in rainfall or temperature, but also by changes in certain kinds of human activities. For their part, societies are not just the victims of the climate system but are involved in the various ways in which the climate system and its impacts might be changing.

Rich and poor societies alike have increasingly come to realize the extent to which human activities (e.g. industrialization processes and land-use practices) and ecological processes can affect the local and global atmospheres as well as be affected by them. In addition, an increasing number of government, individual, and corporate decisions are being made for which a knowledge of climate affairs is required. There is a growing awareness among educators in many disciplines of the need for a better understanding of just how climate variability, change, and extremes can and do affect the environment and the socio-economic and political affairs of people, cultures, and nations.

For their part, social scientists have become much more engaged in research efforts to distinguish between the impacts of physical processes on various socioeconomic sectors of society and those impacts that have actually resulted from decision-making processes. They are also active in trying to identify as well as develop ways to use climate and climate-related information to address a wide range of local to global societal needs.

1.3.4 Problem societies

The phrase "problem societies" refers to climate and climate-related factors that affect the ability of society to interact effectively with the climate system. Accepting the fact that there are many things about the behavior of the atmosphere that we do not yet know or understand, it is also important to note that there is a considerable amount of usable information that we do already know about the interactions between human activities and the climate system. Nevertheless, societies knowingly still engage in activities that increase their vulnerability or reduce their resilience in the face of a varying climate system.

Human activities can alter the physical characteristics of climate from local to global levels. In addition, societal changes can make them more vulnerable to a variable climate. Policy makers at various levels of government knowingly make decisions (explicitly or implicitly) about land use in areas that are prone to climate-related hazards, e.g. deforestation, increasing soil erosion, decrease in soil fertility, destruction of mangroves, over-fishing, chemical emissions to the atmosphere, the drying out of inland seas, and so forth. These decisions set societies up for the impacts of varying and extreme climate and weather conditions, and are the underlying causes of many climate-related problems. For example:

- Tropical deforestation is occurring wherever such forests exist, such as in South America, sub-Saharan Africa and Southeast Asia. Research shows that in the Amazon basin, for example, 50 percent of the rain that falls there is the result of evapotranspiration from the vegetation therein.
- As productive land becomes scarce, people are forced to inhabit increasingly marginal areas for agricultural production or for livestock rearing. As a result of, for example, moving up hillsides and mountain slopes, the cultivation of the soils leads to an increase in soil erosion and to sediment loading of nearby streams, rivers, and reservoirs. In time the land may have to be abandoned, leaving eroded hillsides exposed to the vagaries of nature.
- Land use in arid and semiarid areas can be very destructive, if care is not taken for agricultural and livestock rearing activities. As land is cleared of vegetation to grow crops, it is left vulnerable to wind and water erosion. Irrigated lands need to be drained properly to avoid salinization of the soils or waterlogging.

- In the mid 1970s atmospheric chemists (Rowland and Molina 1974) discovered that chlorofluorocarbons (CFCs), while inert in the lower atmosphere, break down in the stratosphere in the presence of ultraviolet radiation thereby freeing chlorine atoms that combine and recombine with oxygen by breaking down ozone molecules (see Section 5.7). As a result there is a thinning of the ozone layer that protects the Earth's surface from lethal amounts of UV radiation. Once emitted, these chemicals have a lifetime in the atmosphere on the order of many decades. There is still an illegal trade in CFCs.

- The demise of the Aral Sea serves as a good example of environmental degradation that resulted from political decisions. After 1960, the Soviet government expanded cotton cultivation from about 3.5 million ha to 7 or more million ha in its Central Asian Republics. A sharp increase in diversions for irrigation from the region's two major rivers has reduced the surface area of the sea by more than half, and its volume by more than a third. The sea has broken into two parts, and salinity and pollution have made its water unfit for most living things. The sea continues toward total desiccation as a result of policy makers paying little regard to the fragility of the natural environment. In addition winters have apparently become colder and the summers hotter.

- Many countries continue to base their economic growth plans on the continued, if not expanded, use of fossil fuels (coal, oil, and natural gas) that are known to produce heat-trapping carbon dioxide emissions. Much debate has taken place on how and when to reduce such greenhouse gases (GHGs) emissions (i.e. Kyoto Protocol), but the increases in emissions continue.

The long-term changes of concern to policy makers as well as scientific researchers have been in temperature, precipitation, winds, relative humidity, and seasonality. Sea level rise and glacial melt are other major climate change indicators of paramount concern, especially to those living in coastal low-lying areas. Today, the debate is whether human-induced changes to physical forcing factors, which influence the behavior on various time scales of elements of the global climate system, can bring about "deep" climate changes that before humans could only occur naturally.

Societies have difficulty in coping effectively with today's climate anomalies and their impacts on societies and environments. Reasons include but are not limited to the following: scientific uncertainty, a blind faith in the development of new mitigating technology, scientific uncertainty about climate phenomena and about their impacts, the 2–4–6 years political cycle in the USA (the attention span of politicians in various issues relates to their length of term in office), and the mysterious reasons why known ways to cope with anomalies are not used.

With the advent of satellite imagery in the 1960s we have been able to see from space the extent to which human activities on different sides of domestic

(a)

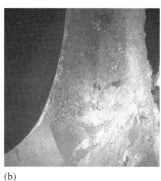

(b)

Figure 1.8 Photographs illustrating the extent of human impacts: (a) dust storm over Texas, USA; (b) The Negev Desert region in the Middle East.

and international political borders can have different consequences for the environment (Figure 1.8). There are some situations where one can see how differences in decisions about land use have led to differences in land degradation that closely follow political jurisdictions.

Atmospheric processes and climate-related impacts do not stop at political borders. Both sides of a border are put in the position of having to cope with their climate-related hazards and using their climate as resources. A good example of a transboundary climate-related impact is the forest fire situation in Indonesia and the resultant regional veil of haze it produces. However, not all societies have the resources to protect their citizens from the hazards or to help their citizens capitalize on the resource potential of their given climates.

Hence, people end up living in areas at risk to climate anomalies and climate-related hazards, mostly people who are poor or politically disenfranchised. Some of the existing natural hazard risks have been increased as a result of government policies; people are allowed if not encouraged to live in flood plains, on coastal chars, in arid areas, in tropical or mangrove forested regions. Some risks have increased because population growth numbers have far outstripped the natural resources needed to support them. Then, the present as well as future populations are at increasingly greater risk to adverse impacts from existing climate conditions, and more so in the face of deep climate change.

Table 1.1 *Selected phrases illustrating "problematic climates"*

Problematic climate

- Present climate data from Africa are problematic because of data collection methods (UNEP/GRIDA)
- Drought and water availability are problematic even without climate change (UNEP/GRIDA)
- There is a problematic funding of pure versus applied science because funding channels and goals are different (INPECO, Clinical Laboratory Automation)
- Complexity of the Earth's dynamic climate system makes long-term prediction problematic
- Climate has problematic stimuli and beneficial stimuli

Problematic climate change

- Dependence on fossil fuels is a problematic situation
- If problematic climate change is occurring...
- CO_2 warming is not problematic
- Climate change is problematic for the next generation rather than the current one (CRU)
- The rate of change from today's climate to a new one is problematic (CRU)
- Policies to maintain the GHGs status quo risks are worsening an already problematic problem
- Detecting changes in observing extreme weather events and attributing them unequivocally to anthropogenic climate change remains problematic (CRU)
- Setting binding GHG emissions in developing countries is extremely problematic (UNEP)

1.3.5 Concluding comments

While every climate can be viewed to varying degrees as a problem climate in the natural science sense, the word "problematic" better captures the contemporary view of what constitutes a problem climate than the one originally defined by Trewartha (1981). The phrases in Table 1.1, selected randomly from different websites provide apt illustrations.

Without doubt, in the 40 years since Trewartha first introduced the notion of problem climates, our knowledge of what we know and what we don't know yet about the climate system has greatly improved. We now realize that all climate regimes, local to global, are problem climates in some respect. Today, it would be appropriate to replace the phrase "problem climates" with "problematic climates," a phrase that directs the focus on the interconnectedness between the atmospheric processes that give us our climates and the human activities that can influence or be influenced by them.

Because of the lack of complete understanding of the behavior of the climate system, early warnings depend on the type of climate anomaly that merits early warning, and they are of varying reliability. So, climate can be problematic (i.e. worrisome) from a natural science perspective and also be problematic (i.e. troublesome) from a societal perspective because its impacts can cause problems (or generate benefits) for human activities and settlements.

As noted at the outset, climate can be viewed as **a resource** to be exploited, **a hazard** to be avoided and even as **a constraint** to economic development. There is yet another often overlooked aspect of climate – climate as **a scapegoat**, meaning that climate anomalies provide decision-makers with handy excuses for socioeconomic or political problems, regardless of whether or not those anomalies really contributed to those adverse impacts.

Thus, it is really important to use all methods available to identify those aspects of the impacts of a climate anomaly that can legitimately be linked to climate and those that can be blamed on society. Only then can policy makers take correct and appropriate action to prepare for or adapt to the adverse impacts of climate on society and of society on climate. A failure to correctly identify the linkages between climate processes and human activities leads to policy responses that do not address the climate-related problems at hand.

Our problem is not that we have to cope with a variable and changing global climate but with the ways societies have chosen to develop their economies with little regard to the impacts on climate. This brings to mind the Pogo cartoon: "*I have met the enemy and he is us!*" It is time to start recognizing problem societies as well as problem climates.

1.4 Examples of general climate websites

The Hadley Centre of the United Kingdom Met Office (www.metoffice.com/research/hadleycentre/) is a fine source for European climate information. Descriptions of the various research activities are provided.

The Bureau of Meteorology of the Commonwealth of Australia (www.bom.gov.au/) has a varied content that provides satellite information and images and basic physical climatology.

The National Center for Environmental Prediction is available through the NOAA website (www.nws.noaa.gov). The site provides comprehensive information concerning models and forecast systems.

The National Climate Data Center (www.ncdc.noaa.gov/oa/ncdc.html) is the basic source of data and climatic information in the United States. The site provides links to many international agencies.

The Max Planck Institute for Meteorology (www.mpimet.mpg.de/) describes numerical models and their value in studying interacting components of the Earth system.

1.5 References

Angström, A., 1935. Teleconnections of climate changes in the present time. *Geographiska Analer*, **17**, 242–258.

Barry, R. G. and Chorley, R. J., 1998. *Atmosphere, Weather, and Climate*, 7th edn. New York: Routledge.

Bonan, G., 2002. *Ecological Climatology Concepts and Applications*. Cambridge: Cambridge University Press.

Burroughs, W., 2002. *Climate: Into the 21st Century*. Cambridge: Cambridge University Press.

Geer, I. W., 1996. *Glossary of Weather and Climate*. Boston: American Meteorological Society, Chapter 1.

Glantz, M. H., 1994. *Drought follows the Plow*. Cambridge: Cambridge University Press.

IPCC, 2001. *Climate Change 2001: The Scientific Basis. Contribution of Working Group I to the Third Assessment Report of the Intergovernmental Panel on Climate Change.* J. Houghton, Y. Ding, D. Griggs, *et al.*, eds. Cambridge: Cambridge University Press.

NRC (National Research Council) 2002. *Abrupt Climate Change: Inevitable Surprises*. Washington, DC: National Academy Press.

Oliver, J. E., 1991. The history, status and future of climatic classification. *Physical Geography*, **12**, 231–251.

Oliver, J. E. and Hidore, J. J., 2002. *Climatology: An Atmospheric Science*. Upper Saddle River, NJ: Prentice Hall.

Rowland, F. S. and Molina, M. J., 1974. Stratospheric sink for chlorofluoromethanes: chlorine atom-catalyzed destruction of ozone. *Nature*, **249**, 810–812.

Ruddiman, W., 2001. *Earth's Climate Past and Future*. New York: W. H. Freeman and Co.

Sellers, W. D., 1965. *Physical Climatology*. Chicago: University of Chicago Press.

Steffen, W., 2001. Toward a new approach to climate impact studies. In L. Bengtsson and C. Hammer, eds., *Geosphere–Biosphere Interactions and Climate*. Cambridge: Cambridge University Press, pp. 273–279.

Trewartha, G., 1981. *The Earth's Problem Climates*, 2nd edn. Madison: University of Wisconsin Press (first edition 1966).

Tribbia, J., 2002. What constitutes 'normal'? In M. H. Glantz, ed., *La Niña and its Impacts: Facts and Speculation*. Tokyo: UN University Press.

Chapter 2
Oscillations and teleconnections

2.1 History and definitions

The concept of atmospheric oscillation began with studies of the Asian monsoon. Following the great 1877 drought in India, the India Meteorological Department was established under the leadership of H. F. Blanford. His task, in part, was to examine whether any monsoon seasonal prediction could be identified. Concentrating upon solar relations and climate, he could not report success. It was, however, Sir Gilbert Walker who, as Director-General of Observatories in the India Meteorological Department, initiated extensive studies of pressure patterns that eventually led to the identification of atmospheric oscillations. After he retired in 1924, Walker observed a see-saw like oscillation of sea level pressures in various parts of the Pacific Ocean (Walker 1923–4). He labeled this the Southern Oscillation. Further studies in the 1920s and 1930s saw identification of North Atlantic and North Pacific oscillations. Not a great deal of attention was accorded this work, and it was not until many years later that the contribution of Walker was recognized and the Walker circulation named in his honor (Bjerknes 1966). It is interesting to note that the statistical methods used by Walker were sophisticated enough to become the "Yule–Walker equations" that refer to properties satisfied by the autocorrelations of an autoregressive process (Katz 2002).

Throughout this text the discussion of regional climates and anomalies will, in part, concern teleconnections. This chapter provides a background to the major oscillations that relate to teleconnections. Of these, ENSO events are by far the best known and, as a result, this topic and its extended influences are given additional emphasis through an essay (Section 2.8).

2.1.1 Oscillations

Any phenomenon that tends to vary above or below a mean value in some sort of periodic way is properly designated as an oscillation. Classical oscillation research states that oscillation occurs when a system is disturbed from a position of stable equilibrium. It may be recognized eventually as a predictable *cycle*, but this term should not be used unless the period has a recognizable regularity. The term oscillation is sometimes used for the swing from one extreme to the other, that is a half cycle.

Fairbridge (1986) noted that a number of types of oscillation are recognized: *damped oscillation*, one with constantly decreasing amplitude; *neutral, persistent* or *undamped oscillation*, maintaining constant amplitude; *unstable oscillation*, growing in amplitude and then breaking down; *stable oscillation*, consistent amplitude with little change; *forced oscillation*, one set up periodically by an external force; *free oscillation*, a motion established externally but which then receives no further external energy.

Atmospheric oscillations can fit any number of these categories. Of singular importance, however, are the planetary atmospheric pressure fields that are considered here. Most of these oscillations are involved with the strength and location of centers of action, the major global highs and lows, and sea surface temperatures. The character of the identified oscillations is mostly derived statistically from long-term pressure observation series.

Oscillations have various periodicities. Some, like the Madden–Julian (Section 2.5), are intraseasonal, occurring at periods of less than 1 year. A quasi-biennial (QB) occurs at approximately 2-year intervals while some periodicities, such as ENSO, have a quasi-cyclic periodicity ranging from 3 to 5 years. At the same time, and as described in Section 2.8, the ENSO phenomena may also be considered as quasi-biennial with a 2 to 2.5 year frequency together with a low-frequency (LF) interannual component. Thus QB ENSO and LF ENSO signals are identified. Decadal oscillations, such as the Pacific Decadal Oscillation (Section 2.3), have been linked to the 11-year sunspot cycle, while bidecadal may be linked to the double sunspot, or Hale, cycle. Multidecadal and longer term oscillations have been identified, some of which may be associated with Earth orbit parameters.

Table 2.1 outlines some of the oscillations that have been identified and which will be described in this chapter.

2.1.2 Teleconnections

Teleconnection is a term used to describe the tendency for atmospheric circulation patterns to be related, either directly or indirectly, over large and spatially non-contiguous areas. Glantz (2001) defines them succinctly as linkages between climate anomalies at some distance from each other.

It seems that a number of late nineteenth and early twentieth century climatologists believed that changes in one location were related to changes at a different location. While the best known of such scholars was Walker, others pointed out, for example, that drought in South Africa seemed to occur at the same time as that in India, suggesting a connection between the hemispheres. However, the word "teleconnection" was not used in a climate context until it appeared in a paper by Ångström (1935).

Table 2.1 *Major oscillations*

Southern Oscillation (SO)

A strongly anti-correlated pressure anomaly over the Indian and South Pacific Oceans. It has a slightly variable period averaging 2.33 years and is often analyzed as part of an ENSO event.

North Atlantic Oscillation (NAO)

Reflects pressure variations and the stability of the Icelandic Low and the Azores–Bermuda High pressure cells. The NAO has a marked influence upon the climates of western Europe.

North Pacific Oscillation (NPO)/Pacific Decadal Oscillation (PDO)

A long-lived phenomenon defined by surface ocean temperatures in the northeast and tropical Pacific Ocean.

Madden–Julian Oscillation (MJO)

An eastward progression of tropical rainfall, which produces anomalous rainfall patterns that may be enhanced or suppressed.

Quasi-biennial Oscillation (QBO)

A low-latitude oscillation that is longer than the dominant annual cycle and results in a periodic reversal of winds in the lower stratosphere.

Pacific North American Oscillation (PNA)

An alternating pattern between pressures in the central Pacific Ocean and centers of action over western Canada and the southeastern United States.

Arctic Oscillation (AO)

An oscillation in which atmospheric pressure, at polar and mid-latitude locations, fluctuates between defined positive and negative phases.

Antarctic Oscillation (AAO)

An oscillation in values of mid- and high-latitude surface pressure systems in the Southern Hemisphere. It is quantified by the Antarctic Oscillation Index.

Teleconnections play an integral part in the study of air–sea interactions and global climate processes. They often provide the missing piece in the understanding of climate patterns, both spatial and temporal, that occur across the world. The identification of connections suggested by teleconnections has become so important that the study forms a subfield of the atmospheric sciences. Table 2.2 provides examples of some teleconnections between two regions.

The study of teleconnections is largely based upon statistical analysis and requires reliable data sources. Of particular importance is the use of empirical orthogonal functions, the EOFs, or the principal components. Using grid point pressure values as a matrix, principal components, eigenvalues and eigenvectors are obtained from the analysis. Through the manipulation of these data, a set of component scores is derived and these can be clustered to form a classification used to identify the patterns under investigation. The analysis is, of course,

Table 2.2   *Examples of teleconnections involving two regions*

Variable 1	Variable 2
Surface pressure in Indonesia	Surface pressure in the eastern Pacific
Precipitation over Australia	Precipitation over India
Ocean temperature in the eastern tropical Pacific	Upper-level air pressure over the northern Rocky Mountains
Ocean temperature in the eastern tropical Pacific	Rainfall in the southeastern United States
Surface temperature in Greenland	Surface temperature in northern Europe
Ocean temperature in the central tropical Pacific	Ocean temperature in the tropical Indian Ocean
Low-level east–west airflow over the Indian Ocean	Low-level east–west airflow over the tropical Pacific Ocean
Ocean temperature in the eastern tropical Pacific	Strength of upper-level westerly wind flow in the North Pacific
Rainfall in northeast Brazil	Ocean surface temperature in the eastern tropical Pacific
Rainfall in the sub-Saharan area of Africa	Surface pressure difference between the Indonesian and tropical east Pacific areas
Upper-level pressure in the subtropical regions of the west Atlantic	Upper-level pressure in the polar region of the North Atlantic
Upper-level pressure in the western subtropical Pacific area	Upper-level pressure in the North Pacific area

Note: Correlations may be either positive or negative for the variables listed. In all cases, the variables are for monthly or seasonally averaged times and for a broad area in the regions listed.

Source: After D. D. Houghton, 1996. Teleconnections, in S. H. Schneider, ed., *Encyclopedia of Weather and Climate*, New York: Oxford University Press, p. 743.

completed using appropriate computer programs; for those wishing to further investigate the EOFs, the outline provided by Barry and Carleton (2001) will prove useful.

A second method of teleconnection analysis is through the computation of correlation maps. In this, for example, the correlation coefficient between sea level pressures at a selected location is compared to that at all grid points north of $20°$ N. Many similar examples are found in teleconnection studies (Peixoto and Oort 1992).

2.1.3 Major identified oscillations and teleconnections

Ongoing research has identified or suggested a number of oscillations and resulting teleconnections. It need be noted that a number of patterns that form part of the major features may be identified. These are often seasonal in nature. For example, in the North Atlantic the East Atlantic Pattern exists from September to April, and over Eurasia the Polar/Eurasian Pattern may be discerned from December to February. An analysis of the many identified patterns is available from the Climate Diagnostics Center (www.cdc.noaa/ClimateIndices).

2.2 The North Atlantic Oscillation (NAO)

The North Atlantic Oscillation is defined as the difference between sea level pressure at two stations representing the centers of actions that occur over Iceland and the Azores. In calculating the pressure difference between the two pressure systems, data from Stykkisholmur in Iceland have been evaluated using various other stations to represent the more southerly center of action. Rogers (1997) used Ponta Delgarda, while Lisbon, Portugal, has been used by Hurrel (1995) and Gibraltar by Jones *et al.* (1997). Generally, US analyses (e.g. NCAR) use the Hurrell index while those in Europe (e.g. East Anglia Climate Research Unit) use that by Jones *et al.*

The acquired pressure differences are used to derive an index that identifies the phase of the oscillation. A positive phase is represented by a stronger than usual subtropical high pressure and a deeper than normal Icelandic Low. Accordingly, the negative phase shows a weak subtropical high and a weak Icelandic Low. The NAO Index is defined using the winter season, December through March.

In its positive phase, the increased pressure difference between the two centers of action results in frequent strong winter storms tracking in a more northerly path across the North Atlantic Ocean. Conversely, the negative phase sees fewer and weaker storms passing in a more southerly track across the North Atlantic. These two modes have a marked influence on the nature of winter climates in both Western Europe and eastern North America.

In Western Europe, a positive NAO Index results in warm, wet winters, for the storm tracks are carried to the north and air from the subtropical highs will prevail. At the same time, the eastern parts of the United States will probably experience mild and possibly wet winters. In contrast, during the negative phase, eastern North America will experience more cold air invasions and so will Western Europe, especially the Mediterranean.

Comparisons between the NAO and the Southern Oscillation (SO) suggest that while the SO is driven by sea surface temperature, NAO is an atmospheric phenomenon. Additionally, the NAO is of much longer time scale, with changes in phase often taking decades. It is thus of particular interest for long-term climatological analysis.

Figure 2.1 shows values of the NAO from the late 1800s to the late 1900s. The cold European winters of the 1940s and the 1960s, which happened to include one of the coldest decade periods on record, each coincided with negative phases. A high positive NAO Index in the 1980s and early 1990s was in periods

Figure 2.1 Values of the NAO from the late 1800s to the late 1900s. The coldest European winters on record (in the 1940s and the 1960s) coincide with negative phases of the NAO.

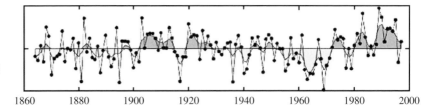

of particularly mild winters across Europe. As noted in Section 2.7, the NAO and the Arctic Oscillation (AO) are intimately linked.

An in-depth analysis of the NAO is presented in a special publication of the American Geophysical Union (Hurrell *et al.* 2003). This work deals not only with the dynamic climatology of the NAO but also provides an account of its ecological and economic consequences.

2.3 The North Pacific Oscillation (NPO)/Pacific Decadal Oscillation (PDO)

Zhang *et al.* (1997) describe the PDO as a long-lived El Niño-like pattern of Pacific climate variability. The term was coined by Mantua *et al.* (1997) in a study investigating the role of sea surface temperature on salmon behavior in the North Pacific Ocean. The index used to assess the PDO is derived from monthly sea surface temperature (SST) anomalies north of 20° N in the Pacific Ocean. Other studies have provided other names including the Pacific Decadal Variation (PDV) and the Interdecadal Pacific Oscillation (IPO).

The PDO is actually represented by a shift of SST that occurs on a 20 to 30 year cycle. It is in its warm or positive phase when the northwest SST anomalies are negative, while the SST anomalies in the eastern tropical Pacific Ocean are positive. The cool or negative phase is the reverse of this, with positive anomalies in the NW and negative in the tropical zone. In effect, it represents a change in the location of cool and warm water masses that impart their influence upon the atmosphere in a variety of ways. This, in turn has impacts upon the western parts of the United States.

Figure 2.2 shows the occurrence of positive and negative phases of the PDO. A number of studies (Mantua *et al.* 1997; Minobe 1997) suggest that there have been just two full PDO cycles in the last 100 years. A cool PDO cycle occurred from 1890 to 1924 and from 1947 to 1976, while the warm cycle was in effect from 1925 to 1946 and from 1977 to the late 1990s. It is thought that a possible change to the cool PDO phase began at that time.

The impact upon climate at the time of the various phases of PDO has been the topic of a number of studies. Of particular interest are those studies relating the PDO to the climate of the American Southwest. In a series of locally distributed articles (also available at www.srh.weather.gov/abq/feature/PDO_NM.htm),

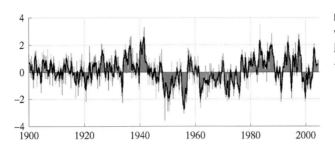

Figure 2.2 Monthly values for the PDO index, January 1900 to August 2004.

C. A. Liles of the NWS in Albuquerque, N. Mex., relates New Mexico rainfall to PDO. He concludes that the PDO cycle in its negative (cool) phase is related to long-term droughts in the region. Given that we may be on the verge of a new negative phase, Liles foresees major drought problems, especially with the burgeoning population of this sunbelt state. A similar result has been derived for Arizona, where it has been found that winter season precipitation is significantly affected by PDO phase. Rainfall was well below normal in the cool phase and above normal in the warm phase.

Causes of the PDO are not known and hence prediction through modeling is not well established. However, because of its multi-year persistence, knowledge of the PDO improves year-to-year and season-to-season forecasting. Recognition of the PDO is of value as a societal function for it illustrates how "normal" climate conditions can vary over a human lifespan.

2.4 The Pacific North American Oscillation (PNA)

The PNA is an alternating pattern between pressures in the central Pacific Ocean and centers of action over western Canada and the southeastern United States (Latif and Barnett 1994). Expressed as an index that is both ocean- and land-based, the PNA is characterized by atmospheric flow in which the west coast of North America is out of phase with the eastern Pacific and southeast United States. It tends to be most pronounced in the winter months. The PNA is associated with a Rossby wave pattern with centers of action over the Pacific and over North America. It refers to the relative amplitudes of the ridge over western North America and the troughs over the central North Pacific and southeastern United States (Leathers *et al.* 1991a).

According to Wallace and Gutzler (1981) the PNA is a "quadripolar" pattern of pressure height anomalies. Anomalies with similar signs are located south of the Aleutian Islands and the southeastern United States and those with opposite signs (to the Aleutian center) are in the vicinity of Hawaii and the intermountain region of Canada. The locations used to derive a PNA Index are $20°$ N, $160°$ W; $45°$ N, $165°$ W; $55°$ N, $115°$ W; $30°$ N, $85°$ W, using the normalized 700 mb height anomalies for a particular season. Other definitions that are used are the result of a statistical pattern analysis.

Figure 2.3 Seasonal mean of the PNA Index. Note the trends of the positive and negative values of the index.

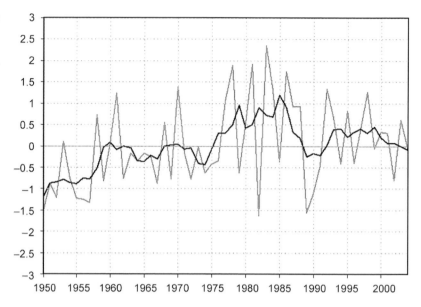

Figure 2.3 shows the seasonal PNA Index for the period 1950 to 2004. When the index is positive, there is a strong Aleutian Low and a strong ridge over western Canada. In the negative phase the pattern is quite different, lacking the strong ridges and troughs with a more zonal flow in effect.

There have been a number of studies linking the PNA to climate anomalies in North America. In a two-part study of the role of the PNA in the United States Leathers *et al.* (1991b) found a number of meaningful relationships. They demonstrated that regional temperatures and precipitation are highly correlated to the PNA Index across the United States especially in winter. Correlations for precipitation were less extensive than those of temperature but clear relationships were obtained. The authors used the PNA Index to corroborate earlier research suggesting a major change in mid-tropospheric circulation in the late 1950s.

An interesting study related winter moisture conditions in the Ohio Valley to the PNA (Coleman and Rogers 1995). It was found that the PNA Index was strongly linked to moisture variability in an area extending from southeast Missouri to Ohio, with a very strong correlation in southern Indiana. They also noted that the mean streamflow of the Ohio River in winter had discharges up to 100% higher in PNA-negative winters compared to PNA-positive winters. The incidence of cyclones in the Great Lakes region of North America was the topic of research by Isard (1999). The study supported the observation that more Great Lakes cyclones originate from the northwest at those months when PNA-positive occurs, while cyclones from the west and southwest occur more often with a PNA-negative index. From these studies, and others like them, it is clearly evident that the PNA plays a significant role in the climates of North America.

2.5 The Madden–Julian Oscillation (MJO)

The MJO is a low-latitude intraseasonal oscillation, meaning that it passes through an identified cycle in a period of 60 to 90 days, quite unlike the annual, biennial or decadal cycles of other oscillations. It is so named for Roland Madden and Paul Julian of NCAR who discovered the wave in the early 1970s. Its identification and possible forecasting is of considerable importance in long-range predictability of tropical and subtropical weather as well as short-term climate variability (see Background Box 2.1).

The MJO is a feature of the tropical atmosphere–ocean system that plays a significant role in precipitation variability. It is characterized by anomalous rainfall conditions, that can be either enhanced or suppressed. The beginning of the cycle, the anomalous rainfall event, usually appears first over the Indian Ocean and Pacific Ocean. It remains identifiable as it moves over the very warm water of the western and central parts of the Pacific Ocean. On meeting the cooler waters of the eastern Pacific Ocean it usually becomes less defined, only to reappear over the tropical Atlantic Ocean and Indian Ocean. As noted this cycle can last two to three months.

Clearly, the higher or lower than normal precipitation of the event is associated with both surface and upper air conditions as they relate to ascending and descending air. A knowledge of the cycle characteristics enables a clearer understanding of tropical rainfall variability and of Pacific Ocean and Atlantic Ocean tropical cyclones. Additionally, such information can be related to the potential impact upon middle-latitude precipitation. In their study of the MJO's role in hurricanes in the Caribbean Sea and Gulf of Mexico, Maloney and Hartmann (2000) found a distinct relationship. They noted that hurricanes are four times more likely to occur when, in the ascending phase, the rising air and surface westerly winds are conducive to formation of the storm.

The Climate Prediction Center (CPC) of NOAA (www.cpc.ncep.noaa.gov/) provides a fascinating account of how a tropical oscillation may impact rainfall in the Pacific northwest of North America and their analytical description is given in Background Box 2.2.

The CPC further notes that there is a coherent simultaneous relationship between the longitudinal position of maximum MJO-related rainfall and the location of extreme west-coast precipitation events. Extreme events in the Pacific northwest are accompanied by enhanced precipitation over the western tropical Pacific and Indonesia (typically centered near 120° E) with suppressed precipitation over the Indian Ocean and the central Pacific.

The MJO is thought to play a significant role in the formation and frequency of hurricanes. If an easterly wave, the initial formation feature of Atlantic hurricanes, meets the cloudy sky phase of the MJO the conditions for hurricane formation improve. The MJO is, of course, moving in the opposite direction to the easterly wave. Unfortunately, although hurricane researchers of the National

---

Background Box 2.1

Monitoring intraseasonal oscillation in the tropics
The major intraseasonal oscillation in the tropics is the Madden–Julian Oscillation, the MJO. Given that most climatological variations that occur on a monthly basis are related to precipitation, the feature plays a significant role in the nature of the regional climate. The MJO is identified by the eastward progression of a large region of either enhanced or suppressed rainfall.

It has proved difficult to accurately model the MJO. Generally, dynamic models do not work well with tropical convective rainfall, while the slow evolution of the MJO further complicates accurate prediction. Because of this, a variety of more conventional means are used to monitor the feature.

Of singular importance is information derived from polar-orbiting and geostationary satellites. This fundamental diagnostic tool uses derived images and data to identify regions in which convectional activity departs from a derived long-term mean.
The global radiosonde network, which provides data twice daily, is used in conjunction with satellite analysis. Radiosondes provide temperature, pressure, moisture values, and wind data at many levels of the atmosphere. The data are incorporated into weather prediction models that are used for weather and climate analysis.

The nature and location of the MJO uses are displayed in time–longitude format to reveal a number of characteristics. The Climate Prediction Center of NOAA lists the following:

1. outgoing long-wave radiation, a satellite derived measure of tropical convective activity and precipitation;
2. velocity potential, a derived quantity that isolates upper level divergence;
3. upper level and lower level wind anomalies;
4. 500 hPa anomalies to represent the atmospheric responses in mid-latitudes.

While the features are used to identify and monitor the MJO, it is also thought that the oscillation plays a role in hurricane formation and ENSO events. Researchers at many atmospheric science centers are actively monitoring the MJO and seeking to assess its role in many teleconnections.

Hurricane Center have been informally using the MJO, the linkage is currently insufficient for accurate modeling and forecasting.

2.6 The Quasi-biennial Oscillation (QBO)

The QBO is a low-latitude oscillation that is longer than the dominant annual cycle and its identification rests with sophisticated computer modeling

---

Background Box 2.2

Evolution of a precipitation event

The typical scenario linking the pattern of tropical rainfall associated with the MJO to extreme precipitation events in the Pacific Northwest features a progressive eastward moving circulation pattern in the tropics and a westward moving circulation pattern in the mid-latitudes of the North Pacific. Typical wintertime weather anomalies preceding heavy precipitation events in the Pacific Northwest are as follows.

7–10 days prior to the heavy precipitation event

Initially, the heavy tropical rainfall associated with the MJO shifts eastward to the western tropical Pacific. A moisture plume extends northeastward from the western tropical Pacific towards the general vicinity of the Hawaiian Islands. A blocking anticyclone is located in the Gulf of Alaska with a strong polar jet stream to its north.

3–5 days prior to the heavy precipitation event

The heavy tropical rainfall shifts eastward towards 180° longitude and begins to diminish. An associated moisture plume extends further to the northeast, often traversing the Hawaiian Islands. The strong blocking high weakens and shifts westward. A split in the North Pacific jet stream develops, characterized by an increase in the amplitude and areal extent of the upper tropospheric westerly zonal winds on the southern flank of the block and a decrease on its northern flank. The tropical and extratropical circulation patterns begin to "phase", allowing a developing mid-latitude trough to tap the moisture plume extending from the deep tropics.

The heavy precipitation event

As the pattern of enhanced tropical rainfall continues to shift further to the east and weaken, the deep tropical moisture plume extends from the subtropical central Pacific into the mid-latitude trough now located off the west coast of North America. The jet stream at upper levels extends across the North Pacific with the mean jet position entering North America in the northwestern United States. Deep low pressure located near the Pacific Northwest coast can bring up to several days of heavy rain and possible flooding. These events are often referred to as "pineapple express" events, so named because a significant amount of the deep tropical moisture traverses the Hawaiian Islands on its way towards western North America.

Source: Climate Predictions Center of NOAA, www.cpc.ncep.noaa.gov/

(Maruyama 1997). Associated with fluctuations of the ITCZ, it is a periodic reversal of winds in the lower stratosphere at elevations between 20 and 30 km.

The alternation of easterly and westerly equatorial upper-air winds has been systematically observed, with increasing precision, since the famous eruption of the Krakatoa volcano in the East Indies in August of 1883. Ash from this volcano reached the lower stratosphere and drifted, in alternate years, either easterly or westerly. The easterly winds were labeled the "Krakatoa Easterlies", named for the volcano. The westerly flow was named the "Berson Westerlies" for the German meteorologist A. Berson, who observed the west–east winds after releasing observation balloons in East Africa (see *Krakatoa Winds* in Fairbridge 1986). The actual discovery of the QBO is credited to R. J. Reed and R. A. Ebdon based upon rawinsonde data obtained at Canton Island (Reed *et al.* 1961). The term Quasi-biennial Oscillation was coined by Angell and Korshover (1964). An extensive and thorough review of the QBO by Baldwin *et al.* (2001) provides an account of both the potential impacts and the theory of the oscillation, and the following account draws, in part, upon that review.

The QBO is a quasi-periodic oscillation of the zonal wind, between easterlies and westerlies that occur in the equatorial/tropical stratosphere. These alternating winds originate at the top of the lower stratosphere, above 30 km or approximately 10 hPa, and propagate downward at a rate of about 1 km each month until they dissipate at the tropical tropopause. The easterlies are generally stronger than the westerlies. The fastest observed oscillation has a period of 20 months, and the slowest 36 months, with an average period of 28 months.

A theory to explain the QBO has been the subject of considerable debate. Lindzen and Holton (1968) suggested that winds of the upper stratosphere, the semi-annual oscillations, were a key; that is the QBO originated in the upper stratosphere. In 1972 the same researchers outlined the role of gravity waves and mixed Rossby waves, each of which provided the necessary easterly and westerly momentum need for the QBO. The details of these are provided in a historical review of QBO theory by Lindzen (1987). The current understanding of the QBO certainly involves gravity waves as shown by computer and numerical models.

The significance of the QBO is still being investigated. It is known to play a role in stratospheric ozone mixing and in the potential for hurricane formation. Promising relationships have been shown to exist between the QBO and Northern Hemisphere winters. As noted, this is well illustrated in the work of Baldwin *et al.* (2001). The QBO is discussed in more detail in an essay given in Section 3.3.

2.7 The Arctic Oscillation (AO) and Antarctic Oscillation (AAO)

In a lecture at the National Center for Atmospheric Research in 2000 (available at tao.atmos.washington.edu/data/annularmodes), J. M. Wallace presented a clear and highly informative summary of the interpretation of the AO and AAO or, generically, the annular modes. One of the several points made in the presentation was that both the identified oscillations are "dynamical twins."

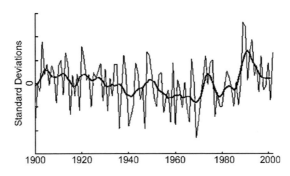

Figure 2.4 A representation of the positive and negative phases of the Arctic Oscillation.

The AO is an oscillation in which atmospheric pressure, at polar and mid-latitude locations, fluctuates between defined positive and negative phases (Thompson and Wallace 1998). It is computed as an index by comparing pressure in the polar region with pressure at 45° N. The negative phase of the index occurs when higher than normal pressure is found over the polar regions and lower than normal at 45° N. Opposite conditions occur in its positive phase. In recent years, the AO has been in a mostly positive phase (Figure 2.4). Shown in the figure are standard deviations from the January–March means of the standardized daily index (www.atmos.colostate.edu). The higher than normal pressure readings in mid-latitudes with lower values in polar regions during this phase have a number of results. Cold outbursts of polar and arctic air are less extensive over North America causing much of the United States east of the Rockies to experience warmer winters. At the same time, cyclonic storms are steered on a more northerly track bringing wetter winters to northwest Europe and Alaska. This more northerly track means that the summer-dry climates of California and Iberia experience less winter precipitation.

The AAO is represented by an oscillation in values of mid- and high-latitude surface pressure systems in the Southern Hemisphere. It is quantified by the Antarctic Oscillation Index (AAOI) which has been presented in a number of ways. In one the monthly zonal mean sea level pressure at 45° S in relation to the same value calculated for 65° S provides the value; in another the EOF of sea level pressure (850 mb height) at 20° S in relation to that at the South Pole is used. The AAO has much in common with the AO pattern in that there is a polar low surrounded by a high-pressure belt to give a dipole pattern.

These annular modes, especially that of the AO, have been used and expressed in other ways. The pattern of the Northern Hemisphere annular mode has been given by the well-known zonal indices that represent the position and strength of the zonal circulation. Additionally, and as noted by Wallace, it is suggested that the AO and the NAO are synonyms and are different names for the same variability, not different patterns of variability. The difference in the use of terms depends upon whether the variability is represented as a regional pattern controlled by the Atlantic sector processes or as an annular mode with strong teleconnections in the Atlantic sector. As noted, the AO is also a representation of what earlier researchers called the Northern Hemisphere Index Cycle.

Further details about the AO and AAO and their impacts are included in Chapter 5. As more data and information become available the role of these oscillations will certainly provide the basis for much continuing research and potentially an area for scientific debate.

2.8 ESSAY: ENSO and related teleconnections

Robert Allan, *Hadley Centre*

2.8.1 Background

Of the natural fluctuations inherent in the global climate system, the El Niño–Southern Oscillation (ENSO) phenomenon is the best known and studied. Interactions between the oceans and the atmosphere across the Indo-Pacific region are at the heart of this irregular climatic feature. Climatic and environmental impacts that result from ENSO activity range from direct influences on countries within and surrounding the Indo-Pacific basin to more indirect near-global manifestations (Glantz 2001; Diaz *et al.* 2001; Trenberth *et al.* 2002; Allan 2003). The latter result from the dynamics of the planetary atmosphere, which often allows ENSO effects to propagate to higher latitudes in both hemispheres.

Scientific research has revealed much about the ENSO phenomenon, its extremes of El Niño and La Niña events, the essential physical processes that power it, its tendency to be locked to the seasonal cycle, and a nature that embodies events which range in magnitude, spatial extent, duration, onset, and cessation (see the overviews of Philander 1992; Allan *et al.* 1996; Glantz 2001). No two individual El Niño or La Niña events are exactly alike and there is still much to be understood about ENSO (Larkin and Harrison 2002).

The structure of the ENSO phenomenon is also shaped by the interplay between its quasi-biennial (QB) (2–2.5 year) and LF interannual (2.5–7 year) components (Allan *et al.* 1996; Ribera and Mann 2003). However, what is now emerging is that natural decadal-multidecadal modes in the climate system (Navarra 1999) can provide lower frequency modulations of ENSO characteristics. A quasi-decadal signal centering on 9–13 years is seen as the cause of "protracted" El Niño and La Niña episodes (Allan *et al.* 2003), while the PDO/IPO (Mantua and Hare 2002) appears to modulate ENSO through its influence on the LF component of the phenomenon. These influences are also manifest through the changes they cause in the nature of the teleconnection patterns linked to ENSO activity. Such interactions also raise the question: is the quasi-decadal signal a separate "ENSO-like" fluctuation that interacts with QB and LF ENSO signals, or is it an integral part of the ENSO phenomenon itself?

Finally, many recent studies have focused on potential modulations of ENSO by global warming resulting from anthropogenic changes to the climate system. This emphasis has been complicated further, because natural low-frequency climate variability must be seen against a background of global warming, which itself may influence natural variability or affect how that variability is observed or interpreted.

2.8.2 ENSO and low-frequency "ENSO-like" climatic variability in mean sea level pressure, surface temperature, and precipitation

The strength and state of the atmospheric component of the ENSO phenomenon have long been monitored by the Southern Oscillion Index (SOI) (see Allan *et al.* 1996 for a history of Southern Oscillation measures and indices). In its contemporary form, the SOI is derived from the normalized difference in monthly atmospheric pressure between Tahiti and Darwin. In Figure 2.5, monthly SOI values from 1866 to 2002 are smoothed with an 11-point running mean, which reveals that El Niño and La Niña events vary in magnitude, onset and cessation times, and in duration. This characteristic is observed in all historical indices of ENSO (e.g. El Niño 3, 3.4 and 4 region sea surface temperatures), indicating that the phenomenon encapsulates a "family" of events.

A more objective way to analyse the nature of ENSO is through the use of spectral and signal detection techniques, such as multi-taper method singular value decomposition (MTM-SVD) (Mann and Park 1999). Figure 2.6 shows the joint local fractional variance (LFV) spectrum generated by an MTM-SVD analysis of surface temperature and mean sea level pressure (MSLP). The temperature fields are based on Hadley Centre monthly gridded sea surface temperatures (HadSST) and the Climatic Research Unit's land surface air temperatures (Jones *et al.* 2001), which together form the HadCRUT data set. MSLP fields are based on the latest version of the Hadley Centre HadSLP data set derived from the initial work of Basnett and Parker (1997). The temperature data are variance-corrected (Jones *et al.* 2001), and gaps in both sets of fields are filled using reduced space optimum interpolation (OI) (Kaplan *et al.* 1997) to give full $65°$N–$35°$S coverage from 1871 to 1998. The resulting surface temperature and MSLP fields are designated as HadCRUTv(OI) and HadSLP(OI).

Figure 2.6 reveals not only QB and LF climatic signals associated with the "classical" interannual ENSO phenomenon over the Indo-Pacific basin, but also the presence of significant decadal-multidecadal signals in the climate system which could influence and modulate ENSO (White and Tourre 2003). Of these low-frequency fluctuations, a quasi-decadal signal operating around

Figure 2.5 Monthly Southern Oscillation Index (SOI) values from January 1866 until September 2002, smoothed with an 11-point running average (except for end points). "El Niño-like" conditions are negative and "La Niña-like" conditions are positive.

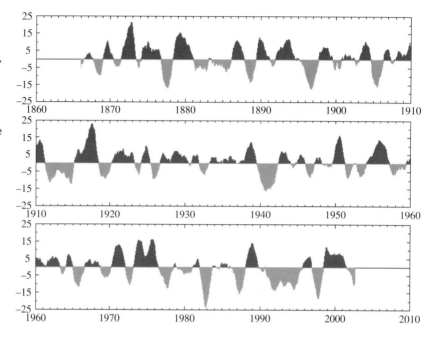

Figure 2.6 Multi-taper frequency-domain singular value decomposition (MTM-SVD) localised variance spectrum (LFV) from a joint analysis of HadCRUTv (OI) and HadSLP (OI) from 1871 to 1998 (relative variance is explained by the first eigenvalue of the SVD as a function of frequency) over the domain 65° N–35° S. The 50, 90 and 99% statistical confidence limits are shown as horizontal lines, and various significant climatic features in the spectrum are pointed out on the diagram.

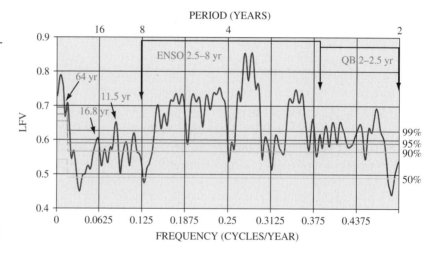

9–13 years is most prominent with another multidecadal peak centered at around 15–20 years. The latter, probably indicative of the PDO/IPO phenomenon, is detailed in the next section. The use of an evolutive LFV spectral option in the MTM-SVD technique permits an analysis of significant climatic phenomena from one epoch to the next. This approach highlights the distinct waxing and waning of QB, LF and quasi-decadal signals in the climate system over the historical record (Allan 2000; White and Tourre 2003). Interestingly, this tendency is manifest across all of the climatic signals examined, and embraces a range of permutations in which robust and weak phases in these signals can be coherent or incoherent at any time. Similar results can be achieved using wavelet analysis methods (Torrence and Webster 1999).

The MTM-SVD signal detection method can also be employed to generate canonical patterns that reveal the spatial structure and temporal evolution of each dominant signal resolved by the LFV spectrum in Figure 2.6. This can be done for any one data field or in a joint mode using two data fields. Experiments using combinations of surface temperature, MSLP, and precipitation data in the joint MTM-SVD mode have shown that they resolve very similar LFV spectral peaks, and thus joint canonical patterns can be derived for each peak. As the precipitation set used in this study is a monthly global land and island rainfall data set, derived originally from Hulme (1992) and updated (1900–1998) on the Climatic Research Unit (CRU), University of East Anglia's website (www.cru.uea.ac.uk/~mikeh/datasets/global/), our joint MTM-SVD canonical analysis results are derived from the 1900 to 1998 period. The initial focus here is on the QB, LF and quasi-decadal signals, followed by an examination of the "ENSO-like" PDO/IPO and its apparent modulation of LF ENSO events (see Bove and O'Brien 2000).

Canonical patterns of surface temperature, MSLP, and precipitation are constructed at QB, LF and quasi-decadal time scales from the joint MTM-SVD analysis, and are shown at 0, 45 and 90 degrees of phase in their evolution (Figures 2.7 to 2.9 (Plates 1 to 3)). Being a canonical pattern, one can complete the sequence to a full half cycle visualizing the 135 degree phase diagrams being in quadrature to the 45 degree panels, and the 180 degree phase diagrams being the complete opposite to the 0 degree panels. As noted in studies cited earlier, the QB and LF signals (Figures 2.7 (Plate 1) and 2.8 (Plate 2)) are dominated by the evolution of warm (El Niño) and cool (La Niña) sea surface temperatures (SSTs) in tongue-like structures extending from the west coast of South America across the equatorial Pacific to the dateline. Both QB and LF ENSO surface temperature sequences also support the findings in studies such as White and Tourre (2003), which indicate that such sequences also involve a significant modulation of the evolution of Indian Ocean SST patterns. Over land in the Northern Hemisphere, surface

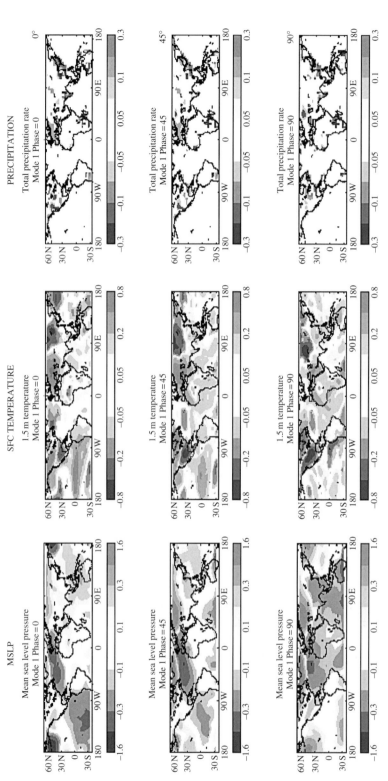

Figure 2.7 Canonical patterns of the spatiotemporal evolution of co-varying 2.2 year LFV peak quasi-biennial (QB) MSLP, surface temperature, and precipitation signals at 0, 45 and 90 degrees of phase over the domain 65° N to 35° S from an MTM-SVD analysis using data from 1900 to 1998. MSLP, surface temperature, and precipitation weights are shown as being redder for positive, and bluer for negative values. For color version see Plate 1.

temperatures on the QB time scale are dominated by a zonal gradient at middle–higher latitudes between Europe and Asia, and a distinctly meridional gradient over the North American continent. An element of the meridional gradient in surface temperatures across North America is still evident in the LF ENSO sequence, but the zonal pattern across Europe and Asia has been replaced by coherent middle–higher latitude warming or cooling across higher latitude Eurasia.

The MSLP evolution of both the QB and LF ENSO signals in Figures 2.7 (Plate 1) and 2.8 (Plate 2) is dominated by the waxing and waning of a distinct SO pattern, with a pronounced Indo-Pacific focus. The SO pattern is strongest and most coherent in the LF ENSO sequence. Higher-latitude MSLP nodes in the Pacific sector are a feature of both QB and LF ENSO signals. A pronounced NAO signal is seen in the QB sequence of MSLP, but any high-latitude AO response that might also be part of this pattern cannot be discerned with the termination of data at $65°$ N (Hurrell *et al*. 2003). The LF ENSO sequence contains a westwardly displaced North Atlantic MSLP feature with perhaps some NAO signatures. Unfortunately, with the termination of the data at $35°$ S, nothing can be said about high latitude Southern Hemisphere phenomena such as the AAO or the Antarctic Circumpolar Wave (ACW) (Simmonds 2003). Linkages between the ACW and ENSO are being explored.

Precipitation responses in the QB and LF ENSO sequences (Figures 2.7 (Plate 1) and 2.8 (Plate 2)) reveal patterns indicative of those defined in numerous studies. However, the sequences display much more information about the evolution and extent of ENSO influences. While "classical" ENSO-induced precipitation impacts are seen over Australia, India, China, Southern Africa, and North America in both sequences, there is evidence of a more widespread influence on European climate, particularly in the QB sequence. This European response has been seen previously in relationships of global rainfall to phases of the SO (Stone *et al*. 1996), and in the specific studies by Fraedrich *et al*. (1992), and Mariotti *et al*. (2002). Over eastern China and Japan, the precipitation signal in both QB and LF ENSO sequences shows more of a wider spatial response than is seen in "classical" studies. As in Allan (2000), and other studies, the LF ENSO precipitation pattern also resolves something of a moderate impact on West Africa, with implications for Sahelian rainfall.

In line with the findings of White and Tourre (2003), the quasi-decadal signal in MSLP and surface temperature (Figure 2.9 (Plate 3)) shows both similarities and differences to the QB and LF ENSO structures (Figures 2.8 (Plate 2) and 2.9 (Plate 3)). SST responses have more of a central to western equatorial Pacific focus with strong links to higher latitudes along the west coast of North America extending to Alaska on quasi-decadal time scales. This Pacific pattern peaks during the mature stage of the quasi-decadal signal

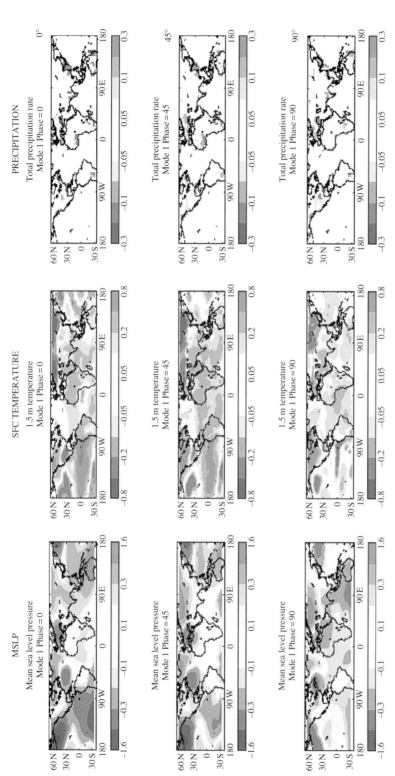

Figure 2.8 Canonical patterns of the spatiotemporal evolution of co-varying 3.6 year LFV peak LF ENSO MSLP, surface temperature, and precipitation signals at 0, 45 and 90 degrees of phase over the domain 65° N to 35° S from an MTM-SVD analysis using data from 1900 to 1998. MSLP, surface temperature, and precipitation weights are shown as being redder for positive, and bluer for negative values. For color version see Plate 2.

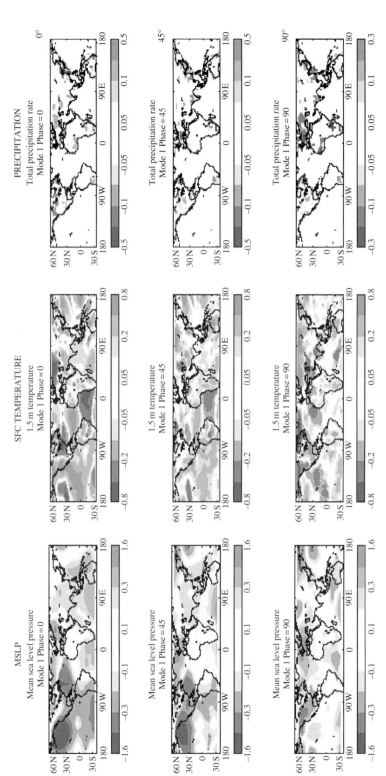

Figure 2.9 Canonical patterns of the spatiotemporal evolution of co-varying 11.6 year LFV peak quasi-decadal MSLP, surface temperature, and precipitation signals at 0, 45 and 90 degrees of phase over the domain 65° N to 35° S from an MTM-SVD analysis using data from 1900 to 1998. MSLP, surface temperature, and precipitation weights are shown as being redder for positive, and bluer for negative values. For color version see Plate 3.

(Figure 2.9 (Plate 3) zero phase). The latter signal also has a stronger imprint on SSTs over the Atlantic sector, as detailed in Chang *et al.* (1997), and additional impacts on Indian Ocean SST structures. The presence of strong, out-of-phase tropical and South Atlantic SSTs is seen only in this sequence. The Eurasian surface temperature pattern is again one of higher latitude responses as seen with the LF ENSO sequence, but is not as clearly defined. Over North America there is evidence of a zonal, rather than a meridional, surface temperature gradient.

In the quasi-decadal band, the MSLP pattern in Figure 2.9 (Plate 3) is suggestive of a westwardly displaced NAO-like structure in the North Atlantic similar to that seen in the LF ENSO sequence in Figure 2.8 (Plate 2). Overall, the SO pattern is much less clearly defined than in the QB or LF ENSO sequences. A strong high-latitude North Pacific MSLP node is seen to dominate the Pacific sector, with poor definition of the "classical" southeastern Pacific node of the SO. There is also evidence of a distinct MSLP node in the South Atlantic, which is not seen in the other sequences of higher frequency signals.

Precipitation patterns in the quasi-decadal sequence (Figure 2.9 (Plate 3)) show marked responses over central-western states of the USA, Labrador in eastern Canada, the Caribbean, southern South America, the UK, western to central Europe, China and Japan, the Indian subcontinent, Sahelian Africa, and central to southern Africa. This encompasses regions experiencing both QB and LF ENSO impacts, and areas where no distinct precipitation responses to ENSO are found. Taken together, MSLP, surface temperature, and precipitation responses suggest that there is a pattern of teleconnections associated with the quasi-decadal signal that has some variations to those linked to QB and LF ENSO events.

The MSLP and surface temperature signals display the type of global propagation of co-varying features that are detailed in White and Tourre (2003). By the very nature of its lower frequency, the quasi-decadal sequence (Figure 2.9 (Plate 3)) evolves more slowly than the LF ENSO sequence (Figure 2.8 (Plate 2)), and in turn this signal evolves more slowly than the QB sequence (Figure 2.7 (Plate 1)). However, there are still questions as to whether this is a real wave-like feature or simply reflects differing responses of each ocean basin to QB, LF, and quasi-decadal signals. These concerns now need to be seen in the light of the evolution and apparent coherent propagation of co-varying MSLP and surface temperature features over both the oceans and land masses.

In the QB sequence (Figure 2.7 (Plate 1)), a coherent MSLP signal evolves and propagates from the southern Pacific into the North Atlantic between the 0 and 90 degree phase snapshots. This feature then forms part of a distinct NAO pattern with an underlying SST tripole pattern in the North Atlantic Ocean as seen in Rodwell and Folland (2002), and other studies. Over the

Indian Ocean, a distinct QB MSLP signal can be seen to evolve and propagate in a northeasterly direction, again with indications of dynamical links to the underlying SST structure. As in White and Tourre (2003), these MSLP structures occur in conjunction with a slow eastward propagation of co-varying surface temperature features, with the main interruption being found in the central to eastern tropical Pacific, where they bifurcate around the tongue of surface temperature of the opposite sign indicative of equatorial Pacific ENSO activity.

A distinct LF ENSO MSLP propagation from the southern Pacific to North Atlantic is again evident in Figure 2.8 (Plate 2), but the signal does not propagate beyond the western North Atlantic in the 90 degree phase diagram. A distinct MSLP signal propagating from the southern Indian Ocean across Australasia and into the northwestern Pacific is also seen again in the LF ENSO sequence. As with the QB surface temperature sequence, the LF sequence displays an eastward traveling signal and its bifurcation around the "classical" warm and cold SST tongues covering the central to eastern tropical Pacific.

The slowest propagation of signals is found in the quasi-decadal sequence, with both the apparent movement of MSLP features from the Pacific to North Atlantic and Indian Ocean across Australasia again being evident. The eastward movement of surface temperature signals is less organized, and the equatorial Pacific tongue structures seen in the QB and LF ENSO sequences have been replaced by more of a tropical central Pacific response linked to higher latitudes along the western coast of North America.

2.8.3 PDO/IPO modes and patterns

Two climatic phenomena operating on inter- to multidecadal time frames with distinct implications for ENSO have received particular attention in recent times, the PDO, defined over the North Pacific by Mantua and Hare (2002), and the IPO, defined over the full Pacific basin and examined in a coupled model by Arblaster et al. (2002). Proxy data have also been used to provide a long temporal record in which to investigate the PDO and its characteristics (Cole et al. 2000; Villalba et al. 2001). The patterns of global SSTs linked to both the PDO and IPO are shown in Figure 2.10a (Plate 4a).

According to the recent work of Folland et al. (2002) the PDO and the IPO appear to be essentially the same, and display signatures at several frequencies (Mantua and Hare 2002) suggestive of a phenomenon ranging across a broad frequency band. Using near-global observational data, Allan (2000) resolved the IPO as occurring in a frequency band of around 20–30 years, producing a close match to the IPO pattern and time series (Arblaster et al. 2002) but also detected an interdecadal signal operating at around 15–20 years (as seen in Figure 2.6). Focusing on the Pacific basin, the lowest frequency signal examined by Tourre et al. (2001) is of a climatic feature manifest in the 12–25 year band with PDO/IPO characteristics. The recent analysis of White and Tourre (2003) concerning QB to interdecadal signals over the global oceans resolved a signal in the 14–22 year band (peaking at 16.7 years in an MTM-SVD LFV spectrum), which they saw as indicative of the PDO/IPO phenomenon. Other work has pointed to the phenomenon being manifest across an even broader frequency range (approximately 20–60 years). If the MTM-SVD LFV spectrum analysis used to construct Figure 2.6 is applied to the data for the 1900 to 1998 period it resolves a significant 32-year PDO/IPO signal, which is more consistent with the temporal nature of the phenomenon.

Studies such as Bove and O'Brien (2000) and Meinke *et al.* (2001) have produced evidence for the modulation of ENSO events, and their climatic impacts, by the PDO/IPO phenomenon over North America, Australia, and Europe. More recently this has been extended into other fields, with studies of specific oceanic and hydrological responses to PDO/IPO extremes (Hildalgo and Dracup 2003). Physically, the PDO/IPO SST pattern (Figure 2.10a (Plate 4a)) has greater SST variance in the northwestern Pacific than in the tropical central to eastern Pacific. In contrast to the dominant phases with zonal equatorial Pacific SST gradients in the QB, LF ENSO or quasi-decadal sequences (Figures 2.7 to 2.9 (Plates 1 to 3)), the peak PDO/IPO SST pattern in Figure 2.10a (Plate 4a) is marked by SSTs of the same sign across the tropical Pacific. It has been suggested that this PDO/IPO SST pattern could change the distribution of diabatic heating, leading to more (less) rainfall in the eastern margins of the west Pacific during the "El Niño-like" ("La Niña-like") phase of the phenomenon. LF ENSO events of either phase occurring during such extremes of the PDO/IPO would undergo modulations of their characteristics and related teleconnection patterns. An example of this is given in Figure 2.10b (Plate 4b) (after the work of Meinke *et al.* 2001) where the probability of exceeding median October to December precipitation, given El Niño conditions in the previous September, differs considerably in some regions of the globe in an "El Niño-like", as opposed to a "La Niña-like", PDO/IPO phase. In this case, western-central Europe and India show the most distinctly opposite precipitation response under contrasting PDO/IPO regimes. Studies have indicated that the QB component of the ENSO phenomenon (Allan 2000; Tourre *et al.*, 2001) may also be modulated by distinctly quasi-decadal fluctuations with "ENSO-like" characteristics that operate in the climate system.

2.8.4 Physical mechanisms

In studies such as Zhang *et al.* (1997), Allan (2000), and White and Tourre (2003), it has been shown that a number of discrete decadal to multidecadal

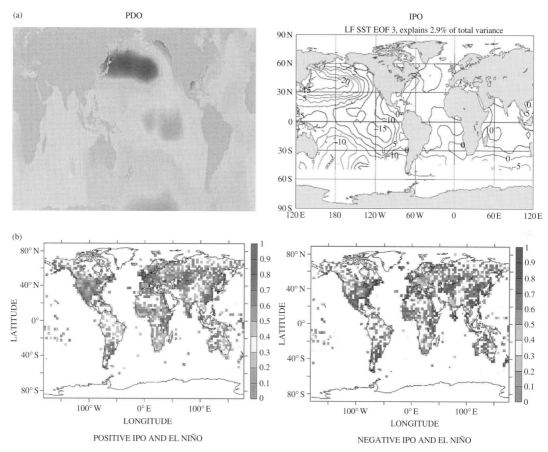

Figure 2.10   (a) Spatial pattern of near-global SSTs indicative of the PDO and IPO phenomena. The PDO diagram is from http://www.iphc.washington.edu/Staff/hare/html/decadal/decadal.html and the IPO panel is the second EOF pattern from the analysis of Folland *et al.* (1999). (b) Probability of exceeding median October–December rainfall over the globe during El Niño events with positive (left panel) and negative (right panel) IPO phase in the previous September. (Derived from the work of Meinke *et al.* 2001) For color version see Plate 4.

modes occur in the climate system. The major question about such modes is whether they arise from stochastic processes or are the result of distinct physical mechanisms. Recently White and Tourre (2003) have postulated that distinct physical processes indicative of the operation of the delayed oscillator mechanism that drives LF ENSO fluctuations, also power higher frequency QB and lower frequency quasi-decadal variability across the Pacific domain. These processes were found to generate progressively slower westward-propagating off-equatorial Rossby waves at increasingly higher latitudes for QB, LF ENSO, and quasi-decadal modes. These waves, along with eastward-propagating equatorial coupled waves generated by the above processes, were found to be primarily responsible for the overall periodicity

Figure 2.11 Schematic of evidence for delayed oscillator physics on biennial (QB), interannual (LF ENSO) and decadal (quasi-decadal) time scales in the Pacific basin. Shading designates Pacific warming for the El Niño or "El Niño-like" phase in all three signals, RW indicates off-equatorial Rossby waves. (From Warren White, JEDA, Scripps Institution of Oceanography, http://coaac.ucsd.edu/PROJECTS/index_current.html)

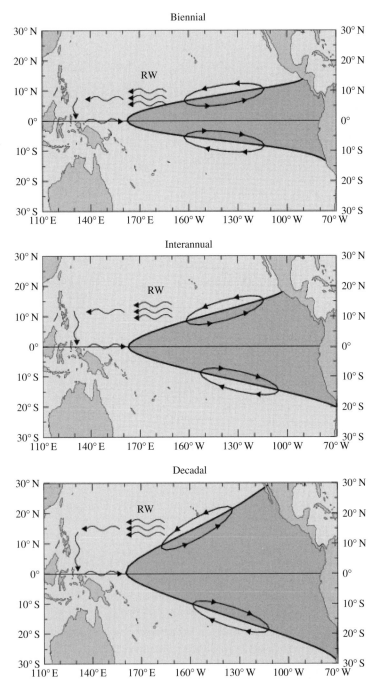

of the modes. A schematic of the basic elements of this mechanism in each band is shown in Figure 2.11. Interestingly, the delayed oscillator mechanism has been found to be the major driving force behind the LF ENSO and decadal "ENSO-like" features observed in the coupled model simulation of Vimont *et al.* (2002). However, there appears to be a temporal restraint on the effectiveness of the negative oceanic feedback at the western boundary of the Pacific in the above model as one goes to lower than quasi-decadal frequencies. Tourre *et al.* (2001) and White *et al.* (2004) have also found evidence that climatic modes occurring at interdecadal time scales could not be generated successfully by delayed oscillator type mechanisms and appear to be the result of advective processes. Examinations of the dynamical solutions to the shallow-water equations on an equatorial beta plane by Jin (2001) confirm the basic findings above, in that delayed oscillator physics were seen to generate realistic QB, LF ENSO and decadal modes.

Given the above, it is not surprising that Mantua and Hare (2002) note that the physical mechanisms underlying the even lower frequency PDO/IPO phenomenon are not clear. Some studies have suggested that "re-emergence" type mechanisms, in which SST anomalies force a seasonal coupling between anomalies in mid-latitude North Pacific winter atmospheric circulation and the following summer equatorial Pacific wind field (Vimont *et al.* 2001), could power such variability. Others see the possibility that advective processes may be the driving force behind this climatic fluctuation (Gu and Philander 1997). An attempt to probe more deeply into the physical nature of the PDO feature is given in Schneider *et al.* (2002).

Allan (2000) has provided evidence of an even lower frequency multidecadal signal in the climate system, which has been equated with a 50–80 year interhemispheric thermal contrast mode thought to be responsible for Sahelian droughts. Further investigations of this mode using paleoenvironmental reconstructions, observational data, and climate model simulations suggest that it may be linked to fluctuations in the global thermohaline circulation (THC) (Delworth and Mann 2000). However, the frequency range defining the THC is a function of the analysis techniques and parameters being used to examine it, and many questions concerning this multidecadal signal have yet to be addressed. For instance, there is some debate as to whether it is a wider manifestation of the Atlantic Multidecadal Oscillation (AMO) recently detailed in Enfield *et al.* (2001), or a distinctly different fluctuation.

2.8.5 "Protracted" ENSO episodes

Figures 2.7 to 2.9 (Plates 1 to 3) also provide an important insight into the nature of so-called "protracted" ENSO episodes (Allan and D'Arrigo 1999; Allan *et al.* 2003), in that QB, LF, and quasi-decadal signal patterns evolve at different rates, and the combination of their various magnitudes and phases has a profound effect on individual episodes. Interactions between these signals can generate a large number of possible climatic sequences. Thus, "protracted" ENSO episodes not only vary in overall duration but also have the potential to wax and wane in magnitude within the lifetime of each individual episode as QB, LF, and quasi-decadal signals interact together. In "protracted" episodes, this is observed as fluctuations between intervals of lower frequency "ENSO-like" and more distinctly higher frequency QB or LF ENSO characteristics. However, in the real world, the interplay between such signals would not occur simply through linear interactions. So there is a need for "protracted" ENSO episode dynamics to be explored further with more sophisticated techniques and using climate models.

Thus, some aspects of "protracted" El Niño and La Niña episodes are not captured in any simple combination of the canonical sequences shown in Figures 2.7 to 2.9 (Plates 1 to 3). As has been detailed in many studies of both individual "classical" LF ENSO events or composites of events, the El Niño and La Niña components of the phenomenon are not exact mirror images of one another. This is also evident when comparing and contrasting composites of variables indicative of "protracted" El Niño and La Niña episodes. In addition, composites also suggest that the quasi-decadal signal may tend to impose more of a broader Pacific SST signature on "protracted" La Niña than on "protracted" El Niño episodes. Even more revealing are the studies of individual episodes, such as the 1990–5 "protracted" El Niño and the 1998–2001 "protracted" La Niña, in Allan *et al.* (2003). Such research is giving new insights into the type of variations that can occur during the course of particular episodes, especially the tendency for El Niño and/or La Niña conditions to wax and wane as episodes evolve.

The wider impacts of "protracted" ENSO episodes are only just beginning to be addressed. Agrawala *et al.* (2001) and Barlow *et al.* (2002) have shown that the recent severe and widespread drought in central and southwestern Asia was a consequence of anomalous climatic patterns generated by the 1998–2001 "protracted" La Niña episode. Potential impacts of the quasi-decadal "ENSO-like" signal on the European sector have been suggested in recent analyses by Allan. Thus, it is apparent that an understanding of "protracted" ENSO episodes is crucial if they are to be incorporated into existing climate prediction schemes.

2.8.6 Global warming and ENSO

Concerns about possible modulations of the ENSO phenomenon by anthropogenic climate warming have been debated widely in the scientific literature. Such work has essentially focused on possible changes in the LF ENSO signal under global warming conditions in state-of-the-art global climate models, and from analyses of historical climate data (Meehl *et al.* 1993; Knutson and Manabe 1998; Collins 2000, Fedorov and Philander 2000; Trenberth *et al.* 2002). In part, this research focus has also been fuelled by evidence of a climate regime shift over at least the Pacific sector around 1976–7 (Gu and Philander 1995; McPhaden and Zhang 2002), and the occurrence of the 1990–5 "protracted" El Niño episode (Gu and Philander 1997; Webster and Palmer 1997).

Recent climatic events and studies have provided further insights into the global warming and ENSO question. The occurrence of the "protracted" La Niña episode of 1998–2001 (Allan *et al.* 2003) provided a counterbalance to concerns that global warming had caused a climatic regime change towards more "El Niño-like" conditions in the Pacific during the first half of the 1990s – a period now known to have been dominated by the 1990–5 "protracted" El Niño episode. A growing research focus on low-frequency climatic variability in general, with an emphasis on phenomena such as the PDO/IPO (Mantua and Hare 2002), and the AMO (Enfield *et al.* 2001), has indicated the importance of understanding these natural climatic fluctuations in any efforts seeking to isolate the global warming signature in the climate system. In fact, studies such as Bratcher and Giese (2002) suggest that natural decadal-multidecadal variability in the Pacific may have played the dominant role in the warming signatures observed to date. They also indicate that current Pacific climate conditions may be in a configuration which is opposite to that seen prior to the 1976–7 regime shift, and that a return to earlier conditions is imminent. Nevertheless, as noted by Lau and Weng (1999), natural and anthropogenic climate signals may well be inextricably bound up and difficult to unravel.

2.8.7 ENSO: summary

The ENSO phenomenon is a major component of the climate system. Through its dynamics and patterns of teleconnections, this climatic fluctuation is able to extend tropical Indo-Pacific influences to higher latitudes in both hemispheres, resulting in near-global climatic impacts. Recent research has indicated that "ENSO-like" fluctuations, operating on decadal to multidecadal time frames, also exist in the climate system and play a significant role in modulating climate. In fact, this may occur frequently through interactions between these "ENSO-like" features and the QB and interannual LF components of ENSO.

There is widespread evidence for the impacts of QB, LF ENSO, and quasi-decadal signals on climate, and their interactions to produce "protracted" ENSO episodes, all within an even broader envelope of lower frequency "ENSO-like" PDO/IPO and AMO climatic fluctuations. The picture emerging is one of interacting phenomena operating on QB to multidecadal time frames, with ENSO to "ENSO-like" characteristics. Not surprisingly, these climatic features display similarities and differences in their physical patterns and structures. Theoretical and modeling studies suggest that they are powered by dynamical processes inherent in the climate system, with the QB, LF ENSO, and quasi-decadal features all showing evidence of the operation of delayed action oscillator physics in their nature. However, the physics underlying lower frequency PDO/IPO and AMO fluctuations has not yet been explained.

Possible modulation of ENSO by the enhanced greenhouse effect remains an area of considerable debate in the scientific community. Although this debate will continue, there is growing evidence that natural "ENSO-like" variability on decadal to multidecadal time scales plays a very significant role in observed climatic patterns.

Acknowledgments

This essay is British Crown Copyright. I would like to thank David Parker, Hadley Centre, Met Office, for his comments and suggestions on the text.

2.9 Examples of oscillations and teleconnections websites

(Note: many significant sites are given in the text of this chapter)

The Climate Prediction Center is linked to www.nws.noaa.gov. Teleconnections provide the main topic and the site provides a comprehensive account of the indices used.

Interannual forecast/climate dynamics is located at www.cdc.noaa.gov/seasonal/dynamics.html. Details and links to many sites including ENSO, index values and monsoons are provided.

2.10 References

Agrawala, S., Barlow, M., Cullen, H. and Lyon, B., 2001. *The Drought and Humanitarian Crisis in Central and Southwest Asia: A Climate Perspective*. IRI Special Report 01-11, Lamont-Doherty Earth Observatory of Columbia University, Palisades, New York, 20pp.

Allan, R. J., 2000. ENSO and climatic variability in the last 150 years. In H. F. Diaz and V. Markgraf, eds., *El Niño and the Southern Oscillation: Multiscale Variability and its Impacts on Natural Ecosystems and Society*. Cambridge: Cambridge University Press, pp. 3–55.

2003. El Niño – A world perspective and what it means in Australia. In P. Attiwell and B. Wilson, eds., *Ecology: An Australian Perspective*. Melbourne, Australia: Oxford University Press.

Allan, R. J. and D'Arrigo, R. D., 1999. 'Persistent' ENSO Sequences: How unusual was the 1990–1995 El Niño? *Holocene*, **9**, 101–118.

Allan, R. J., Lindesay, J. A. and Reason, C. J. C., 1996. Multidecadal variability in the climate system over the Indian Ocean region during the austral summer. *Journal of Climate*, **8**, 1853–1873.

Allan, R. J., Reason, C. J. C., Lindesay, J. A. and Ansell T. J., 2003. 'Protracted' ENSO episodes and their impacts in the Indian Ocean region. *Deep-Sea Research II*. Special Issue on the Indian Ocean, **50**, 2331–2347.

Angell, J. K. and Korshover, J., 1964. Quasi-biennial variability in temperature, total ozone, and tropopause height. *Journal of Atmospheric Science*, **21**, 479–492.

Ångström, A., 1935. Teleconnections of climate changes in the present time. *Geografiska Analer*, **17**, 242–258.

Arblaster, J. M., Meehl, G. A. and Moore, A. M., 2002. Interdecadal modulation of Australian rainfall in the PCM. *Climate Dynamics*, **18**, 519–531.

Baldwin, M. P., Gray, L. J., Dunkerton, T. J., *et al.*, 2001. The quasi-biennial oscillation. *Reviews of Geophysics*, **39**, 179–229.

Barlow, M., Cullen, H. and Lyon, B., 2002. Drought in Central and Southwest Asia: La Niña, the warm pool, and Indian Ocean precipitation. *Journal of Climate*, **15**, 697–700.

Barry, R. G. and Carleton, A. M., 2001. *Synoptic and Dynamic Climatology*. London, New York: Routledge, Chapter 1.

Basnett, T. A. and Parker, D. E., 1997. *Development of the Global Mean Sea Level Pressure Data Set GMSLP2*. Climate Research Technical Note CRTN79, Hadley Centre, Meteorological Office, Bracknell, UK, 16pp.

Bjerknes, J., 1966. A possible response of the atmospheric Hadley circulation to equatorial anomalies of ocean temperature. *Tellus*, **8**, 820–829.

Bove, M. C. and O'Brien, J. J., 2000. *PDO Modification of US ENSO Climate Impacts*. COAPS Tech. Rep. **00–03**, 103pp. (Available from Center for Ocean-Atmospheric Prediction Studies, The Florida State University, Tallahassee, FL 32306–2840, USA.)

Bratcher, A. J. and Giese, B. S., 2002. Tropical Pacific decadal variability and global warming. *Geophysical Research Letters*, **29**, 1918, doi:10.1029/2002GL015191.

Chang, P., Ji, L. and Li, H., 1997. A decadal climate variation in the tropical Atlantic Ocean from thermodynamic air-sea interactions. *Nature*, **385**, 516–518.

Cole, J. E., Dunbar, R. B., McClanahan, T. R. and Muthiga, N. A., 2000. Tropical Pacific forcing of decadal SST variability in the western Indian Ocean over the past two centuries. *Science*, **287**, 617–619.

Coleman, J. S. M. and Rogers, J. C., 1995. Ohio River Valley winter moisture condition associated with the Pacific-North American teleconnection pattern. *Journal of Climate*, **16**, 969–981.

Collins, M., 2000. The El Niño-Southern Oscillation in the second Hadley Centre coupled model and its response to greenhouse warming. *Journal of Climate*, **13**, 1299–1312.

Delworth, T. L. and Mann, M. E., 2000. Observed and simulated multidecadal variability in the Northern Hemisphere. *Climate Dynamics*, **16**, 661–676.

Diaz, H. F., Hoerling, M. P. and Eischeid, J. K., 2001. ENSO variability, teleconnections and climate change. *International Journal of Climatology*, **21**, 1845–1862.

Enfield, D. B., Mestas-Nuñez, A. M. and Trimble, P. J., 2001. The Atlantic multidecadal oscillation and its relation to rainfall and river flows in the continental U.S. *Geophysical Research Letters*, **28**, 2077–2080.

Fairbridge R. W., 1986. Oscillations. In J. E. Oliver and R. W. Fairbridge, eds., *The Encyclopedia of Climatology*. New York: Van Nostrand Reinhold.

Fedorov, A. V. and Philander, S. G., 2000. Is El Niño changing? *Science*, **288**, 1997.

Folland, C. K., Parker, D. E., Colman, A. and Washington, R., 1999. Large scale modes of ocean surface temperature since the late nineteenth century. In A. Navarra, ed., *Beyond El Nino: Decadal and Interdecadal Climate Variability*. Berlin: Springer-Verlag, p. 374.

Folland, C. K., Renwick, J. A., Salinger, M. J. and Mullan, A. B., 2002. Relative influences of the interdecadal Pacific Oscillation and ENSO on the South Pacific Convergence Zone. *Geophysical Research Letters*, **29**(13), 211–214.

Fraedrich, K., Muller, K. and Kuglin, R., 1992. Northern Hemisphere circulation regimes during the extremes of the El Niño/Southern Oscillation. *Tellus*, **44A**, 33–40.

Glantz, M. H., 2001. *Currents of Change: El Niño and La Niña Impacts on Climate and Society*, 2nd edn. Cambridge: Cambridge University Press.

Gu, D. and Philander, S. G. H., 1995. Secular changes of annual and interannual variability in the tropics during the past century. *Journal of Climate*, **8**, 864–876.

    1997. Interdecadal climate fluctuations that depend on exchanges between the tropics and extratropics. *Science*, **275**, 805–807.

Hildalgo, H. G. and Dracup, J. A., 2003. ENSO and PDO effects on hydroclimatic variations of the Upper Colorado River Basin. *Journal of Hydrometeorology*, **4**, 5–23.

Hulme, M., 1992. A 1951–80 global land precipitation climatology for the evaluation of General Circulation Models. *Climate Dynamics*, **7**, 57–72.

Hurrell, J. W., 1995. Decadal trends in the North Atlantic Oscillation and relationships to regional temperature and precipitation. *Science*, **269**, 676–679.

Hurrell, J. W., Kushnir, Y., Visbeck, M. and Ottersen, G., 2003. An overview of the North Atlantic Oscillation. In J. W. Hurrell, Y. Kushnir, G. Ottersen and M. Visbeck, eds., *The North Atlantic Oscillation: Climate Significance and Environmental Impact*, Geophysical Monograph Series, **134**, pp. 1–35.

Isard, S. A., 1999. Zones of origin for Great Lakes cyclones in North America, 1899–1996. *Monthly Weather Review*, **128**, 474–485.

Jin, F.-F., 2001. Low-frequency modes of tropical ocean dynamics. *Journal of Climate*, **14**, 3874–3881.

Jones, P. D., Jonsson, T. and Wheeler, D., 1997. Extension to the North Atlantic Oscillation using early instrumental pressure observations from Gibraltar and south-west Iceland. *International Journal of Climatology*, **17**, 1433–1450.

Jones, P. D., Osborn, T. J., Briffa, K. R., *et al.*, 2001. Adjusting for sampling density in grid-box land and ocean surface temperature time series. *Journal of Geophysical Research*, **106**, 3371–3380.

Kaplan, A., Kushnir, Y., Cane, M. A. and Blumenthal, M. D., 1997. Reduced space optimal interpolation for historical datasets: 136 years of Atlantic sea surface temperatures. *Journal of Geophysical Research*, **102**, 27,835–27,860.

Katz, R. W., 2002. Sir Gilbert Walker and a connection between El Niño and statistics. *Statistical Science*, **17**, 97–112.

Knutson, T. R. and Manabe, S., 1998. Model assessment of decadal variability and trends in the tropical Pacific Ocean. *Journal of Climate*, **11**, 2273–2296.

Larkin, N. K. and Harrison, D. E., 2002. ENSO warm (El Niño) and cold (La Niña) event life cycles: Ocean surface anomaly patterns, their symmetries, asymmetries, and implications. *Journal of Climate*, **15**, 1118–1140.

Latif, M. and Barnett, T. P., 1994. Causes of decadal climate variability over the North Pacific and North America. *Science*, **266**, 634–637.

Lau, K.-M. and Weng, H., 1999. Interannual, decadal-interdecadal, and global warming signals in sea surface temperature during 1955–97. *Journal of Climate*, **12**, 1257–1267.

Leathers, D. J., Yarnal, B. and Palecki, M. A., 1991a. The Pacific/North American teleconnection pattern and United States climate. Part I: Regional temperature and precipitation associations. *Journal of Climate*, **4**, 517–528.

1991b. The Pacific/North American teleconnection pattern and United States climate. Part II: Temporal characteristics and index specification. *Journal of Climate*, **4**, 707–716.

Lindzen, R. S., 1987. The development of the theory of the QBO (Personal recollections). *Bulletin of the American Meteorological Society*, **68**, 329–337.

Lindzen, R. S. and Holton, J. R., 1968. A theory of quasi-biennial oscillation. *Journal of Atmospheric Science, 26, 1095–1107.*

Malony E. D. and Hartmann, D. L., 2000. Modulation of hurricane activity in the Gulf of Mexico by the Madden-Julien Oscillation. *Science*, **284**, 2002–2004.

Mann, M. E. and Park, J., 1999. Oscillatory spatiotemporal signal detection in climate studies: A multiple-taper spectral domain approach. *Advances in Geophysics*, **41**, 1–131.

Mantua, N. J. and Hare, S. R., 2002. The Pacific Decadal Oscillation. *Journal of Oceanography*, **58**, 35–44.

Mantua, N. I., Hare, S. R., Zhang, Y., Wallace, I. M. and Francis, R. C., 1997. A Pacific decadal climate oscillation with impacts on salmon. *Bulletin of the American Meteorological Society*, **78**, 1069–1079.

Mariotti, A., Zeng, N. and Lau, K.-M., 2002. Euro-Mediterranean rainfall variability and ENSO. *CLIVAR Exchanges*, **7**, 3–5.

Maruyama, T., 1997. The Quasi-Biennial Oscillation (QBO) and equatorial waves – A historical review. *Meteorology and Geophysics*, **48**, 1–17.

McPhaden, M. J. and Zhang, D., 2002. Slowdown of the meridional overturning circulation in the upper Pacific Ocean. *Nature*, **415**, 603–608

Meehl, G. A., Branstator, G. W. and Washington, W. M., 1993. Tropical Pacific interannual variability and CO_2 climate change. *Journal of Climate*, **6**, 42–63.

Meinke, H., Power, S., Allan, R. and de Voil, P., 2001. *Can Decadal Climate Variability (DCV) be Predicted?* Project Reference Number **QPI44**, Final Report to Land & Water Australia, 31pp.

Minobe, S., 1997. A 50–70 year climatic oscillation over the North Pacific and North America. *Geophysical Research Letters*, **24**, 683–686.

Navarra A. (ed.), 1999. *Beyond El Niño: Decadal and Interdecadal Climate Variability.* Berlin: Springer-Verlag, 374pp.

Peixoto, J. P. and Oort, A. H., 1992. *Physics of Climate*. American Institute of Physics.

Philander, S. G. H., 1992. El Niño. *Oceanus*, **35**, 56–61.

Reed, R. J., Campbell, W. J., Rasmussen, L. A. and Rogers, D. G., 1961. Evidence of a downward-propagating annual wind reversal in the equatorial stratosphere. *Journal of Geophysical Research*, **66**, 813–818.

Ribera, P. and Mann, M. E., 2003. ENSO related variability in the Southern Hemisphere, 1948–2000. *Geophysical Research Letters*, **30** (1), 1006.

Rodwell, M. J. and Folland, C. K., 2002. Atlantic air-sea interaction and seasonal predictability. *Quarterly Journal Royal Meteorological Society*, **128**, 1413–1443.

Rogers, J. C., 1997. North Atlantic storm track variability and its association to the North Atlantic Oscillation and climate variability of Northern Europe. *Journal of Climate*, **10**, 1635–1647.

Schneider, N., Miller, A. J. and Pierce, D. W., 2002. Anatomy of North Pacific decadal variability. *Journal of Climate*, **15**, 586–605.

Simmonds, I., 2003. Modes of atmospheric variability over the Southern Ocean. *Journal of Geophysical Research*, **108**, 8078.

Stone, R. C., Hammer, G. L. and Marcussen, T., 1996. Predictions of global rainfall probabilities using the phases of the Southern Oscillation Index. *Nature*, **384**, 252–255.

Thompson, D. W. J. and Wallace, J. M., 1998. The Arctic Oscillation signature in the wintertime geopotential height and temperature fields. *Geophysical Research Letters*, **25**(9), 1297–1300.

Torrence, C. and Webster, P. J., 1999. Interdecadal changes in the ENSO-monsoon system. *Journal of Climate*, **12**, 2679–2690.

Tourre, Y. M., Rajagopalan, B., Kushnir, Y., Barlow, M. and White, W. B., 2001. Patterns of coherent decadal and interdecadal climate signals in the Pacific basin during the 20th century. *Geophysical Research Letters*, **28**, 2069–2072.

Trenberth, K. E., Caron, J. M., Stepaniak, D. P. and Worley, S., 2002. The evolution of ENSO and global atmospheric surface temperatures. *Journal of Geophysical Research*, **107**, 10.1029/2000JD000298.

Villalba, R., D'Arrigo, R. D., Cook, E., Wiles, G. and Jacoby, G., 2001. Decadal-scale climatic variability along the extra-tropical western coast of the Americas over past centuries inferred from tree-ring records. In V. Markgraf, ed., *Interhemispheric Climate Linkages*, Cambridge: Cambridge University Press, pp. 155–172.

Vimont, D. J., Battisti, D. S. and Hirst, A. C., 2001. Footprinting: A seasonal connection between the tropics and mid-latitudes. *Geophysical Research Letters*, **28**, 3923–3926.

Vimont, D. J., Battisti, D. S. and Hirst, A. C., 2002. Pacific interannual and interdecadal equatorial variability in a 1000-year simulation of the CSIRO coupled general circulation model. *Journal of Climate*, **15**, 160–178.

Walker, G. T., 1923–4. World weather, I and II. *Indian Meteorol. Dept. Mem.*, **24**(4), 9.

Wallace, J. M. and Gutzler, D. S., 1981. Teleconnections in the 500 mb geopotential height field during the Northern Hemisphere winter. *Monthly Weather Review*, **109**, 784–812.

Webster, P. J. and Palmer, T. N., 1997. The past and the future of El Niño. *Nature*, **390**, 562–564.

White, W. B. and Tourre, Y. M., 2003. Global SST/SLP modes/waves during the 20th century. *Geophysical Research Letters*, **30**, 1651.

White, W. B., Annis, J. L. and Allan, R. J., 2004. Modulation of global biennial and interannual climate signals by decadal and interdecadal signals during the 20th century. *Journal of Climate*, **17**, 3109–3124.

Zhang, Y., Wallace, J. M. and Battisti, D. S., 1997. ENSO-like interdecadal variability: 1900–93. *Journal of Climate*, **10**, 1004–1020.

Chapter 3
Tropical climates

3.1 Introduction

For many years, from early exploration to perhaps the middle of the twentieth century, the weather and climate of the tropical world were considered among the most easily explained of all world climate systems. The daily rainfall of the equatorial zone, the constancy of the trade winds, and the unrelenting aridity of the tropical deserts gave an impression of benign and unchanging conditions. Such is far from the case. Within the tropics are some of the most interesting and difficult to explain phenomena of the world's climates.

The Earth's tropical regions are, in terms of geographic location, the area between the Tropic of Cancer (23.5° N), and the Tropic of Capricorn (23.5° S). Such a definition is not suitable, however, for identification of the climatic regimes and over the years a number definitions have been suggested. In his classification of world climate in 1896, Supan proposed that locations with annual average temperature greater than 20 °C might be considered as the tropics. In his widely used 1918 classification, Köppen classed tropical climates as having the average temperature of each month greater than 18 °C. Using this criterion, the wet tropical climates occupy some 36% of the Earth's surface. If the tropical deserts are added to this class, then the tropics comprise almost 50% of the surface area of the world.

Additional thermal boundaries have been proposed by other authors, but in reality the tropical climates of the world may be considered as those latitudes that lie between, and partly include, the subtropical high-pressure regions that are centered at about 30°–35° N and S. The equatorial zone is a subset of the tropical regions, occurring perhaps 10° north and south of the equator.

3.2 The climatic controls

As with all Earth climates, the most basic control is the energy balance. Within the tropics, although varying slightly with the seasons, intensity of solar radiation is high all year and there is very little variation in the length of the day from one part of the year to the next. The photoperiod, the relative lengths of day and

Figure 3.1 The relationship between annual and diurnal energy cycles in the tropics. (After Oliver and Hidore 2002)

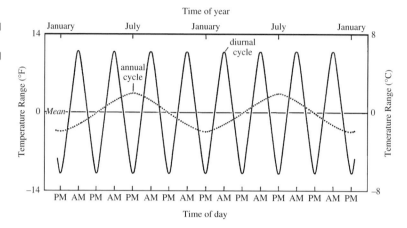

night to which plants must adjust, varies between 11 and 13 hours from winter to summer. Though solar radiation is relatively high all year, it is sometimes not as high as it is in mid-latitudes, particularly in summer. Cloud cover in wet equatorial lands usually reflects more than half of the incident total solar radiation.

In relation to the energy inputs, annual temperatures average about the same throughout the moist tropical regions and the annual range in temperature depends on the length of the dry season. Where there is no dry season, the annual range in mean monthly temperatures may be as little as one or two degrees. Where there is dry weather in the winter months the mean temperatures decrease to increase the annual temperature range.

In equatorial climates the primary energy flux is diurnal (Figure 3.1) and the variation in temperature from day to night is greater than the variation from season to season. Tropical regions are not places of continuous high temperatures. Nights can be rather cool so that the diurnal radiation and temperature cycles are more important than the annual temperature cycle as a regulator of life cycles.

Given these characteristics, it is not surprising that rainfall rather than temperature determines the seasons and it is the amount and timing of rainfall that form the chief criterion for distinguishing the various climates. The largest portion of tropical environments has a marked seasonal regime of rainfall that governs the biological productivity of the system. The remaining areas are deserts, where rainfall is incidental throughout the year. It is, in fact, the seasonal moisture pattern that distinguishes the major tropical environments – the rain forest, the savannas, and the desert – from each other.

Essentially, the dominant dynamic controls of tropical climates are the inter-tropical convergence zone (ITC or ITCZ) and the subtropical high pressure systems. A further aspect over large areas of the tropical world concerns the seasonal change of winds to give the monsoon climates.

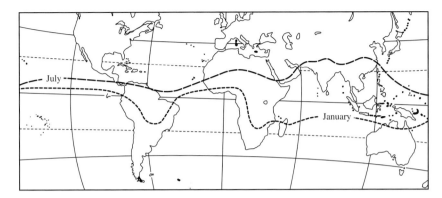

Figure 3.2 Average extreme positions of the ITCZ. (After Oliver and Hidore 2002)

3.2.1 The intertropical convergence zone

The intertropical convergence zone (ITCZ), an area once designated as the doldrums, is an east–west oriented low pressure region near the equator. It is characterized by low surface pressure, and convergence of air masses giving rise to cloudiness and rainfall. The ITCZ is best developed in the tropical oceans and is the most prominent climatic feature in the tropics. It plays principal roles in tropical climate by interacting with the planetary scale circulations of the atmosphere and oceans.

The position of the ITCZ varies seasonally (Figure 3.2), occurring in the hemisphere of the most intense solar radiation. As illustrated in the figure, the seasonal migration over land is greater than over the oceans. The feature is obliterated during the onshore monsoon period over the Indian Ocean. The seasonal migration of the ITCZ is of high significance to the amount of rainfall that occurs in the semiarid regions that bound the wet tropics. As indicated in the account of desertification in the Sahel (Section 3.4.1), failure of the ITCZ to migrate to its "normal" position can lead to drought with resulting social disruptions.

The ITCZ is not a simple band of clouds, and double ITCZs often occur. In this, an ITCZ is found on each side of the equator. In his review of this phenomenon, Zhang (2001) notes that the most noticeable double ITCZs are usually found over the eastern Pacific during spring while over the western and central Pacific double ITCZs occur from June to September. It is only during ENSO events that the eastern Pacific double ITCZ fails to materialize.

For most of the time, the ITCZ south of the equator is weaker than that which is north. This occurs when the northeast and southeast trade winds converge over the warm waters and intense convection results in the formation of clouds that produce heavy precipitation. The south ITCZ is formed when the southern trade winds blow over the cold upwelling waters near the equator; these decrease vertical mixing and slowing of surface winds (Liu and Xie 2002). The south ITCZ does not have the cloud creation capability of the north ITCZ, hence its weaker signal.

Figure 3.3 Models of tropical meridional circulation. (a) Hadley Cells on two side of the equator; (b) two equatorial cells separating the two Hadley Cells; (c) and (d) a single equatorial cell separating the Hadley Cells. (Based upon various sources including Asnani 1968)

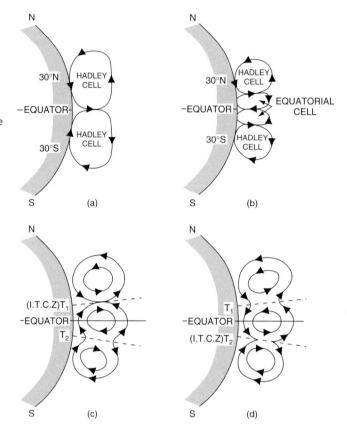

Over the years various models have been proposed to explain the ITCZ. It is commonly viewed as the ascending branch between the Hadley Cells (Figure 3.3a). Fletcher (1945) proposed a twin equatorial cell model (Figure 3.3b) while Asnani (1968) suggested another model with a single equatorial cell separating the two Hadley Cells (Figures 3.3c and 3.3d). In the single cell, the air flows toward the equator in the lower layers and flows away from the equator aloft, producing upward motion on one side and downward motion on the other. Near the equator, subsidence occurs.

There have also been numerous theories and numerical modeling studies for the explanation of the ITCZ. Charney (1971) considered that the position of the ITCZ depends on Ekman pumping efficiency and moisture availability. Schneider and Lindzen (1977) discovered that zonally symmetric convective heating was crucial in determining the location of the ITCZ. Tomas et al. (1999) stated that a zonally symmetric cross-equatorial pressure gradient would result in a convergent flow at the ITCZ latitudes.

Some modeling studies reveal that the ITCZ follows the SST maximum. Goswami et al. (1984) discovered that a steady ITCZ occurred over the SST maximum. Hess *et al.* (1993) and Numaguti (1993) found that the ITCZ location following SST maximum depended heavily on the cumulus parameterization used.

Numerical modeling has been used also to examine a variety of relationships. Included among the many studies are an examination of the relationship between the ITCZ and atmospheric internal dynamics (Waliser and Somerville 1994), energy transport by transient waves (Kirtman and Schneider 2000), and radiative convective instability (Raymond 2000). Needless to say, an appreciable amount of research continues.

Of particular interest to researchers are the oscillations that occur in the tropical realm. The Quasi-biennial Oscillation (QBO), the periodic reversal of winds in the lower tropical stratosphere, is discussed later in this chapter in the essay by Randall Cerveny (Section 3.3). Less has been written about the Madden–Julian Oscillation (MJO), a description of which is given in Background Box 2.1.

3.2.2 The South Pacific Convergence Zone

Extending SE from the ITCZ in Southeast Asia (New Guinea) to around $30°$ S, $120°$ W is a persistent convective cloud band called the South Pacific Convergence Zone (SPCZ). Details of the SPCZ are described in an excellent review by Vincent (1994). The SPCZ separates the mainly E–NE airflows on the western side of the South Pacific Subtropical High Pressure (STHP), from the SE airflows in the mid-latitudes. It is an area of low-level convergence, leading to moist air confluence and convection. The SPCZ is most evident, and active, in summer, as a result of the propagation of monsoon convective activity originating in India in the previous NH summer. Its persistence is also assisted by location over the western Pacific pool of warm SST. While the northern and western sections of the SPCZ are strongly tied to the ITCZ and tropical convection, evidence suggests the southern and eastern parts are instrumental in steering storm tracks in the central Pacific toward higher latitudes.

Fluctuations in the strength and location of the SPCZ are linked to several time scales. Aside from seasonal variations, the 30–60 day variations of the MJO (Section 2.5) are important influences. On an interannual basis, the primary influence is ENSO (Section 2.8). During an El Niño event, the SPCZ shifts to the northeast of its average location, and to the southwest during La Niña. The SPCZ is also affected by wave activity in the circumpolar vortex (CPV) (see Chapter 4). These variations have strong influences on the climate variations, especially precipitation and temperature, in New Zealand and the rest of the southwest Pacific.

3.2.3 The subtropical high pressure cells and trade winds

The subtropical high pressure systems are located between latitudes $20°$ and $40°$ in both hemispheres. Where not modified by continental climates, they are

characterized by light winds and fine, clear weather. These highs were sometimes known in times past as the horse latitudes or the Calms of Cancer and
Capricorn. According to the Oxford English Dictionary, the origin of the name
horse latitudes is uncertain. However, it has been suggested that sailing ships
carrying horses to the West Indies, if becalmed unduly in the Sargasso Sea,
occasionally had to jettison their live cargo as fodder ran out.

With an east–west oriented surface pressure system, the STHP systems
dominate the circulation patterns over the oceans. The location of these systems
varies throughout the year, being closer to the equator during the winter season
and most distant during the summer. The intensities of the STHP systems, as
represented by sea level pressures, are also variable, and may be declining in
some portions while strengthening in others (Jones 1991; Inoue and Bigg 1995).
They are also asymmetric, with the highest surface pressures in their eastern
portions. It is in these eastern areas that subsidence occurs; in the western
portions rising air is more common and it is here that precipitation is more likely
to occur. This is explained in part by the vertical structure of the STHP systems,
which is illustrated in Figure 3.4. Cores of maximum high pressure are at the
surface in the east, but at higher elevations on the western sides of the systems.

Figure 3.4 Isobaric patterns of subtropical high pressure cells at upper levels (top) and close to the surface (center). The lower figure shows a cross section of orbital movements. (After McGregor and Nieuwolt 1998)

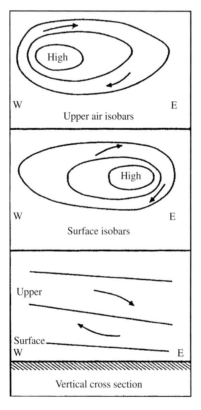

Upper air isobars

Surface isobars

Vertical cross section

The classical explanation for causation and location of the STHPs suggests that air rising at the equator flows poleward aloft and converges at about 20° to 30° N and S. At those latitudes, convergence of air aloft creates a high pressure at the surface. However, in order for the STHPs to be as persistent as they are, the winds aloft, the antitrades, would need to be constant. It just so happens that the antitrades are frequently absent, so that this process alone could not result in the STHPs.

Another potential cause relies upon the differential heating and cooling of air in the cells. In the poleward flows of the Hadley Cells, continuous longwave radiation to space causes upper air cooling. The more distant from the equator, the more the air will become cooler and denser. Following this, a higher pressure will occur at all levels of the system resulting in the creation of the surface high pressure systems. The cold water of cold currents on the eastern sides of the cells (and the cold air over continents, when the cells expand in winter) would enhance this effect.

The STHPs have also been attributed to middle-latitude dynamics. At the boundary between polar and tropical air (polar front), changes in Coriolis force would cause anticyclonic cells to move equatorwards while low pressure systems would move polewards. These cold equatorward-moving anticyclones would reinforce and regenerate the STHPs. At the very least, the infusion of the cold anticyclonic air would provide an explanation of short-term changes in intensity of the STHPs.

Clearly each of these explanations is insufficient in itself. McIlveen (1992, p. 417) expertly summarizes the interaction of these processes and his wording is used here. "During the development of a particular subtropical high-pressure cell, convergence aloft obviously exceeds divergence below. However, during long quiet periods in which the intensity of an anticyclone changes little, divergence and convergence must balance closely in a dynamic equilibrium. No simple argument explains the continuing presence of the high surface pressure, despite the unwise claims of some descriptive texts. For example it is not the direct consequence of the downward motion of air in the main mass of air, since this would imply a gross failure of the hydrostatic approximation, which actually must be especially accurate in the quiet conditions prevailing there. It seems that dynamic constraints such as the Coriolis effect maintain the excess of atmospheric mass in the central parts of the anticyclone, once it has been established by the initial excess of convergence. The relative warmth of subsiding air means that air pressure falls more slowly with increasing height than is the case in the surrounding cooler air masses, with the result that the upward doming of isobaric surfaces increases with height, often into the upper troposphere."

Easterly winds of the tropical oceans, both north and south of the equator, have long been known as trade winds. The word "trade" is probably derived from an obsolete form, meaning "track", and refers to the persistent direction of these winds. The trade winds are a highly significant part of global atmospheric circulation, controlling a large portion of the Earth's climate regions. The trade winds form the equatorward limb of the Hadley circulation in each hemisphere. They flow out of the subtropical high pressure cells as northeasterlies in the Northern Hemisphere, and southeasterlies in the Southern Hemisphere.

As they travel across the oceans, the trade winds accumulate latent and sensible heat. This is transported to the zone of the ITCZ where it contributes toward massive convection which, in turn, fuels the Hadley circulation. Over continents, the annual cycle of heating and cooling disrupts the patterns of the trades.

Throughout much of the trade wind regions, a persistent temperature inversion occurs. Known as the trade wind inversion, it is a highly stable layer that effectively caps vertical motion, and hence deep convection seldom occurs in the trade wind zone. The trade wind inversion is not the upper limit of the trades, but separates the well-mixed, humid marine air from arid upper air. In fact, four layers in the trade zone may be identified: the subcloud layer, cloud layer, inversion layer, and free atmosphere.

The climate of the trade wind zone is remarkably consistent, in large part due to the persistence of the trade winds and the trade wind inversion as illustrated by weather in Hawaii and other similarly positioned oceanic island groups (Giambelluca and Schroeder 1998). The inversion limits rainfall in the region near Hawaii to around 700 mm per year. However, because of orographic lifting of trade winds on the windward slopes, some spots in Hawaii receive spectacular amounts. Mount Waialeale on Kauai, one of the wettest places on Earth, averages around 11 m of rainfall annually.

3.2.4 The monsoons

In a later chapter the monsoons are used as an illustration of climatic modeling. Accordingly, at this juncture only basic characteristics of the monsoon climates will be discussed.

Monsoon, a term derived from an Arabic word meaning season, refers to the conditions in which the mean surface wind reverses its direction from summer to winter. Most regions lying between 35° N and 25° S and between 30° W and 170° E experience a monsoon effect. They are best developed in south and east Asia.

Generally, the summer monsoons of both the hemispheres are wet and winter monsoons are dry. The Asian summer monsoon consists of the Indian monsoon and the east Asian monsoon, both of which are responsible for abundant summer rainfall. West Africa also experiences a wind reversal from southwesterly in summer to northeasterly in winter. As a result, on the west coast of Africa the heaviest rainfall occurs from June to August.

The east Asian winter monsoon and the north Australian summer monsoon are intermingled because the dry winter air of the Northern Hemisphere flows across the equator towards the Southern Hemisphere continents, picking up moisture from the warm tropical oceans to become the wet monsoon over north Australia. The southwestern part of North America is also monsoonal. Here the surface zonal wind changes from an easterly in January to westerly in July and there is an increase in rainfall over large areas of southwestern North America and southern Mexico from June to July. Recent data suggest a summer monsoon circulation over the subtropical South American highlands with distinct rainfall summer increases over the central Andes and the southern parts and north coast of Brazil.

The complexities of the monsoon circulation systems, as already noted, are described in Chapter 9. Further excellent descriptions may be found in the works of Das (Das 1986, 2002).

3.3 ESSAY: The Quasi-biennial Oscillation and tropical climate variations

Randall Cerveny, *Arizona State University*

The Quasi-biennial Oscillation (QBO) is a tropical atmospheric phenomenon evidenced by a reversal in the stratospheric wind direction near the equator approximately every 28 months. It is an unusual phenomenon that researchers now postulate arises from an interaction of propagating waves in a rotating stratified atmosphere. The interactions of these waves produce oscillatory perturbations that fundamentally modify the extratropical tropospheric circulation and consequently their climates and that of the globe as a whole. A wide variety of climate records around the world demonstrate periodicities that appear to be well correlated with the QBO. With regard to this essay, much research has recently been published indicating that the QBO also exerts a strong influence on the tropics of the Earth – and even other planets! For a brief history and discussion of QBO observations and theory, please see the discussion in Chapter 2. In this essay, I address the impact of the QBO on climate variability in the tropics, including the controls on seasonal hurricane and tropical storm development, the Southern Oscillation and even the planetary length of day.

3.3.1 Planetary concerns

Although terrestrial observations of features associated with the QBO have been made for the last 120 years (dating back to the eruption of Krakatoa in the East Indies), only recently have scientists identified QBO-like phenomena on other planets, such as the planet Jupiter. Recent information from a Jovian space probe (Flasar *et al.* 2004) identified intense ($140\,\mathrm{m\,s^{-1}}$) high-altitude equatorial winds from measurements of the infrared spectra of Jupiter's stratosphere (Figure 3.5). This jet apparently has many of the

Figure 3.5 Observed zonal (east–west) winds over the planet Jupiter, averaged over longitude, as functions of pressure as the vertical coordinate and latitude as the horizontal. Winds are expressed in units of meters per second. (Adapted from Flasar *et al.* 2004)

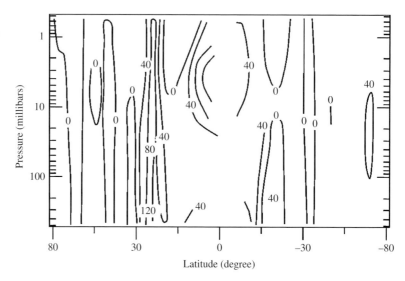

characteristics found in the terrestrial QBO. Additionally, as with terrestrial models, simulations of planetary atmospheres, such as that of Jupiter, have produced unstable zonal jets.

Strong linkages also exist between the QBO and solar activity. For example, while the QBO is primarily driven by gravity wave interaction, researchers identified as early as the 1980s that a linkage exists between the QBO and sunspot activity. Although previous research stated that the north polar stratosphere during winter tended to be colder during the west phase of the QBO, Labitzke discovered that, at solar maximum, the polar stratosphere was warm if the QBO was in its west phase (Labitzke 1987). Subsequent work by Labitzke with van Loon showed that the solar cycle and QBO were linked such as in the case of west-phase QBO years (Labitzke and van Loon 1988). There was a strong correlation with warmer winters when the Sun was active and colder winters when the Sun was less active. Although their results were statistically significant, they could not identify a strong physical explanation for the linkage.

Recent studies have refined that initial work by suggesting that a strong connection exists between solar ultraviolet irradiance and the QBO. Small changes in solar radiative forcing – as seen through modeling studies – can serve as pacemakers to the period and phrase of the QBO. For example, a recent study indicated that average UV irradiance tends to be higher for the east QBO phase and lower for the west phase.

A key impact of this stratospheric oscillation in equatorial winds is the generation and propagation of planetary scale waves that move poleward. These planetary scale waves strongly influence the extratropical tropospheric atmosphere. Burroughs stated that the QBO is "the most widely observed feature in [climate] records, and must clearly be regarded as a real feature of almost all meteorological records" (Burroughs, 1992, p. 60). For example, early climatic research identified a statistical link between the QBO and one of the longest monthly measured European temperature series dating back to 1660. Other researchers have shown that such statistical correlations are likely linked to variations in pressure (such as the Arctic Oscillation) created by planetary waves propagated by the QBO.

3.3.2 Tropical influences: surface climate

Because the QBO is centered in the equatorial stratospheric tropics, one would assume that it should have an effect on the underlying tropospheric atmosphere. However, one factor potentially limits that interaction; the QBO does not directly penetrate significantly into the troposphere so any effects from the QBO are likely to be indirect. For example, although most measurements of the tropical tropospheric temperature do not show a strong oscillation at the QBO periodicity, some studies have suggested an indirect link between tropical temperatures and the QBO via a linkage between tropical stratospheric (QBO) and lower tropospheric winds.

However, given that caveat of a lack of strong physical association, analysis has shown some climatic indicators of the QBO operating in the tropics and subtropics. For example, past records of South Asian interannual monsoon rainfall do display some characteristics of the QBO. Another study has identified that one of the most prevalent vector-borne diseases in Australia – Ross River virus – demonstrated a reoccurrence periodicity that matches the QBO (Done *et al.* 2002). The likely physical mechanism involves the influence of the QBO on Australian summer rainfall. However, even the identification of a statistical linkage is particularly useful in that the QBO may be used as a long-term predictor of the incidence of Ross River virus by public health authorities.

Other research has shown that precipitation in the tropics displays a strong QBO periodicity. The summer monsoonal rainfall in India, for example, has displayed periodicities that suggest an influence by the QBO. Research has also linked variability in rainfall regimes of other parts of the world to the stratospheric QBO variations. Scientists have identified in African rainfall data that negative anomalies in rainfall extend over the central Sahel during the eastern phase of the QBO.

Many studies have suggested that correlations between tropical precipitation and the QBO might be created by the enhanced deep convection present during the QBO westward wind shear phase and suppressed when there is eastward shear in the tropical stratosphere. Such a linkage was recently evaluated using datasets involving measures of tropical convection (i.e. outgoing longwave radiation, OLR; and an index of highly reflective clouds, HRC). Researchers identified oscillations in both equatorial OLR and HRC values that appear to coincide with the fundamental periods of the QBO. Their findings indicate that the QBO modulation of the lower stratospheric vertical wind gradient may result in cloud tops more likely being "sheared off" in some years than in others.

Such linkages between tropical deep convection and the QBO led hurricane expert William Gray and colleagues to suggest that the timing of El Niño–Southern Oscillation (ENSO) may be modulated by the QBO. ENSO is the set of phenomena associated with oscillating shifts in water (El Niño/La Niña) and atmosphere (the Southern Oscillation) in the Pacific Ocean. Gray and colleagues theorized that since ENSO is strongly linked to the creation of deep convective activity in the tropical Pacific Ocean, the relative phase (east/west) of the QBO should play a large role in modulating the character of ENSO. During the easterly phase of the QBO, the vertical wind shear favors deep convective activity near the equator while inhibiting such activity in the subtropics. Conversely, during the westerly shear phase of the QBO, deep equatorial convection is suppressed while subtropical monsoonal convection is enhanced. This may be part of the physical linkage between the QBO and Indonesia sea-level pressure oscillations that appear to display some QBO periodicity (Gray, 1984).

However, the direct linkage between the QBO and the tropics is unclear because the tropical troposphere itself does have an identifiable quasi-biennial oscillation. That tropical QBO is apparently unrelated to the stratospheric QBO. Instead it appears that the "tropospheric QBO" is directly linked to the Southern Oscillation. This tropospheric biennial variability is less regular than the stratospheric QBO, more asymmetric in its longitudinal structure and has its greatest amplitude over Indonesia.

3.3.3 Tropical influences: ozone and aerosols

One easily identifiable – but not surprising – influence of the stratospheric QBO is seen in the interannual variability of subtropical stratospheric ozone (Randel and Wu 1996). Annual variations matching those seen in the QBO are evident in ozone concentrations (Figure 3.6). The likely mechanism in accounting for the QBO influence on ozone involves the synchronicity of the maximum westerly vertical wind shear and the maximum diabatic cooling, which fundamentally leads to the descent of air parcels and their constituents. One interesting aspect of the ozone space/time variability is the dichotomy (evident equatorward and poleward of 15° latitude) in the signal between the tropics and high latitudes. Generally, this variability coincides with ascent

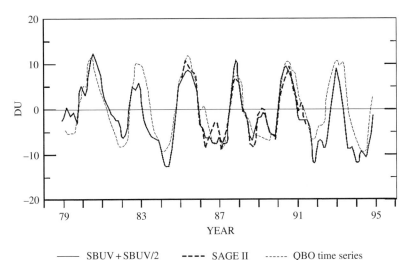

Figure 3.6 Time series of equatorial ozone anomalies in Dobson units (solid line) with a reference QBO wind time series (dashed line) based in part on Singapore stratospheric winds. (Adapted from Randel and Wu 1996: permission from American Meteorological Society)

(descent) in the subtropics associated with the westerly (easterly) wind shear phase. However, in addition to the QBO, ozone also displays an annual cycle that modulates the fundamental periodicity of ozone to some extent.

Given the initial identification of one of the QBO phases by plume movement from the volcano Krakatoa, it is not surprising that QBO variability is evident in records of trace atmospheric substances such as volcanic aerosols, as well as water vapor and methane. Researchers have illustrated markedly different aerosol spatial distributions from equatorial volcanic eruptions as a function of the phase of the QBO. Similar findings have been reported for water vapor, methane, and nitrates concentrations.

3.3.4 Tropical influences: tropical cyclones

One of the strongest apparent linkages between the QBO and the tropical atmosphere is the association between stratospheric winds and tropical cyclone activity in the Northern Hemispheric Atlantic Ocean. William Gray revealed one of the first analyses of a relationship between tropical storm frequency (in this case, the number of hurricane days per season) and the QBO in 1984 (Figure 3.7). Analyzing a variety of hurricane parameters, he determined that the QBO exerts the strongest influence on the minimum sea surface pressure anomalies of hurricanes such that overall summer sea level pressure in the Caribbean averages 0.21 millibars lower under the western phase of the QBO as opposed to the eastern phase. Subsequently, Gray and his colleagues incorporated the QBO as a primary variable in their statistical scheme for forecasting seasonal tropical storm activity. Fundamentally, they have

Figure 3.7 Relationship between 30-millibar (mb) stratospheric wind direction (West phase – black bars; East phase – white bars) and seasonal number of hurricane days from 1949 to 1982. Years with no observations are those in which the 30 mb zonal wind is changing direction or is very weak. (Adapted from Gray 1984: permission from American Meterological Society)

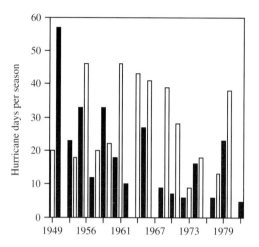

identified that strong Atlantic hurricanes tend to occur more frequently when the QBO is transitioning into or is already in its westerly phase. Conversely, when the QBO is in its easterly phase, the Atlantic hurricane season tends to be milder. Even today the QBO remains one of the main statistical variables in Gray's seasonal predictions of tropical storm activity in the Atlantic Basin.

Two major physical explanations have been postulated to account for a QBO-tropical cyclone. The first suggests that lower stratospheric vertical wind shear restricts the penetrative convection associated with strong storms. Following this line of thinking, the easterly phase of the QBO would have inhibited tropical cyclone activity because of the extra lower stratospheric wind ventilation and increased tropospheric to stratospheric wind shear.

The second explanation involves the effect of the QBO on the position of critical levels of propagation for tropical easterly waves. Shapiro (1989) noted that slower tropical waves have their propagation speeds comparable to winds at the 85 hPa level while stronger system (hurricanes and tropical storms) move at speeds closer to 50 hPa wind speeds. These differences are accented during different phases of the QBO. During the east phase of the QBO, the relative advection (the rate of airflow into the moving tropical system) is smaller than during the west phase. Consequently, the shearing associated with the east phase would advect air away from the developing system and thereby inhibit its intensification. The reverse would hold true for the west phase of the QBO. Ultimately, no matter the physical explanation, the statistical relationship between the QBO and Atlantic tropical cyclone activity has led to the QBO's incorporation into Atlantic seasonal tropical cyclone algorithms.

Unfortunately, the influence of the QBO on typhoon activity in the Pacific basin is less clear. Some climatologies and prediction schemes of North

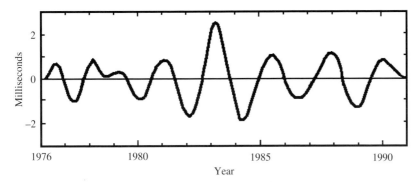

Figure 3.8 Variations in a monthly time series of length-of-day (computed as deviations in milliseconds) as filtered recursively in the quasi-bennial (18–35 months) band. (Adapted from Dickey *et al.* 1994)

Pacific tropical cyclone activity have not identified any strong QBO signal in the occurrence or intensity of typhoons and tropical storms. However, a recent case study of the 1994 Typhoon Orchid revealed that deep convection from the typhoon induced a gravity wave in the lower stratosphere.

3.3.5 Tropical influences: unusual linkages

One of the most unusual influences of the QBO is its identifiable impact on the terrestrial length-of-day. Researchers have identified the QBO as exerting a possible effect on length-of-day because of its contribution to changes in the angular momentum of the atmosphere (Dickey 1994). Several researchers have computed the axial angular momentum linked to alternating zonal winds of QBO and discovered that the QBO, when operating in conjunction with momentum shifts generated by ENSO, has a small but discernable impact on the terrestrial length-of-day (Figure 3.8). The QBO effect at biennial periods is, according to these researchers, one of the clearest influences evident on the length-of-day. But, some researchers have noted that day length variability is also modulated by ENSO (producing triennial–quadrennial variability) and by mantle–core interactions (producing six-year periodicity).

A second somewhat unusual influence of the QBO appears to be on the strength of the semidiurnal pressure waves associated with normal thermal heating and cooling. Study of the long-term variation in these semidiurnal pressure waves indicates some QBO variability. The existence of a QBO signal in these long-term pressure records also suggests that the QBO is not a transitory phenomenon, but rather a fundamental and persistent characteristic of the equatorial atmosphere.

3.3.6 The future of QBO research

Many climatologists who evaluate periodicities in various types of climatic data around the world regard the QBO as a ubiquitous signal embedded in those data. The dominant presence of QBO signals in some types of climate data has led to marked improvements in forecasting, particularly with regard to the tropics and subtropics. For example, measures of the QBO are considered primary variables in seasonal forecasting equations developed for Atlantic tropical cyclone activity. In a similar fashion, study of global spatial and temporal variations in ozone variability must address the mechanisms of the QBO.

However, specification of the actual physical mechanisms by which the QBO influences climate around the world remains elusive. While we have achieved a good – albeit not perfect – understanding of how the QBO operates, there remain many avenues of research on the QBO. In particular, the incorporation of a realistic QBO into numerical general circulation models (GCMs) remains unachieved.

In addition, continued identification of a QBO signal in existing climate data must continue. Because of the relative ubiquity of the QBO in climate records around the world, some researchers have, unfortunately, tended to disregard the identification of quasi-biennial signals as "unimportant" or stated that QBO influences have been sufficiently addressed. However, the usefulness of the QBO in Atlantic tropical cyclone seasonal forecast algorithms demonstrates that the QBO is not only a climatically important phenomenon but also one whose variations can affect the structure and development of human society. We need to continue to address the QBO's causes and influences.

To accomplish that task, more and varied climate records – particularly for the tropics – must be established. Sadly, the tropics remain a region of relatively sparse climate observations. Consequently, although new and continued climate monitoring should be undertaken in the tropics, the study of long-term paleoclimatic proxy records of sufficient resolution to record QBO fluctuations should also be analyzed and evaluated. These types of surrogate climate records might include tree-ring chronologies, equatorial glacial ice-cores or similar annual resolution records. Study of the QBO should continue to be an area of investigation for researchers around the world.

3.4 Human activities and problem climates in the tropics

3.4.1 Desertification

The multi-year drought in the early 1970s in the Sahelian zone, a region extending from West Africa to the Horn of Africa, led to increases in death and morbidity of humans and livestock and extensive environmental deterioration. It brought to the fore the process of desertification. Satellite images and photographs of the landscape and human suffering were widely seen in all types of media. The relative roles of climate variability and human impacts on the environment in this type of setting led to many articles and professional papers. It is an argument still not totally resolved and the United Nations recognizes both in its definition of desertification. The definition adopted by the 1992 United Nations Conference on Environment and Development states that desertification is land degradation in arid, semiarid and dry subhumid areas resulting from various factors including climate variations and human activities. The essay by Michael Glantz (Section 1.3) provides a lucid analysis of the roles of climate and human activities in such events. Although desertification occurs outside of the tropics, it is certainly of enormous concern within the tropical realm.

At the Earth Summit (the UN Conference on Environment and Development, or UNCED) in Rio de Janeiro in 1992, delegates called on the UN General Assembly to establish an Intergovernmental Negotiating Committee to prepare, by June 1994, a convention to combat desertification, particularly in Africa. In December 1992, the General Assembly agreed and adopted the resolution. This led to the formation of the United Nations Convention to Combat Desertification (UNCCD) in 1996.

The Conference of Parties, the Convention's supreme governing body, held its first session in October 1997 in Rome, Italy, when the UNCCD stated, "The Convention aims to promote effective action through innovative local programs and supportive international partnerships. The treaty acknowledges that the struggle to protect drylands will be a long one – there will be no quick fix. This is because the causes of desertification are many and complex, ranging from international trade patterns to unsustainable land management practices. Real and difficult changes will have to be made, both at the international and the local levels." (www.unccd.int)

It is evident that there is no "quick fix", for in a UN publication (*The UNCCD After 10 Years*) Kofi Annan, the UN Secretary-General, writes "Current estimates are that the livelihoods of more than one billion people are at risk from desertification, and that, as a consequence, 135 million people may be in danger of being driven from their land . . . Indeed, recognizing the urgent need to address the far-reaching implications of this problem, the UN General Assembly has declared 2006 the International Year of Deserts and Desertification." For some countries, such as the People's Republic of China, it drew the attention of national policy makers toward arid lands research and helped to elevate such research to a sustained national priority.

There is no agreement on where desertification can take place. Many researchers identify arid, semiarid, and sometimes subhumid regions as the areas in which desertification can occur or where the risks of desertification are highest. Others assert that desertification can only occur along the desert fringes. Following Le Houérou (1975), some researchers suggest that desertification can occur only in the 50–300 mm isohyet zone. This interesting statement suggests that the role of climate is of high significance.

When examining climate over a given time period, it is important to recognize that different terms need to be used. Climate variability refers to the natural variations that appear in the atmospheric statistics for a designated period of time, usually on the order of months to years, and can occur in any climate type. However, in dealing with desertification it should be noted that during the annual dry season, characteristic of semiarid regions, the climate is desertlike. Improper use of the land during this period leads to land degradation that has little to do with climate variability. Thus, both seasonal dry periods and short-term variations in climate when combined with improper land-use practices, can give the appearance of the impact of longer term regional climate changes when none may have occurred at all. Climate fluctuations are variations in climate conditions that occur on the order of decades. Such may have occurred in the West African Sahel when drought conditions extended over multi-year growing seasons.

The question of relative responsibility for the desertification, whether it be climate variability or human activity, has been approached by Kelly and Hulme (1993), who related a satellite-derived index of active vegetation to rainfall variations. Figure 3.9 shows how the extent of the Sahara relates to the rainfall deficit. When removing the rainfall effects from the relationship, variations in the Sahara still exist; this may be due to the cumulative effect of dry years delaying the recovery of vegetation or directly related to human degradation of the environment.

The records of rainfall in the Sahel indicate that the drought of the 1970s and 1980s was severe and long lasting. Only recently has it eased. By 2003 a wet period appears to have returned, for in 2003 Burkina Faso recorded record rainfall amounts. However, how well the data represent the actual conditions

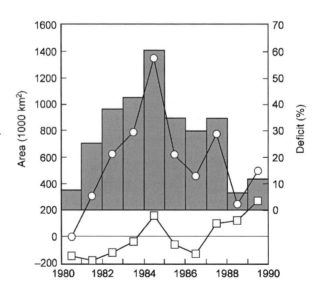

Figure 3.9 A rainfall deficit index (bars) is compared with a satellited index (circles) to obtain the residual trend showing the extent of the Sahara when rainfall effects are removed. This is shown by the squares. (After Tiempo Climate Cyberlibrary, *Tiempo* Issue 9)

has been questioned. In a 2004 study, Chappell and Agnew explored the contention that, since the major droughts of the 1970s, mean annual summer rainfall has declined. They suggest that previously published work did not take into account "the location of the rainfall station network each year in relation to the spatial and temporal heterogeneity of West African Sahel rainfall." They went on to suggest that the identified decline in rainfall was an artifact of changing location and number of rainfall stations.

Clearly, there is much research remaining to be completed in the study of the Sahel. In his review of the relationship between surface and atmosphere in the Sahel, Taylor (2001) provides a synopsis of a climatic situation that will occupy many researchers for years to come.

3.4.2 Amazon deforestation

The Amazon Basin of Brazil occupies some 5 million square kilometers, some 4 million of which is forested. It is an area about which much has been written, especially since the Brazilian government's efforts to colonize the region and the deforestation that has resulted. Essentially, the construction of highways into the region (the Belem–Brazilia road was paved in 1974, the Cuiba–Porto Vehlo in 1984, and the Rondonia to Acre in 1986) permitted settlement that was stimulated by programs to attract migrants from other parts of the country. Subsistence farming on the poor soil of cleared forest areas was a failure and cattle pasturing became the main land use in cleared areas. This was greatly aided by the "right of possession" for whoever clears a forested area; after logging, pasture is the easiest way to occupy an extensive area and large land tracts are often owned by a few cattle ranchers. Brazil now has the largest commercial cattle herd in the world and is a major exporter of beef and beef products. Beyond cattle ranching, the planting of soybeans is also a major contributor to deforestation. Social scientists find many problems with the events that have occurred and various solutions have been suggested (Fearnside 1986; Anderson 1990; Amelung and Diehl 1992; Hall 1997).

As Figure 3.10 shows, the amount of deforestation is remarkable. Through its National Institute for Space Research (INPE), the Brazilian government monitors the amount of deforestation through remote sensing. As shown in the essay in Section 3.5, remote sensing is of considerable importance in monitoring tropical environments. It is largely through remote sensing that much of the data concerning changes in the forest are derived.

Climate change resulting from deforestation can be considered at a number of levels. First there is the global effect. A response to the burning of the forests over a huge area is the addition of carbon dioxide to the atmosphere, with the potential for enhancing the global greenhouse effect; then there is the removal of the forest which serves as an oxygen source for the planet. As shown in a number of publications, both of these concerns are difficult to evaluate and model and the

Figure 3.10 Annual deforestation rates in the Brazilian Amazon. (Data from Brazilian Ministry of Science and Technology)

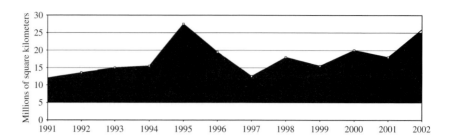

results are not, at this time, totally reliable. As Shukla *et al.* (1990) note, climatic fluctuations over northern middle latitudes are large, and forced model perturbations make it difficult to draw any definitive conclusions.

At the more local scale it is clear that a number of climate impacts must occur when a rainforest is replaced by a savanna grassland. There will be increased surface albedos leading to a modification of day–night temperatures and changing surface properties that will decrease both roughness and infiltration rates. A reduction in precipitation is anticipated and may well lengthen the dry seasons that are experienced in some of the forest climates. The impacts have been shown to occur in models as illustrated in the research of Dickinson and Kennedy (1992).

A number of research organizations have been responsible for obtaining and modeling climate in Amazonia. The type of research completed as part of ABRACOS (Anglo-Brazilian Amazonian Climate Observation Study) is illustrated in the assessment of Amazonian deforestation on climate. In their study, Gash and Nobre (1997) review measurements and modeling completed by the group and demonstrate that energy balance differences between forest and clearings give higher temperatures in the clearings. Where substantial deforestation has occurred the higher sensible heat fluxes in the cleared areas produce deeper convective boundary layers. This leads to variations in cloud cover and meso-scale circulations.

The Large-Scale Biosphere–Atmosphere Experiment (LBA) in Amazonia is an international research program led by Brazil. It was designed to generate new information needed to understand the "climatological, ecological, biogeochemical, and hydrological functioning of Amazonia, the impact of land use changes on these functions and the interactions between Amazonia and the Earth system" (http:daac.ornl.gov/lba).

The LBA has many components and is perhaps the most comprehensive source for current Amazonia scientific research. Analyses of largely unknown properties of the climate are providing some interesting results. For example, it has been shown (Roberts *et al.* 2001) that cloud condensation nuclei are surprisingly low. Thus the increase that will occur because of human activity in the region will have a stronger impact on climate than it might in other continental areas. In the study of "smoking clouds" over the Amazon (Andreae *et al.* 2004) the point is made that the burning of the forests is responsible for some 75% of Brazil's greenhouse gas emissions.

As might be expected, modeling and remote sensing both play an important part in assessing the impacts of deforestation climate. An early modeling study is exemplified by the use of a coupled model of atmosphere and biosphere that showed that, when forests were degraded, there was a significant increase in surface temperature and decrease in precipitation and evapotranspiration (Shukla *et al.* 1990). A research project using geosynchronous visible and infrared satellite data revealed differences in local circulation and that dry season rainfall occurrence is larger over deforested and non-forested (savanna) regions than over areas of dense forest (Negri *et al.* 2004). The following essay provides an in-depth look at a remote sensing project in the Amazon.

3.5 ESSAY: Remote sensing of Amazonia deforestation and vegetation regrowth: inputs to climate change research

Paul Mausel[1], Dengsheng Lu[2] and Nelson Dias[3]
[1]*Indiana State University,* [2]*Indiana University,*
[3]*Universidade de Taubaté*

Moist tropical forest deforestation and vegetation succession are topics of great interest throughout the world. The area with the most abundant moist mature forests in the world is Amazonia (5 000 000 sq. km). One of the largest and most respected research groups studying Amazonia deforestation and succession is the Large Scale Biosphere–Atmosphere Experiment (LBA) that is a research initiative led by scientists from NASA, INPE (Brazilian Space Agency), and the European Union (EU). It is designed to create new knowledge needed to understand the climatological, ecological, biogeochemical, and hydrological functioning of Amazonia, the impact of land use change on these functions, and the interactions between Amazonia and the Earth system (lba.cptec.inpe.br/lba/indexi.html). Collectively, there are approximately 130 research groups and 155 projects engaged in seven categories of LBA research interest (Figure 3.11).

This essay focuses most on land use and ecological aspects of LBA interest with an emphasis on classification and analysis of (1) deforestation, (2) vegetation secondary succession, (3) land use/land cover (LULC) change, and (4) biomass modeling in several Amazonia research sites. The essay authors are participants in the LBA project "Human and Physical Dimensions of LULC Change in Amazonia." This project does not focus on atmospheric phenomena, but elements of its research can provide important data for modeling carbon sequestration and release of carbon (CO_2) into the atmosphere. LULC research has connections with carbon storage and exchange, atmospheric chemistry, and the physical climate system (Figure 3.11). Classifying, analyzing, and modeling LULC in Amazonia through remote sensing provides valuable data for a variety

Figure 3.11 LBA-ECO science working groups and their interactions.

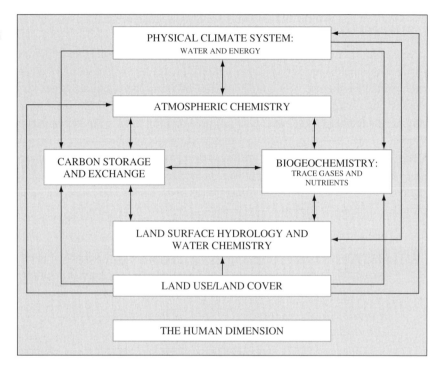

of users, including those concerned with biogeochemistry/trace gases and atmospheric chemistry, and physical climate.

3.5.1 Remote sensing of Amazonia

Introduction

Collectively, a majority of satellite sensors used in LULC research acquire data in wavelengths from blue light (0.4 µm) through thermal infrared radiation (12.5 µm) with spatial resolutions with pixel sizes ranging from approximately 1.0 m to 12 km+. These satellite sensors have several spectral configurations and spatial resolutions. They are available to provide spectral data for analysis with: (1) high spatial resolution of 1–5 m (e.g. IKONOS, Quickbird-2); (2) medium spatial resolution of 10–30 m (e.g. SPOT series, Landsat TM/ETM+, and ASTER); (3) low spatial resolution of 200–1100 m (e.g. MODIS, MISR); and (4) very low spatial resolution of >1.1–12 km+ (e.g. GOES, Meteosat series).

A major problem in most types of Amazonia spectral data acquisition is weather, because most areas are only relatively cloud free 6–12 weeks/year. Radar sensors have the advantage of acquiring data irrespective of weather or light conditions, but these data are often costly and somewhat more limited in applications, thus their use is much less than that of sensors using the 0.4–12.5 µm wavelengths. A few Earth resources radar satellites exist and collectively they have spatial resolutions from 3 m to 38+ km that use many parts of the microwave spectrum (L, S, C, X, K, Q, W bands with wavelengths >30 cm to <1 cm respectively). Most radar systems used for LULC research have high or medium spatial resolution in the 10–100 m range (e.g. ERS series, Radarsat series). Low spatial resolution satellite radar sensors also exist with spatial resolutions most typically from 4 km to 38 km (e.g. TRMM/TMI). Technical specifications and uses of major satellite sensors that collect electromagnetic energy (visible, infrared, and microwave) can be found at the NASA website: geo.arc.nasa.gov/sge/health/sensor/cfsensor.html.

High-resolution sensors provide detail that can acquire information about parts of an individual tree. Medium-resolution sensors provide data to develop information about LULC features at a subregional scale. Low-resolution sensors are used to develop information about Earth surface features at a regional scale, and most of the very low-resolution sensors are used to acquire continental and global scale data that are frequently focused on atmospheric phenomena (e.g. gases, pollutants, clouds) and other large Earth surface features (e.g. oceans, ice). Meteorologists, climatologists, and oceanographers are the biggest users of very low-resolution spectral data.

Land use/land cover (LULC) and biomass changes in Amazonia
Road building, political incentives/programs, and commercial developments, that began in the late 1960s and 1970s, started to change Amazonia (Moran 1981). Moist mature forests were cut and/or burned in large quantities, particularly near the roads that supported movement of goods and people (Moran 1981). Almost one-ninth of Amazonia was deforested by 1988 (Skole and Tucker 1993), but the exact percentage that currently has been deforested varies from authority to authority due to different definitions of the original mature forest base and deforestation statistics. Common recent estimates of Brazilian Amazonia deforestation range from 12% to 18%. Deforestation from 1988 to the present averages approximately 0.45% annually according to Brazil's official National Institute for Space Research (INPE) figures from the 2004 website (rainforests.mongabay.com/amazon/2004_deforestation.html). This deforestation has converted large areas of mature forest into a mosaic of patches of agricultural lands, pasture, and different stages of forest succession, with about 20–50% of the deforested area in different succession stages (Lucas *et al.* 2000).

Tropical moist mature forests have accumulated large amounts of biomass. Removal of these forests to make way for alternate land uses initially results in large quantities of CO_2 release to the atmosphere through burning and decomposition. The most common initial land use following deforestation is grass for cattle grazing and/or crops. A major release of CO_2 occurs as forests are replaced by forms of low biomass vegetation. Commonly, after a few years of crop or grazing, increased sequestration of carbon from CO_2 occurs when shrubs, small trees, and dense grasses become established as initial secondary succession (SS1). Passage of another 5–12 years usually results in trees again becoming dominant with a relatively dense cover resulting in the establishment of an intermediate secondary succession forest (SS2). Change in land use from SS1 to SS2 results in further sequestration of carbon derived from CO_2. If SS2 is left undisturbed for an additional 10–15 years, an advanced SS forest that has an appearance similar to the original moist mature forest often appears (SS3). The SS3 forests have developed a multi-canopy and have many trees almost as tall as the original moist mature forest, but their species complexity and total biomass is less than the original forest. Even after 30–50 years of re-growth in a good physical environment (and much longer than 100 years or never in poorer nutrient environments) succession vegetation has less biomass than the original forest.

The LULC sequence from mature forest to crop/pasture to SS1 to SS2 to SS3 to mature forest conditions again is often interrupted. At any SS stage, human activity can intercede by reintroducing crops, grazing economies, and agroforestry. The reintroduction of crops and grazing in a SS forest creates a net increase in CO_2 released and a net decrease in sequestered carbon. The introduction of agroforestry in a SS forest makes little change in carbon gain or loss to the atmosphere.

Remote sensing of LULC changes: implications to carbon sequestration

The most important LULC classes in Amazonia to study biomass changes, from highest sequestered carbon/high biomass to lowest, are moist mature forest/rainforest, SS3, SS2, SS1, agriculture (grazing/crop), bare or nearly bare (e.g. bare soil, road, very sparse grass), and water. Accurate classification of these LULC features in Amazonia using remote sensing was historically difficult to achieve. Spectral data acquired from low-resolution satellite sensors can classify huge areas, but primarily identify mature forest, succession vegetation, bare, and water, often with accuracy problems. Low-resolution sensors provide a general insight into the location and extent of deforestation/afforestation over large areas, but they do not provide data suitable to identify LULC changes in the detail required to assess the amount and trends of carbon sequestration and CO_2 released to the atmosphere or to help explain rates and nature of SS growth.

High-resolution satellite and airborne sensors can acquire spectral data at 1-meter resolution or less. However, their use is costly and too much analysis is needed to provide subregional information useful for LULC change analysis relevant to the carbon cycle and determining driving forces that affect processes that influence the nature and rate of vegetation succession.

A medium-resolution sensor such as Landsat TM (30 m) theoretically can accurately classify the seven basic feature classes previously identified. However, as recently as 1993, TM data could not consistently and accurately differentiate the three primary SS classes. During the period 1993–6, three studies were conducted that focused on differentiation of multiple classes of succession features, intensively using supervised and unsupervised (clustering) classification techniques on TM data, in spectral and spatial formats, integrated with a large field observation and measurement database. This approach provided a consistently high level of classification accuracy suitable for use in estimating the amount, rate, and direction of LULC changes.

The initial research using the integrated analysis of TM data and detailed field data was conducted near Altamira, Brazil (Mausel *et al.* 1993). Based on analysis of test samples in Altamira (Figure 3.12), accuracy in identification

Figure 3.12 Amazonia study area.

of moist mature forest and water exceeded 97%, accuracy of bare or very nearly bare features was 90%, agriculture feature accuracy was 89%, and accuracy of the three SS classes ranged from 81% to 88%. Two other studies were conducted independently in Ponta de Pedras (Brondízio *et al.* 1996) and Tome Acu (Batistella 1999) implementing the methods applied in Altamira. These two studies had good classification accuracy that validated using TM data combined with detailed field data to accurately develop subregional LULC maps. Recent research focusing on SS classification has resulted in improving accuracy in identification of SS stages by using forest stand structure data, subpixel information, advanced classification approaches, and other types of spectral data (Lu *et al.* 2004; Vieira *et al.* 2003).

Good LULC accuracy using TM data made it economically feasible to classify full TM scenes (185 km × 185 km) to assess LULC conditions. Multiple dates of LULC classifications can be integrated to provide temporal perspectives that, when analyzed, result in identifying LULC change patterns. One of the early change detection experiments that traced temporal patterns for 1985, 1988, and 1991 was near Altamira, Brazil. The spectral data from a small subscene (12.0 km × 5.5 km) of a TM scene near Altamira was used (Figure 3.13 (Plate 5)) to develop LULC temporal trajectories, pixel by pixel, for a three-date (1985–91) period using computer image processing techniques (Mausel *et al.* 1993). Table 3.1 summarizes the three date changes in the subscene study area and Table 3.2 provides LULC change statistics for the larger TM scene based on an analysis identical to that implemented in the subscene.

The 1985 to 1991 change analysis focused on creating a large number of three-date classes of interest such as very slow secondary succession, very fast secondary succession, normal secondary succession, and no change-rainforest (see Mausel *et al.* 1993 for the change image). For example, a class (3.6% of the small study area) was identified that represented very fast secondary succession. Remote sensing analyses supported by field data indicate that pixels in this class primarily were abandoned fields of grass or crop in 1985 that was very advanced SS1 in 1988, and was very advanced SS2 by 1991. In six years, there was a substantial forest regrowth. The reason why there was such rapid growth was a result of a combination of two or more favorable physical characteristics that included rich soil, good access to water, fast growing succession vegetation species, and good access to seed sources. This LBA research group was most interested in studying LULC changes to provide insights into processes causing the changes. However, scientists interested in tracing CO_2/sequestered carbon from its terrestrial origins to the atmosphere could use the LULC change data to provide insights into changes in sequestration or release of CO_2. In this very fast succession example it is evident that the pixel areas in this class had lost a majority of

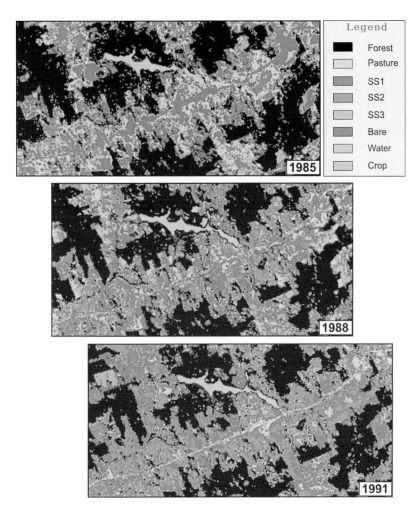

Legend

- ■ Forest
- ▢ Pasture
- ▨ SS1
- ▨ SS2
- ▨ SS3
- ▨ Bare
- ▢ Water
- ▨ Crop

Figure 3.13 Land use/ land cover classifications of a small (12.5 km × 5.5 km) study area near Altamira, Brazil, using computer analyzed August 1985, 1988, and 1991 Landsat TM. For color version see Plate 5.

their sequestered carbon when a moist mature forest was cut/burned a few years prior to 1985. However, after a short phase of crop or grazing, the return to forest was very fast; thus between 1985 and 1991 a rapid return of sequestered carbon derived from CO_2 occurred.

Tables 3.1 and 3.2 give the percentage of LULC by class, for the small study area and the large TM scene respectively, for each of the three years of remote sensing data used. Analysis of these tables shows the rate at which some features are losing biomass while others, through succession, are regaining lost biomass. The tables can provide qualitative assessments of overall gain or loss of sequestered carbon during a designated period of study. The exact C/CO_2 net gain or loss of a given area requires an association of specific biomass gains and losses to LULC features.

Table 3.1   *Land cover changes in the Altamira subsite study area: 1985–91*

Land cover class	Percentage land cover		
	1985	1988	1991
Mature forest	44.32	40.71	39.21
Pasture/grass	18.71	10.88	5.71
Initial SS (SS1)	10.68	13.78	14.67
Intermediate SS (SS2)	6.43	18.82	26.42
Advanced SS (SS3)	1.33	6.91	10.65
Bare/nearly bare	14.74	2.72	0.93
Crop/agriculture	2.38	4.78	1.01
Water/wetland	1.41	1.41	1.41

Modified from *Geocarto International* (1993), Vol. 8, No. 4, p. 67.

Table 3.2   *Land cover changes in the full Altamira study area: 1985–91*

Land cover class	Percentage land cover		
	1985	1988	1991
Mature forest	64.01	58.73	56.99
Pasture/grass	10.82	7.92	3.06
Initial SS (SS1)	5.58	8.58	10.91
Intermediate SS (SS2)	4.19	9.97	15.47
Advanced SS (SS3)	0.91	3.96	5.93
Bare/nearly bare	7.59	1.71	1.33
Crop/agriculture	1.16	2.82	0.64
Water/ wetland	5.74	6.31	5.68

Modified from *Geocarto International* (1993), Vol. 8, No. 4, p. 69.

Quantitative insights into net gains or losses of carbon in the sequestered or gaseous states are becoming possible to derive through biomass modeling studies combining field and spectral data. Research is being conducted to use moderate resolution spectral data and field sample data to inform lower resolution sensor data to acquire accurate regional LULC information. Neural network classification is being implemented in this scaling-up research that promises to provide greater quantification of carbon sequestration/CO_2 release associated with remote sensing-derived LULC changes over large areas.

3.5.2 Biomass modeling using remote sensing data

Total biomass includes aboveground (AGB) and belowground living mass, but due to difficulty in measuring belowground biomass, most research has focused on the former. Approaches to measuring AGB are based on (1) field data, (2) remote sensing, and (3) geographic information systems (GIS). Traditional field measurement techniques are time-consuming, labor intensive, and difficult to implement. Representative field samples are required as input in developing AGB estimation models using remote sensing and/or GIS approaches. GIS-based methods using ancillary data are also difficult to implement because of problems in appropriate data acquisition, knowledge about relationships between AGB and ancillary data, and impacts of environmental conditions on AGB accumulation. The advantages of remote sensing (e.g. repetitive nature of data collection, synoptic view, digital format, and high correlations between spectral bands and vegetation parameters) make it the primary data source for large-area AGB estimation (Foody *et al.* 2003).

Few studies have used low spatial resolution data or have used high spatial resolution data in AGB estimation. Medium spatial resolution optical sensor data, such as Landsat TM (Nelson *et al.* 2000; Foody *et al.* 2003) and SAR (radar) data (Santos *et al.* 2003) are most often used for AGB estimation, which is based on statistical relationships between AGB and TM or SAR responses. AGB can be (1) directly estimated using remotely sensed data implementing analytical approaches, such as multiple regression, K nearest-neighbor (Fazakas *et al.* 1999), and neural network (Foody *et al.* 2003), and (2) indirectly estimated from canopy parameters, such as crown diameter, which are derived from remote sensed data using multiple regression analysis or canopy reflectance models (Popescu *et al.* 2003).

TM spectral data used for AGB estimation have been explored in tropical regions (Foody *et al.* 2003). Spectral responses alone often are not sufficient to accurately estimate AGB because of the complexity of forest stand structure and environmental conditions. A combination of spectral responses and textures is usually helpful in improving AGB estimation. The variables used for AGB estimation vary, depending on the characteristics of the study areas. Different forest stand structures related to soil conditions and land use history significantly influence AGB estimation. Large differences in biomass density and associated forest stand structure exist between succession and mature forests that result in significantly different performance in AGB estimation. Use of Landsat TM imagery is more successful for AGB estimation in succession forests than in mature forests because of a less complex stand structure than found in mature forests. Examples of biomass modeling research conducted in eastern Amazonia study sites using TM data showed that models developed for succession forests had R^2 values ranging from

0.71 to 0.78, for the better models, to R^2 values ranging from 0.16 to 0.18 for models developed in mature forests (Lu *et al.* 2005a).

AGB modeling of SS features using remotely sensed data is sufficiently accurate for providing good qualitative estimates of biomass and is becoming increasingly important in providing quantitative measures. Nevertheless, additional research is needed in accuracy assessment to better assess the degree to which AGB estimates can be used in models that focus on carbon sequestration and potential for CO_2/C exchanges. More research is needed in integrating remotely sensed data and ancillary data using advanced models, such as neural networks, to more accurately estimate mature forest AGB. Selection of suitable remote sensing variables, reducing uncertainties caused by AGB field data collection and the position errors between sample plots and image pixels, are critical in improving AGB accuracy. AGB estimation and validation of results is difficult; however, progress has been made in biomass modeling, particularly for SS features, and many of the results have value for scientists with CO_2 and sequestered carbon interests.

3.5.3 Linking atmospheric research with LULC processes

Progress has been made toward improving understanding of linkages between LULC and aspects of Amazonia local and regional climate and its effects on global climate. LBA atmospheric research has raised questions and has provided some answers about humid tropical climate processes, forcings, and consequences associated with LULC. This research primarily addresses: surface fluxes and boundary layer growth; convection, clouds and rainfall; and climate modeling.

Artaxo *et al.* (2002) indicate that cloud structure over Amazonia is unique among terrestrial ecosystems. Anthropogenic emissions have converted unperturbed clouds into perturbed clouds with smaller cloud droplets and a much higher number of cloud condensation nuclei (CCN). Concentrations of aerosols and CCN are largely determined by the nature of Amazonia LULC. Vegetation directly emits aerosol particles but they are also formed from oxidation of gaseous volatile organic compounds (VOC). Increased CCN results in reducing droplet size and forces precipitation to form at higher elevations through ice processes, which, in turn, affect local precipitation patterns. This process occurs predominantly during the dry season related to higher intensity of biomass burning.

Amazonia is a major direct source of organic aerosols to the global atmosphere that, coupled with vegetation and soil ammonia emissions, generate large-scale atmospheric effects. The tropical troposphere is responsible for about 70% of the global atmospheric oxidation of long-lived gases including CH_4, CO, HCFCs and CH_3Br. Results from the LBA program indicate that:

the Amazon Basin consistently is a large net carbon sink in the undisturbed rainforest; nitrogen emitted by forest soils is subject to chemical cycling within the canopy space resulting in vegetation re-sequestration of much of the soil-derived NO_x; forest vegetation is both a sink and a source of VOC; and concentrations of aerosol and CCN are seasonal, with a pronounced maximum in the dry (burning) season (Andreae *et al.* 2002).

Results obtained from the atmospheric mesoscale wet season (WETAMC) campaign indicate that deforestation may be of secondary importance during the wet season compared to the convection system in affecting wind regime and cloud formation. This experiment also indicates that mesoscale convective systems are initiated over high terrain where deforestation may be enhancing rainfall (Silva Dias *et al.* 2002). Changes in canopy structural properties exert substantial control on rates of photosynthesis and forest respiration (Vourlitis *et al.* 2002). Forest conversion to pasture and selective logging can alter canopy structure, microclimate, soil water availability, and disturbance (e.g. fire) regimes (Fearnaside 2000). Partition of precipitation water in deforested areas increases runoff and decreases soil storage and evapotranspiration thereby increasing the magnitude and/or duration of drought during dry and transition periods that further limits transitional forest net ecosystem CO_2 exchange (NEE) (Vourlitis *et al.* 2002).

Amazonia is a large global store of carbon and both forest clearing and agricultural development are often cited as large net sources of CO_2 to the atmosphere. Studies of NEE indicate that intact tropical rainforest and savanna can accumulate 1 to $6\,t\,C\,ha^{-1}\,y^{-1}$ (Malhi *et al.* 1999), waterlogged valley forested areas can accumulate up to $8\,t\,C\,ha^{-1}\,y^{-1}$ (Araujo *et al.* 2002), and even $36\,t\,C\,ha^{-1}\,y^{-1}$ was estimated on an eastern Amazonia site (Carswell *et al.* 2002). Nepstad *et al.* (2002) conducted an experiment on the impact of partial precipitation throughfall on carbon cycling and found that the net accumulation of carbon in mature Amazon forests may be very sensitive to small reductions in rainfall. They also found that soil water reductions were also sufficient to increase soil emissions of N_2O and soil consumption of CH_4.

Global Circulation Models (GCM) are structured for global scale analysis. Research has been conducted with the intent of developing circulation models adapted to microscale and mesoscale simulations that could depict land–atmosphere interactions at the level of detail needed for Amazonia, where convective clouds and precipitation play an important role. Fine-scale process-based multilayer canopy models (soil–plant–atmosphere) have been used to predict hourly gross photosynthesis and respiration that can be tested against collected tower eddy covariance flux measurements. Aggregation of these models with coarser scale daily whole-canopy carbon fixation and transpiration models and even a coarser ecosystem biogeochemistry

(MBL-GEM) model, can allow quantification and improve our understanding of the changes in C storage, C exchange, and nutrient dynamics (Williams *et al.* 2002). Field results indicate that NEE fluxes during deforestation cause the switch from net sinks to net sources with emission totals reaching $1.6\,\mathrm{kg\,C\,m^{-2}\,y^{-1}}$, a flux magnitude equivalent to 10 years of undisturbed CO_2 sink fluxes in Amazonia (Potter *et al.* 2001).

3.5.4 Conclusions

LULC features and changes in feature patterns can be determined with useable accuracy for relatively large areas using analysis of medium-resolution (e.g. TM) remotely sensed data. The information derived through remote sensing of LULC contains insights to C/CO_2 exchanges primarily in qualitative and semi-quantitative forms. However, modeling utilizing spectral data, field data, and a host of improved remote sensing and GIS techniques is beginning to estimate ABG with improving accuracy that soon will be become an important quantitative data source for scientists engaged in studying components of atmosphere and ecosystem. Links exist between remote sensing scientists whose focus is on exploring LULC and LULC changes and scientists with atmospheric/climatological interests. These links need to be strengthened through better understanding of where terrestrial and atmospheric interests overlap in order to most effectively address issues of mutual concern.

3.6 Chapter summary

The tropical climates of the world are providing the climatologist with an endless set of research questions. In this chapter a number of components have been emphasized. First, in his highly informative essay, Cerveny provides the essence of the complexities associated with low-latitude teleconnections. Thereafter, the discussion relates to the relative roles of climate variability and human activities in tropical climates. In the case of the desert boundaries, it is clear that over time there have been changes induced by long-lasting drought. At the same time, it is shown that degradation of the surface plays a highly significant role in desertification. The various arguments presented are placed within the global framework as represented by the United Nations. Continued research and meaningful discussion are clearly called for.

The same is true of the other example presented, that of deforestation in the Amazon. The impacts here are a result of government policies, human migration resulting from drought and poverty in other regions, and the ready availability of an export market for a product. Needless to say, the changing of any forested region, especially one in the tropics, has long-term consequences in local and regional climates. These changes may, too, influence the global climate in ways not totally understood. There is little doubt, however, that the type of research described in the essay by Mausel and colleagues will lead to a greater understanding.

3.7 Examples of tropical climates websites

The climate of the tropical Pacific is found at www.nws.noaa.gov with a link to the CPC. Information provided includes online data, climatic impacts of El Niño and La Niña and various island/ocean climatologies/forecasts.

The USGS site, pubs.usgs.gov/gip/deserts, provides information about desertification. A descriptive account is accompanied by appropriate photographs.

The National Hurricane Center/Tropical Prediction Center is linked to www.nws.noaa.gov. It provides a wealth of detail on current and historic hurricane information.

Southern Hemisphere storm information and a weekly tropical climatology note is available at the Australian site www.bom.gov.au/climate/tropnote/tropnote.shtml.

3.8 References

Amelung, T. and Diehl, M., 1992. *Deforestation of the Tropical Rain Forests: Economic Causes and Impact on Development*. Tubingen, Germany: J. C. B. Mohr.

Anderson, A., 1990. *Alternatives to Deforestation: Steps Toward a Sustainable Use of the Amazon Rain Forest*. New York: Columbia University Press.

Andreae, M. O., Artaxo, P., Brandao, C., *et al.*, 2002. Biogeochemical cycling of carbon, water, energy, trace gases, and aerosols in Amazonia: The LBA-EUSTACH experiments. *Journal of Geophysical Research*, **107**, LBA 33–1 to 33–25.

Andreae, M. O., Rosenfeld, D., Artaxo, P., *et al.*, 2004. Smoking clouds over the Amazon. *Science*, **303**, 1337–1342.

Araujo, A. C., Nobre, A. D., Kruijt, B., *et al.*, 2002. Comparative measurements of carbon dioxide fluxes from two nearby towers in a central Amazonian rainforest: The Manaus LBA site. *Journal of Geophysical Research*, **107**, LBA 58–1 to 58–20.

Artaxo, P., Martins, J. V., Yamasoe, M. A., *et al.*, 2002. Physical and chemical properties of aerosols in the wet and dry seasons in Rondonia, Amazonia. *Journal of Geophysical Research*, **107**, LBA 49–1 to 49–14.

Asnani, G. C., 1968. The equatorial cell in the general circulation. *Journal of Atmospheric Sciences*, **25**, 133–134.

Batistella, M., 1999. Exploratory comparison between maximum likelihood and spatial-spectral classifiers using Landsat TM bands and principal component analysis for selected areas in Tome Açu, Brazilian Amazon. In *GIS Brasil 99* – V Congresso E Feira Para Usuários De Geoprocessamento Da América Latina, 1999, Salvador/Brasil. Anais. Curitiba, Brasil: Sagres Editora.

Brondízio, E. S., Moran, E. F., Mausel, P. and Wu, Y., 1996. Land cover in the Amazon estuary: linking of the Thematic Mapper with botanical and historical data. *Photogrammetric Engineering and Remote Sensing*, **62**, 921–929.

Carswell, F. E., Costa, A. L., Palheta, M., *et al.*, 2002. Seasonality in CO_2 and H_2O flux at an eastern Amazonian rain forest. *Journal of Geophysical Research*, **107**, LBA 43–1 to 43–16.

Chappell, A. and Agnew, C. T., 2004. Modelling climate change in the West African Sahel (1931–90) as an artifact of changing station location. *International Journal of Climatology*, **24**, 547–554.

Charney, J. G., 1971. Tropical cyclogenesis and the formation of the Intertropical Convergence Zone. In W. H. Reid, ed., *Mathematical Problems of Geophysical Fluid Dynamics*. Lectures in Applied Mathematics, Vol. 13, American Mathematical Society, pp. 355–368.

Das, P. K., 1986. *Monsoons*. World Meteorological Organization, WMO No. 613.
2002. *The Monsoons*. India: National Book Trust.

Dickey, J. O., Marcus, S. L., Hide, R., Eubanks, T. M. and Doggs, D. H., 1994. Angular-momentum exchange among the solid Earth, atmosphere, and oceans – a case study of the 1982–1983 El Nino Event. *Journal of Geophysical Research – Solid Earth*, **99**, 23921–23937.

Dickinson, R. E. and Kennedy, P., 1992. Impacts on regional climate of Amazon deforestation. *Geophysical Research Letters*, **19**, 1947–1950.

Done, S. J., Holbrook, N. J. and Beggs, P. J., 2002. The Quasi-Biennial Oscillation and Ross River virus incidence in Queensland, Australia. *International Journal of Biometeorology*, **46**, 202–207.

Fazakas, Z., Nilsson, M. and Olsson, H., 1999. Regional forest biomass and wood volume estimation using satellite data and ancillary data. *Agricultural and Forest Meteorology*, **98–99**, 417–425.

Fearnside, P. M., 1986. *Human Carrying Capacity of the Brazilian Rain Forest*. New York: Columbia University Press.
2000. Global warming and tropical land use change: Greenhouse gas emissions from biomass burning, decomposition, and soils in forest conversion, shifting cultivation, and secondary vegetation. *Climatic Change*, **46**, 115–158.

Flasar, F. M., Kunde, V. G., Achterberg, R. K., *et al.*, 2004. An intense stratospheric jet on Jupiter. *Nature*, **427**, 132–135.

Fletcher, R. D., 1945. The general circulation of the tropical and equatorial atmosphere. *Journal of Meteorology*, **2**, 167–174.

Foody, G. M., Boyd, D. S. and Cutler, M. E. J., 2003. Predictive relations of tropical forest biomass from Landsat TM data and their transferability between regions. *Remote Sensing of Environment*, **85**, 463–474.

Gash, J. H. C. and Nobre, C. A., 1997. Climatic effects of Amazonian deforestation: some results from ABRACOS. *Bulletin American Meteorological Society Society*, **78**, 823–833.

Giambelluca, T. W. and Schroeder, T. A., 1998. Climate. In S. P. Juvik, J. O. Juvik and T. R. Paradise, eds., *Atlas of Hawai'i*. Honolulu: University of Hawai'i Press.

Goswami, B. N., Shukla, J., Schneider, E. K. and Sud, Y. C., 1984. Study of the dynamics of the Intertropical Convergence Zone with a symmetric version of the GLAS climate model. *Journal of Atmospheric Sciences*, **41**, 5–19.

Gray, W., 1984. Atlantic season hurricane frequency. Part I: El Nino and 30 mb quasi-biennial oscillation influences. *Monthly Weather Review*, **112**, 1649–1668.

Hall, A., 1997. *Sustaining Amazonia*. New York: St. Martins Press.

Hess, P. G., Battisti, D. S. and Rasch, P. J., 1993. Maintenance of the Intertropical Convergence Zones and the large scale tropical circulation on a water covered earth. *Journal of Atmospheric Sciences*, **50**, 691–713.

Inoue, M. and Bigg, G. R., 1995. Trends in wind and sea level pressure in the tropical Pacific Ocean for the period 1950–1979. *International Journal of Climatology*, **15**, 35–52.

Jones, P. D., 1991. Southern Hemisphere sea level pressure data; an analysis and reconstruction back to 1951 and 1911. *International Journal of Climatology*, **11**, 585–608.

Kelly, M. and Hulme, M., 1993. Exploring the links between desertification and climate change. *Environment*, July/August, **35**(6), 4–15.

Kirtman, B. P. and Schneider, E. K., 2000. Spontaneously generated tropical atmospheric general circulation. *Journal of Atmospheric Sciences*, **57**, 2080–2093.

Labitzke, K., 1987. Sunspots, the QBO and the stratospheric temperature in the North Polar Region. *Geophysical Research Letters*, **14**, 535–537.

Labitzke, K. and van Loon, H., 1988. Association between the 11-year solar cycle, the QBO and the atmosphere. Part I: The troposphere and stratosphere in the Northern Hemisphere in winter. *Journal of Atmospheric and Terrestrial Physics*, **50**, 197–206.

Le Houérou, H. N., 1975. The nature and causes of desertification. In *Proceedings of the IGU Meeting on Desertification*. Cambridge: Cambridge University Press.

Liu, W. T. and Xie, X., 2002. Double intertropical convergence zones – A new look using scatterometer. *Geophysical Research Letters*, **29**, 291–294.

Lu, D., Mausel, P., Batistella, M. and Moran, E., 2004. Comparison of land-cover classification methods in the Brazilian Amazon basin. *Photogrammetric Engineering and Remote Sensing*, **70**, 723–731.

Lu, D., Moran, E., Mausel, P. and Brondizio, E., 2005a. Comparison of aboveground biomass across Amazon sites. In E. Moran and E. Ostrom, eds., *Seeing the Forest and the Trees: Human-Environment Interactions in Forest Ecosystems*. Cambridge, Mass.: MIT Press.

Lu, D., Batistella, M. and Moran, E., 2005b. Satellite estimation of aboveground biomass and impacts of forest stand structure. *Photogrammetric Engineering and Remote Sensing*, **71**, 967.

Lucas, R. M., Honzák, M., Curran, P. J., *et al.*, 2000. Mapping the regional extent of tropical forest regeneration stages in the Brazilian Legal Amazon using NOAA AVHRR data. *International Journal of Remote Sensing*, **21**, 2855–2881.

Malhi, Y., Baldocchi, D. D. and Jarvis, P. G., 1999. The carbon balance of tropical, temperate, and boreal forests. *Plant, Cell and Environment* **22**, 715–740.

Mausel, P., Wu, Y., Li, Y., Moran, E. F. and Brondízio, E. S., 1993. Spectral identification of succession stages following deforestation in the Amazon. *Geocarto International*, **8**, 61–72.

McGregor, G. R. and Nieuwolt, S., 1998. *Tropical Climatology*, 2nd edn. New York: Wiley.

McIlveen, R., 1992. *Fundamentals of Weather and Climate*. London: Chapman & Hall.

Moran, E. F., 1981. *Developing the Amazon*. Bloomington, Ind.: Indiana University Press.

Negri, A. J., Adler, R. F., Xu, L. and Surratt, J., 2004. The impact of Amazonia deforestation in dry season rainfall. *Journal of Climate*, **17**, 1306–1391.

Nelson, R., Kimes, D. S., Salas, W. A. and Routhier, M., 2000. Secondary forest age and tropical forest biomass estimation using Thematic Mapper imagery. *Bioscience*, **50**, 419–431.

Nepstad D. C., Moutinho, P., Dias-Filho, M. B., *et al.*, 2002. The effects of partial throughfall exclusion on canopy processes, aboveground production, and biogeochemistry of an Amazon forest. *Journal of Geophysical Research*, **107**, LBA 39–1 to 39–18.

Numaguti, A., 1993. Dynamic and energy balance of the Hadley circulation and the tropical precipitation zones: significance of the distribution of evaporation. *Journal of Atmospheric Sciences*, **50**, 1874–1887.

Oliver, J. E. and Hidore, J. J., 2002. *Climatology: An Atmospheric Science*. Upper Saddle River, N.J.: Prentice Hall.

Popescu, S. C., Wynne, R. H. and Nelson, R. F., 2003. Measuring individual tree crown diameter with lidar and assessing its influence on estimating forest volume and biomass. *Canadian Journal of Remote Sensing*, **29**, 564–577.

Potter, C., Davidison, E., Nepstad, D. and de Carvalho, C. R., 2001. Ecosystem modeling and dynamic effects of deforestation on trace gas fluxes in Amazon tropical forests. *Forest Ecology and Management*, **152**, 97–117.

Randel, W. J. and Wu, F., 1996. Insolation of the ozone QBO in SAGE II data by singular decomposition. *Journal of Atmospheric Sciences*, **53**, 2546–2559.

Raymond, D. J., 2000. The Hadley circulation as a radiative-convective instability. *Journal of Atmospheric Sciences*, **57**, 1286–1297.

Roberts, G. C., Andreae, M. O., Zhou, J. and Artaxo, P., 2001. Cloud condensation nuclei in the Amazon Basin: Marine conditions over a continent? *Geophysical Letters*, **28**, 2807–2810.

Santos, J. R., Freitas, C. C., Araujo, L. S., *et al.*, 2003. Airborne P-band SAR applied to the aboveground biomass studies in the Brazilian tropical rainforest. *Remote Sensing of Environment*, **87**, 482–493.

Schneider, E. K. and Lindzen R. S., 1977. Axial symmetric steady-state models of the basic state for instability and climate studies. Part I: Linearized calculations. *Journal of Atmospheric Science*, **34**, 263–279.

Shapiro, L. J., 1989. The relationship of the quasi-biennial oscillation to Atlantic tropical storm activity. *Monthly Weather Review* **117**, 1545–1552.

Shukla, J., Nobre, C. and Sellers, P., 1990. Amazon deforestation and climate change. *Science*, **247**, 1322–1325.

Silva Dias, M. A. F., Rutledge, S., Kabat, P., *et al.*, 2002. Cloud and rain processes in a biosphere-atmosphere interaction context in the Amazon Region. *Journal of Geophysical Research*, **107**, LBA 39–1 to 39–18.

Skole, D. L. and Tucker, C. J., 1993. Tropical deforestation, fragmented habitat, and diversely affected habitat in the Brazilian Amazon: 1978–1988. *Science*, **260**, 1905–1910.

Taylor, C. M., 2001. Feedbacks between the land surface and the atmosphere in the Sahel. *Arid Lands Newsletter*, **49**.

Tomas, R. A., Holton, J. R. and Webster, P. J., 1999. The influence of cross-equatorial pressure gradients on the location of near-equatorial convection. *Quarterly Journal of the Royal Meteorological Society*, **125**, 1107–1127.

Vieira, I. C. G., de Almeida, A. S., Davidson, E. A., *et al.*, 2003. Classifying successional forests using Landsat spectral properties and ecological characteristics in eastern Amazonia. *Remote Sensing of Environment*, **87**, 470–481.

Vincent, D. G., 1994. The South Pacific Convergence Zone (SPCZ): A review. *Monthly Weather Review*, **122**, 1949–1970.

Vourlitis, G., Priante Filho, N., Hayashi, M. M. S., *et al.*, 2002. The role of seasonal variations in meteorology on the net CO_2 exchange of a Brazilian transitional tropical forest. In *II Scientific Conference LBA* (Large scale biosphere-atmosphere experiment in Amazonia), Manaus, 9–11 July, 2002.

Waliser, D. E. and Somerville, R. C. J., 1994. Preferred latitudes of the intertropical convergence zone. *Journal of Atmospheric Sciences*, **51**, 1619–1639.

Williams, M., Shimabukuro, Y. E., Hebert, D. A., 2002. Heterogeneity of soils and vegetation
    in an eastern Amazonian rain forest: Implications for scaling up biomass and production.
    Ecosystems, **5**(7), 692–704.

Zhang, C., 2001. Double ITCZs. *Journal of Geophysical Research*, **106**, 11785–11792.
    http://orca.rsmas.miami.edu/~czhang/Research/ITCZ/research.itcz

Chapter 4
Middle-latitude climates

4.1 Introduction

In geographical terms, the middle-latitude climate zone is generally located between the poleward edges of the subtropical high pressure systems (approximately 35° N and S), and the beginnings of the polar circulations (approximate 60° N and S). For many years the middle-latitude climates were referred to as climates of the temperate zone. This proves a definitive misnomer, for while the zone contains some of the most equable of climates, it also has some of the most extreme. In the Northern Hemisphere (NH), the middle-latitude climates have some of the highest measure of continentality, an index that is essentially a measure of seasonal extremes.

This chapter examines various aspects of middle-latitude climates. It begins with an essay providing details of the development and significance of reanalysis. This recent analytical method draws upon a state-of-the-art data assimilation system to reprocess all past atmospheric environmental observations, combining them with short forecasts in order to derive the best estimate of the state and evolution of the environment. Since the statistical combination of the forecast and observations is denoted in operational applications as "analysis," the new method is usually known as reanalysis. Thereafter, an account of selected aspects of middle-latitude climates of the NH is followed by a discussion of the much-neglected climates of the Southern Hemisphere (SH).

4.2 Data availability

Those studying the climate of the middle latitudes in the NH have been relatively fortunate in terms of data availability. It is there that temperature and precipitation have been measured in reasonably accurate form for several hundred years. The longest range from some 300 years in Central England (Manley 1953) to about 200 years for a few locations in Europe and the eastern United States. Most of the stations that make up today's observational network have been kept for little more than 100 years.

Until the advent of weather satellites in 1966, when the Applications Technology Satellite (ATS1) was launched, SH and ocean data were sparse.

Archives for the oceans began in 1854 when the major maritime nations began a regular program of recording atmospheric and oceanic data. Such data were recorded mostly in the popular sea lanes, and for large oceanic areas only limited data were available.

As shown in Table 4.1, today's climatologists have available a number of outstanding data sources. All are readily accessible, many from the World Wide Web. However, extreme caution must be used when analyzing the data, for without a good knowledge of statistical techniques, many fundamental errors can be made. This is particularly true when correlations are completed, for some conclusions may be reached when a statistical relation appears between parameters with no identified scientific physical relationship.

Together with satellite imagery, perhaps the most significant development for climatology has been the availability of rapid computations through computer analysis. Many research papers that are published in the current literature use reanalysis, the subject of the essay that follows.

4.3 ESSAY: Reanalysis

Brian Giles, *University of Birmingham*

Reanalysis is the most recent in a long line of techniques used to improve our understanding of the climatology of planet Earth. Climatology is variously described as the synthesis of average and extreme weather conditions (Durst 1951), or the long-term aspects of meteorological processes (Lamb 1972). As Sorre (1934) put it climate is "l'ensemble des phénomènes météorologiques qui caractérise la condition moyenne de l'atmosphère en chaque lieu de la terre."[1] But each of these phrases requires refinement, not least in the way in which the meteorological elements are measured. Landsberg (1987) described it as "a statistical collective," which means that the data used must be compatible. After the first scientific weather instruments were invented in the seventeenth century (Chapter 3 in Khrgian 1970) scientists in Europe exchanged weather data, began to produce means and extremes, ensured their instruments were compatible with each other, and rationalized their units. This was helped during the nineteenth century by the development of national meteorological institutes and services (Chapters 8 and 9 in Khrgian 1970).

4.3.1 The background

In the twentieth century, four periods can be distinguished: the first 40 years were the "surface" period – the state of the art of climatology at the

---

[1] The ensemble of meteorological phenomena that characterize the average condition of the atmosphere in each part of the Earth.

Table 4.1   *Selected members of the World Data Center network that are directly related to research in climatology*

World Data Centers	Affiliation and location	Data emphasis
Atmospheric trace gases	Carbon Dioxide Information Center, Oak Ridge National Laboratory, Oak Ridge, Tennessee	Data related to atmospheric trace gases that affect and contribute to the Earth's energy budget
Glaciology – USA	National Snow and Ice Data Center, Boulder, Colorado	Snow cover, snow pack, sea-ice extent, sea-ice thickness, images and pictures of historic glacial extent
Glaciology, Geocryology – China	Lanzhou Institute of Glaciology and Geocryology, Chinese Academy of Sciences, Lanzhou, China	Glacial Atlas of China, snow cover, glacial extent and variation, periglacial data and hydrologic data
Glaciology – UK	The Royal Society and the Scott Polar Research Institute, University of Cambridge	Data related to glaciers, periglacial processes, satellite imagery, snow and ice chemistry
Airglow – Japan	National Astronomical Observatory, Tokyo, Japan	Airglow data, solar radiation data
Meteorology – China	Climate Data and Applications Office, Beijing, China	Real-time synoptic data, historical surface climate data, dendrochronology data, glacial data, and atmospheric chemistry data
Meteorology – Russia	Federal Service of Russia for Hydrometeorology and Monitoring of the Environment, Obninsk, Russia	Both surface observations and gridded surface and upper air meteorology data, marine ship observation data, and aerology data
Meteorology – USA	National Climate Data Center, Ashville, North Carolina	Archives of data from many national and international research projects and experiments including data from the IGY 1957–8 and International Quiet Sun Year, 1964–5 among many others. Synoptic surface and upper air climate data Global Historic Climate Data Network, Ozone Data for the World since 1965
Paleo-climatology – USA	National Geophysical Data Center (NGDC), Boulder, Colorado	Dendrochronology data, ice-core data, sea-floor sediment cores, coral data, proxy data on climatic forcing, including volcanic aerosol data, ice volume, atmospheric composition, etc. Numerical model simulation experiments data, climate reconstructions and maps

(From Simpson 2005)

beginning of the Second World War in 1939 was shown in *Climate and Man* (US Department of Agriculture 1941); the 1940s and 1950s were the "upper balloon period"; the 1960s and 1970s were the "modern rawinsonde" period and the 1980s heralded the "satellite" period.

Until the advent of satellites most of the reporting stations were unevenly based on the continents and there was a sparsity of data for the ocean areas, which constitute about 71% of the Earth's surface. In the second half of the twentieth century more data were collected over the ocean areas as well as over some of the more inhospitable land areas, more sophisticated measuring instruments (including rawinsondes and satellites) were developed, but also there was a realization that perhaps not all these data were comparable. Various international efforts were made to make this diverse amount of data internally compatible. It became apparent in the last quarter of the century that the data could not be used objectively in some climatological analyses, especially over the question of climate variability. Problems included urbanization, comparatively sparse amounts of data over the poles and oceans, changes in meteorological practices, and changes in station sites. The development of computers and computer models required that data be evenly distributed both horizontally and vertically rather than in the uneven manner that typified the observing station network. Two strategies were developed to overcome these problems and both relied on producing "gridded" data. The first strategy was the development of "standardization" or "homo-genization" of climate data. An early example of this was the careful way in which Manley (1953) constructed his first time series (1698–1952) of monthly mean temperatures for Central England (CET). He used overlapping sequences of observations from documented and carefully located stations. Although Manley never produced daily values of CET, this was done in the British Meteorological Office in the 1970s and early 1980s. But as Jones (1987) points out, there were by then three different monthly series and five different daily series; which should be considered the definitive one? Parker *et al.* (1992) published their CET daily series 1772–1991 and Jones (1994) a NH land temperature anomaly series for 1851–1993. Each of these papers briefly describes the methodology for producing a homogeneous temperature series that makes allowances for site changes whether locational or environmental. A detailed account of the methodology can be found in the three-part paper on Swedish temperature homogenization by Alexandersson and Moberg (1997), Moberg and Alexandersson (1997) and Moberg and Bergström (1997). For other examples of this strategy see the works of Trenberth and Paolino (1980), Easterling and Peterson (1995), and Jones and Bradley (1995). The homogen-ization of rawinsonde data is described in Lanzante *et al.* (2003).

At the same time, the second half of the twentieth century saw the rapid development of synoptic climatology (Barry and Perry 1973). This discipline focuses on the problems of analyzing surface and upper air circulation patterns and their related weather conditions. It grew out of the techniques of meteorological synoptic analysis (Petterssen 1956; Saucier 1955) based primarily on pressure fields and their high and low pressure centers. Despite the methodology being standard and the data being the same, the analyzed charts often varied because of differences of interpretation. This is also true of the modern computer generated charts because of the programs that are used. Hence the second strategy in the 1990s was the production of assimilated data sets of both atmospheric (and later oceanographic) fields that were needed by the climatological community and their computers. Assimilation refers to the technique whereby current data is introduced iteratively into atmospheric general circulation models (AGCM) that were developed in the 1970s and 1980s by the major operational weather prediction centers originally for producing weather forecasts. Later similar models were developed for climate modeling (see McGuffie and Henderson-Sellers 2005).

The analyses depended on systems that quality control, check for and correct inconsistencies in the raw data, and finally produce gridded data. The outputs of these global analyses have been used in global climate models (see Gates *et al*. 1996). Unfortunately the systems and the analyses produced were continually evolving and so were strictly not compatible. One answer to this was to produce "control" data sets for a period of years such as the NASA/DAO (National Aeronautics and Space Administration/Data Assimilation Office of the Goddard Laboratory for Atmospheres) assimilated data set 1985–9 (Schubert *et al*. 1993). Examples of other data sets that are available include those at the British Atmospheric Data Centre (badc.nerc. ac.uk/data/), the University of East Anglia, UK, Climate Research Unit (www.cru.uea.ac.uk/cru/data) and the Intergovernmental Panel on Climate Change Data Distribution Centre (IPCC DDC) (ipcc-ddc.cru.uea.ac.uk/asres/ scenario_home.html). The second answer was to produce retroactive records of global analyses of atmospheric fields using a frozen state-of-the-art analysis system – the reanalysis projects.

4.3.2 Development

The concept of the first reanalysis project was discussed at a National Meteorological Center (NMC) workshop in April 1991. A problem had arisen because there seemed to be some "climate changes" in the NMC climate assimilation system that were not real. It was discovered that these changes were really due to operational changes and so were analogous to the problem that homogenization had solved with station or point data. In the case of the NMC assimilation system, upper air observation times changed in June 1957 from 6 hourly intervals beginning at 0300 UTC to the current 6 hourly intervals beginning at 0000 UTC. Also the statistical techniques used in the assimilation model were continually being improved. The past data also included changes in the observation systems and changes in the number of stations (or raw observation points). The workshop had two choices: either to pick a subset of data that remained stable throughout the chosen period, or to use observational data available at a given time.

The first choice would give the most stable climate and is analogous to the use of 30-year means that the World Meteorological Organization (WMO) recommend to describe the climate of a place. The second choice would provide the most accurate analysis (on the basis that more is better!) for the period of the reanalysis. The workshop opted for the second choice and so the National Centers for Environmental Prediction/National Center for Atmospheric Research (NCEP/NCAR) 40-Year Reanalysis project was born and would cover the period 1957–96. The data used in NCEP/NCAR-40 consisted of global rawinsonde data obtained from NCEP (1962 onwards), the US Air Force (1948–70), China (1954–62), the former USSR (1961–78), Japan and several other countries; COADS (Comprehensive Ocean-Atmosphere Data Set, Woodruffe *et al.* 1987); aircraft and constant level balloon data from a variety of sources, mainly post-1962; surface land synoptic data since 1949; satellite sounder data since 1969; and satellite cloud drift winds using geostationary meteorological satellites. Once all these data were assimilated the reanalysis output came in two main formats: the binary universal format representation (BUFR) of the WMO and the gridded binary format (GRIB).

The data output is given for 17 standard pressure levels on a 2.5° latitude × 2.5° longitude grid at 0000, 0600, 1200 and 1800 UTC each day. In addition, a variety of monthly means of atmospheric fields are produced. Full details of NCEP/NCAR-40 are given in Kalnay *et al.* (1996).

Whilst NCEP/NCAR-40 was being developed it was intended to extend the reanalysis backward in time to incorporate (the mainly NH) upper air data for the 1948–57 decade. This hope became a reality in 2001 when the NCEP/NCAR 50-Year Reanalysis (1948–98) was described by Kistler *et al.* (2001). Information and data for both of these reanalyses can be accessed online at www.cdc.noaa.gov/cdc/reanalysis/reanalysis.shtml and at dss.ucar.edu/pub/reanalysis/rean_model.html.

Both these reanalyses are used widely by the climatological community in current research. As an example, out of 118 articles published in the *International Journal of Climatology* in 2003, eleven used these NCEP/NCAR reanalyses. Generally they compared the reanalysis data with other contemporary data: in the Arctic Ocean with sea level pressure measured by the Soviet drifting camps (1950–90) (Cullather and Lynch 2003); synoptic weather classifications in Midwestern USA using "raw" radiosonde data series (1946–96) (Schoof and Pryor 2003); global tropospheric temperatures derived from satellite microwave sounding units and global surface air temperatures (1979–90) (Sturaro 2003).

The second United States reanalysis project developed from the NASA/ DAO assimilated data set 1985–9, mentioned above, when it was expanded to provide a 17-year reanalysis covering the period 1979–97. It was designated as the NCEP/DOE (Department of Energy) AMIP-II Reanalysis or Reanalysis 2 (note that by 2004 the organization had been renamed GMAO or the Global Modeling and Assimilation Office of the Goddard Space Flight Center). AMIP (Atmospheric Model Intercomparison Project) was inaugurated in 1990 and is part of the World Climate Research Programme (see www-pcmdi.llnl.gov/projects/amip) with offices at Lawrence Livermore National Laboratory, California, USA. Reanalysis 2 has improved parameterizations of the physical properties and also fixed the errors discovered in the original assimilation data set. It has been running in real-time since 2001 and now covers the period from 1979 to present. Details may be found on wesley.wwb.noaa.gov/reanalysis2/index.html.

In February 1993 the European Centre for Medium-Range Weather Forecasts (ECMWF) began a 15-year reanalysis programme (ERA-15) to produce an assimilated data set for the period 1979–93. This project used its own archive of observations received from the WMO, COADS, the Hadley Centre (UK) sea ice and sea surface temperature (SST) data set as well as the NCEP SST analyses, data from the First GARP (Global Atmospheric Research Program) Global Experiment (FGGE) of 1978–9, the Alpine Experiment (ALPEX) of 1982, Japanese and Australian data, TOGA (Tropical Ocean Global Atmosphere) data, and the vertical sounder data from the TIROS satellites. ECMWF had its own data assimilation system and the project was completed in September 1996 (see www.ecmwf.int/ research/era/ERA-15/Project/). Although this was a useful experiment, the documentation acknowledges that the reanalysis contained some problems particularly with respect to changes in the amount and coverage of the radiosonde data. In 2000 a new reanalysis was begun by ECMWF known as ERA-40 to cover the period 1957–2001. This was completed in 2003. It used the same data sources as ERA-15 but with the addition of a variety of satellite radiances (from 1972 onwards), cloud photographs and cloud motion winds, particularly those of the European Meteosat geostationary satellite from 1982–8. The output is 6-hourly analyses on a grid with spacing of about 125 km in the horizontal and with 60 levels in the vertical between the surface and 65 km. (Simmons and Gibson 2000). The current availability of ERA-40 analyses can be checked online at www.ecmwf.int/research/era/ Data_Services/section1.html. During the process of validation some deficiencies in the reanalyses have been found so there was a rerun covering the latter part of the ERA-40 period. An example of the use is given by Fink *et al.* (2004), who compare the reanalysis data with the raw data obtained in GATE (GARP Atlantic Tropical Experiment) in 1974. They show that the subjectively analyzed wind fields presented in the original studies compare favorably with the ERA-40 horizontal wind fields although the vertical motion field shows the largest differences from previously published results. This is almost certainly due to the fact that the original ship dispositions produced a much denser upper air network than that produced by the reanalysis.

The Japanese Re-Analysis 25 years (JRA-25) is another major reanalysis and is a joint project of the Japanese Meteorological Agency (JMA) and the Japanese Central Research Institute of the Electric Power Industry. It covers the period 1979–2004. The project began in 2001 with data preparation and development of the assimilation procedures. The JMA will combine its own historical archive of observational data with the merged databases of the NCEP and ECMWF archives. The output is based on a 2.5° latitude × 2.5° longitude grid and will consist of pressure level analyses, a variety of atmospheric variables and land surface variables. The assimilation system will use the most recent JMA operational models. The reanalysis began in 2003 and was due for completion in 2005. Some comparisons with the NCEP/NCAR and ERA reanalyses have already been carried out and the results can be found within the JRA-25 website at www.jreap.org/indexe.html. There are also proposals for a JRA-30 and JRA-50. The former would cover the satellite period and the latter would use only conventional observations. These were outlined at a workshop in Colorado in 2003 (see below).

Other projects have been based in universities. The Climate Research Unit of the University of East Anglia, UK, has been an important center for climate research since it was formed in 1972. Its various members have produced a series of databases of both sets of stations and gridded series (see Jones 1994; Jones and Moberg 2003; Konnen *et al.* 2003; Hanson *et al.* 2004). Within the United States many universities are involved and their research input is usually publicized on their individual websites. The American Meteorological Society website also provides numerous links as well as showing research via its various journals (ametsoc.org/AMS/).

Several workshops on reanalysis have already taken place. The WMO World Climate Research Programme (WCRP) sponsored one in the USA in October 1997 (WCRP 1998) and one in the UK in 1999 (WCRP 2000). The latter consisted of 109 papers covering the development and use of the various reanalysis projects. A third workshop was held under the auspices of ECMWF at Reading, UK, in November 2001 and the 34 papers are available at www.ecmwf.int/publications/library/ecpublications/proceedings/ERA40-reanalysis_workshop/index.html. In August 2003 there was a workshop on ongoing reanalysis of the climate system held in Boulder, Colorado, USA. Amongst other things a white paper was presented outlining a US national programme on three new reanalysis initiatives: R1979 – reanalysis for the satellite era using TOVS/SSU (TIROS Operational Vertical Sounder/Stratospheric Sounding Unit) data and aimed at the best possible spatial analyses rather than the best temporal analyses; R1950 – reanalysis for the era of upper air observations; R1850 – reanalysis for the era of surface observations, which would provide globally consistent gridded analyses of the key surface fields. The final workshop report can be found at www.joss.ucar.edu/joss_psg/meetings/climatesystem as can the text of the white paper.

4.3.3 Summary

This essay has attempted to describe and explain one of the most recent and exciting developments in the field of climatology. It has briefly traced the subject from the early days of simple instrumental measurements, through the analysis of the data that they produced to the current situation where the demands of the current computers require data that is evenly spread spatially across and above the Earth's surface. As a consequence of the computer revolution in meteorology and climatology in the second half of the twentieth century, the uneven station distribution that had developed in the previous 150 years, almost entirely on 26 percent of the land surface (very few being sited on Antarctica or Greenland), was deemed unsuitable. In addition, because of changes in location and the local environment, station data were not always comparable. The first step was to statistically manipulate the data so that local inconsistencies were ironed out and this led to homogenized data. These techniques were developed to such an extent that data sets covering the whole globe were produced. The final step was to again manipulate the data so that their uneven distribution over the Earth's surface was removed and they were replaced, via various interpolation techniques, by gridded values in both horizontal and vertical planes. This was and is the aim of the various reanalysis projects. But what should always be remembered is that data are only as good as the original measurement and ground truth will always be required to check the outputs of the computers in the reanalysis projects.

4.4 Using reanalysis

The enormous value of reanalysis is illustrated in the number of research papers that use the data. The wide variety of applicability and utility of reanalysis can be aptly demonstrated by looking at but a few example from among the many that exist. The following examples show how reanalysis can be used for different time intervals, in relation to other data sources, and for various climatic elements.

A study using NCEP/NCAR reanalysis data for varying time intervals was provided by Gulev *et al.* (2001). They used 6-hourly NCEP/NCAR reanalysis data for the period from 1958 to 1999 together with software providing improved accuracy in cyclone identification to study the climatology of NH winter cyclone activity. The findings of this work showed that: (1) there are secular and decadal-scale changes in cyclone frequency, intensity, lifetime, and deepening rates; (2) in the Arctic and in the western Pacific there is an increase in the number of cyclones, while over the Gulf Stream/North Atlantic Drift and in the subpolar Pacific Ocean there is a decreasing trend; (3) both Atlantic and Pacific cyclone activity is positively correlated to the NAO and PNA. The NAO relationship also reflects the 1970s shift associated with European storm tracks. The PNA is largely linked to the eastern Pacific cyclone frequencies, and, especially over the last 20 years, cyclone activity over the Gulf of Mexico region and the eastern North American coast.

The NCEP/NCAR reanalysis is also frequently used in conjunction with other data sources to examine many aspects of the surface–atmosphere interaction. A study by Washington *et al.* (2003) used reanalysis together with both ground station data and that derived from satellite, in this case the Total Ozone Monitoring Spectrometer (TOMS). The objective of the study was to better understand terrestrial sources and the transport mechanisms responsible for the production and distribution of atmospheric dust, detailed knowledge of which is important in reducing uncertainties in the modeling of past and future climates. The study showed the Sahara to be the primary key source region while other key regions were the Middle East, Taklamakan, southwest Asia, central Australia, the Etosha and Mkgadikgadi basins of Southern Africa, the Salar de Uyuni (Bolivia), and the Great Basin (United States). The study also outlined some discrepancies that occurred between surface and satellite observations.

Reanalysis can be used in a number of ways for various climatic elements. Using the Climate Prediction Center (CPC) Merged Analysis of Precipitation (CMAP) product together with the Goddard Earth Observing System (GEOS) reanalysis and the National Center for Environmental Prediction (NCEP) sea surface temperature (SST) data Zhou and Lau (2002) examined the principal modes of summer rainfall over South America for the period 1979–95. The leading modes of rainfall variation identified annual, decadal, and long-term variability. The first mode is highly correlated with El Niño–Southern Oscillation (ENSO), showing a regional rainfall anomaly pattern largely consistent with previous results. This mode captures the summer season inter-annual variability in northeast Brazil and its connection with excessive rainfall over southern Brazil and Ecuador. The modes associated with long-term variability show that since 1980 there has been a decrease of rainfall from the northwest coast to the southeast subtropical region and a southwards shift of the Atlantic ITCZ. This variation leads to increased rainfall over northern and eastern Brazil.

One final example illustrating the use of the NCEP/NCAR reanalysis data deals with the climatology of the spring monsoon in South China (Yan 2002). The results indicate that the spring monsoon season in South China occurs in April and May, a finding supported by both seasonal and interannual variation of circulation and precipitation patterns. The interannual variation of the spring monsoon rainfall in South China relates primarily to the anomalous circulation over the North Pacific Ocean. This is related to the westerly jet stream over North Asia, the polar vortex, and sea surface temperature anomalies in the Pacific. It is shown that changes in the Asian tropical atmospheric circulation have little influence on the spring monsoon in South China.

4.5 The Northern Hemisphere

4.5.1 Middle-latitude circulation: evolution of a concept

In 1735 George Hadley provided a hypothesis that was to become the starting point of succeeding models of the general circulation of the atmosphere. In an attempt to explain the trade winds observed and reported by mariners, Hadley suggested that air heated at the equator would rise and, at some high elevation, spread toward the poles. The return flow near the surface would complete the cell and hence explain the trades. Accordingly, a convective cell (a Hadley Cell) was identified in each hemisphere. Subsequent observations of surface pressure and wind patterns negated the idea of a single huge cell in each hemisphere and saw the creation of the three-cell model in which three cells (Hadley, Ferrel, and Polar) were identified in each hemisphere.

General circulation models from the 1800s are not markedly different from the general circulation models used in many textbooks of today. Superficially, the three-cell model appears to explain observed surface phenomena; equatorial low pressure, trade winds, the subtropical high pressure, mid-latitude surface westerlies, subpolar lows, and polar easterlies. However, the three-cell model does not consider many variables including seasonal variations.

First, consider data applicable to meridional circulation (Figure 4.1). Winter meridional circulation in the NH shows a definitive transfer between the equator and about 30° N, the Hadley Cell, but poleward of 30° N there is no clearly defined pattern (Palmén and Newton 1969). The meridional circulation of the Ferrel Cell is some 80% weaker than the Hadley Cell, while a polar meridional circulation is not evident at all. In the extratropical areas of the Ferrel and Polar Cells, zonal mass transfer is, in fact, an order of magnitude larger than meridional transfer during the winter season. During summer NH meridional circulation patterns are even weaker.

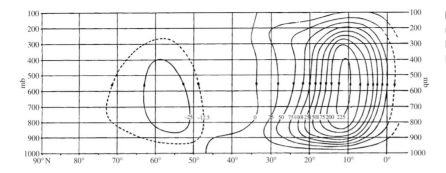

Figure 4.1 Winter meridional circulation in the Northern Hemisphere. (After Palmén and Newton 1969)

Upper-air westerlies

Carl Gustoff Rossby *et al.* (1939) provided a physical or process-based explanation for the wave-like features (now called Rossby waves) that characterize the upper-level westerlies. The main reason for the existence of a general atmospheric circulation is the need for redistribution of energy over the globe. Rossby, and many subsequent researchers, demonstrated that the need for equator–pole energy transfer and the strength of the wind flow aloft combine to produce waves in the zonal westerlies. The physics of air movement through ridges and troughs of Rossby waves results in upper-level convergence and divergence in selected portions of the waves.

If the positions of longwave troughs and ridges during a given season are predisposed toward specific locations and orientations, then surface pressure systems or the centers of action can be related to the Rossby wave pattern and position. The surface thermal properties of the Earth, specifically the temperature differences between continental and oceanic areas, play a significant role in anchoring the position of certain Rossby wave features. In winter, for example, air movements from over relatively warmer ocean waters toward an intensely cold landmass result in a tendency for anticyclonic curves in the flow (a ridge); the Western North American Ridge is a dominant feature of the winter climate of North America. In contrast, air flowing off a cold landmass and over a warm ocean current will be given a cyclonic torque; major troughs are climatologically preferred winter season features for the east coasts of both North America and Asia. A seasonal analysis of the frequency distribution of trough and ridge axes at 45° N (Harman 1991) is shown in Figure 4.2. The most clearly defined modes are seen in the January distribution where well-defined and persistent ridges occur at 90° E (continental Asia), 20° W (North Atlantic), and 120–140° W (Eastern Pacific–Western North American Ridge). Airflow away from a ridge and toward the next downstream trough produces subsident motion and surface high pressure. When this sinking motion occurs above a cold and snow-covered landmass, major "centers of action" are produced. The North American (Canadian) and Asian (Siberian) highs are positioned just downstream from major ridges.

Figure 4.2 Percentage
frequency distribution of
mean monthly 500-mb
trough and ridge axes at
45° N by 10° deg longitude
increments for (a) January
and (b) July. (From Harman
1991)

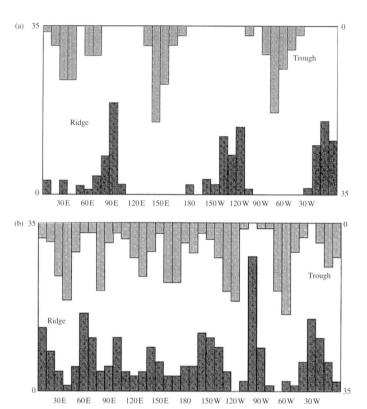

Similarly, the preferred locations of the winter season trough axes (60° W,
150° E and 30° E) are associated with surface low pressure positioned just
downstream. Two of these trough axes (60° W and 150° E) are located above
the east coasts of major continents where relatively warm, poleward flowing
ocean currents provide additional energy to fuel cyclonic storms. The Icelandic
and Aleutian "centers of action" are produced in this manner. Wind patterns and
thermal advection associated with the surface cyclones again help reinforce the
preferred location of ridges and troughs in the upper tropospheric westerlies.
Winds on the backside of a cyclone bring cold polar air southward reinforcing
the trough; while southerly surface flow to the east of the low helps advect warm
air poleward reinforcing the next, downstream, ridge.

The frequency distributions of ridge and trough locations for the remaining
months (Harman 1991) show less clustering of ridge and trough axes with
greatest variability in longwave positions occurring in July (Figure 4.2). This
redistribution of ridge and trough locations is indicative of the breakdown of the
strong hemispheric thermal gradient of winter through the transition months
(spring and fall) to the weak thermal regime of summer (July), as illustrated in
Figure 4.3. The orientation of the westerlies (including ridge–trough strength

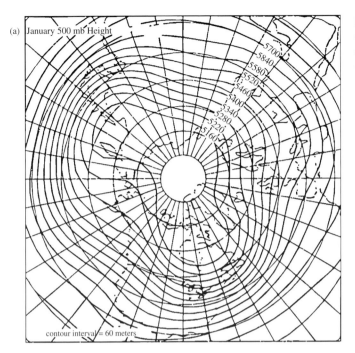

(a) January 500 mb Height

5700
5840
5580
5520
5460
5400
5340
5280
5220
5160

contour interval = 60 meters

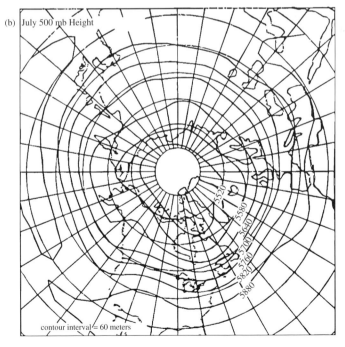

(b) July 500 mb Height

5520
5580
5640
5700
5760
5820
5880

contour interval = 60 meters

Figure 4.3 Mean height of the 500-mb contours in (a) January and (b) July. (From Harman 1991)

and persistence), the location and strength of the centers of action, and seasonal circulation variations are thus related to the seasonally changing geography of the planetary energy budget (Harrington and Oliver 2003).

It is clear that the positions of the ridges and troughs are the keys to seasonal weather and, more recently, these have been related to atmospheric teleconnections. Of particular note in this respect is the PNA. In an earlier discussion of this feature (Section 2.4) the significance of location and extent of upper-air flows were shown to be the key to the use of the PNA in explaining the climates of North America.

4.5.2 Precipitation trends and patterns

Given the amount of published research relating to global warming, there is a vast literature dealing with potential changes in circulation patterns and temperatures in the NH. Research concerning precipitation patterns and trends, while abundant, is less widely known, so this area of research is selected as the topic for this section. Of interest in such research are the problems of data reliability and the fact that a number of studies have related precipitation trends and variability to teleconnections indices. There is also, of course, the question of whether precipitation amounts have increased in recent decades and, if so, whether the increase related to an increase in extreme events.

Among the best-known studies of future climates, including assessments of precipitation, are those provided by the Intergovernmental Panel on Climate Change (IPCC, www.ipcc.ch; Houghton *et al*. 1995). The work of the IPCC is guided by the mandate given to it by its parent organizations the WMO and UNEP. The IPCC has three working groups: Working Group I assesses scientific aspects of climate systems and climate change; Working Group II studies the vulnerability of human and natural systems to climate change and options for adapting to them; Working Group III assesses options for limiting greenhouse gas emissions and economic issues. Also, a task force studies national greenhouse gas inventories.

Precipitation trends
In reporting future precipitation trends there is invariably a caution concerning GCM and other simulations, for much uncertainty exists. Nonetheless, in Europe, for example, the IPCC reports that most of the continent will experience an increase in precipitation, especially in higher latitudes. The Mediterranean regions and parts of Eastern Europe may experience declines. The European Climate Assessment (ECA), which was initiated to analyze temperature and data precipitation in Europe, provides a similar assessment (Klein Tank 2002).

A study considering global precipitation trends was completed by New *et al*. (2001). This research found that, over the land areas, from 1901 to the mid 1950s, precipitation remained below the century long mean. From 1950 until

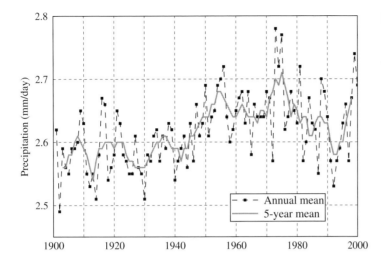

Figure 4.4 Global annual mean and five-year running mean precipitation change. (Source: NASA:GISS)

the 1970s, the precipitation was above the mean; it then declined until the mid 1990s. Hence, for the twentieth century, this study found variations, but essentially from about 1915 there has been no net change. Given that the NH is the land hemisphere, such a finding would seem significant to areas such as agriculture and insurance (see Section 8.4).

The graphical expression of global precipitation trends is shown in Figure 4.4. The NH data are graphed in Figure 4.5, which, for comparison, also shows other world areas. These graphs were derived from the CRU TS 2.0 data set that comprises 1200 monthly grids of observed climate for the period 1901–2000. Available from the Tyndall Centre for Climate Change Research (www.cru.uea. ac.uk/~timm/grid/CRU_TS_2_0_text.html) the data set covers the global land surface at 0.5° resolution. The data are also readily available from other sources.

In another study of global precipitation variations for the twentieth century, Dai *et al.* (1997) used statistical analysis and found the leading EOF to be ENSO-related precipitation variations in low latitudes. The second EOF demonstrated an increasing global linear trend for the world; the EOF trend is seen as a long-term increase in North America, mid- to high-latitude Eurasia and, in the SH, in Argentina and Australia. Interestingly, these researchers found that the NAO accounts for some 10% of winter variance over regions surrounding the North Atlantic Ocean. During high-index NAO winters, northern Europe, the eastern United States and the Mediterranean experience above normal precipitation. Below normal occurs in southern Europe and eastern Canada.

Precipitation patterns

Figure 4.6 provides the twentieth century precipitation trend for the United States based upon the CRU TS data-base. While this does provide an overall

Figure 4.5 Annual (black line) and five-year running (gray line) mean precipitation bands that cover 30%, 40% and 30% of the global area. (Source: NASA:GISS)

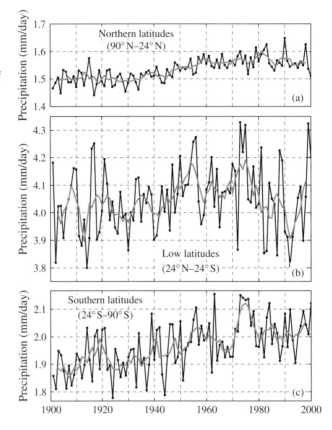

Figure 4.6 Annual and five-year running mean precipitation in the contiguous United States. (Source: NASA:GISS)

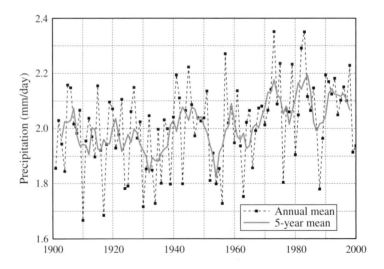

image, studies of identified regions show that area differences do occur. In North America, there has been a gradual increase in precipitation since the 1920s for large parts of the country. Increases of more than 20% have occurred in north-western and eastern Canada, and in parts of the US Gulf coast. Decreases have occurred over much of the High Plains and California (Karl and Knight 1998). However, in a study of 105 years of data, from 1895 to 1999, Garbrecht and Rossel (2002) showed that in the central and south Great Plains above normal precipitation occurred in the last two decades of the twentieth century. During this 20-year span, there was a reduction in the number of dry years, and an increase in the number of wet years.

Yan (2002), in a study of temporal and spatial patterns of seasonal precipitation variability in China, identified four distinct precipitation regions through cluster analysis. In examining trends and variability of each region, none were found to have significant increasing or decreasing trends for the period 1951–99. In relating the role of large-scale circulation patterns in precipitation variability in the identified regions, Yan found that the Arctic and Polar-Eurasia Oscillations were positively associated with winter precipitation. Other teleconnections indices, including the Southern Oscillation, appeared to have negligible effects.

Apart from precipitation trends and spatial patterns, considerable interest has been afforded extreme precipitation events. The IPCC (Cusbasch et al. 2001) has noted that there is likely to be an increase in heavy and extreme precipitation events in those regions where the trend shows increasing precipitation. This, as noted earlier (Figure 4.5), is largely in the mid to high latitudes of the NH. Kunkel et al. (2003) designed a study to test whether extreme precipitation events were increasing in the United States using daily precipitation observations for the period 1895–2000. They revealed that heavy precipitation frequencies in the late nineteenth and early twentieth centuries were very similar to the frequency that occurred in the 1980s and 1990s. The values reached a minimum in the 1920s and 1930s. This finding indicates that for this study area, at least, there has not been an increasing trend in extreme precipitation events when long-term data series are used. Had the longer time series containing earlier observations not been used, then a trend might have been discerned. Clearly, the significance of the data series used is of singular importance.

In a similar study, Michaels et al. (2004) focused on trends of precipitation on the 10 wettest days of the year locations in the contiguous United States. These trends were then compared with trends in overall precipitation. They found that for the region as a whole the trends of the two sets are not significantly different. On a regional scale, only in the northeast and southeast does the rate of increase of wettest days exceed the total; for the rest of the country, precipitation on the wettest days is increasing at a lesser rate than the total precipitation increases.

The results obtained by Kanae et al. (2004) of changes in hourly precipitation at Tokyo also show that extreme precipitation events have not increased in recent years. Using a data set beginning in 1890, the researchers noted that strong/frequent hourly precipitation occurred more in the 1940s, and although the 1990s had a considerable number of extreme hourly events, they did not match the earlier period. The authors did note that precipitation events caused by tropical cyclones or Baiu fronts differed in the two periods.

From the foregoing, it is fairly clear that there are differences of opinion in ongoing research. At the same time, the role of teleconnections in precipitation variability has been considered in a number of ways. In the studies cited above, teleconnections were used to help explain variances. But also to be considered is the role of precipitation, in the form of snow, as a possible forcing mechanism in atmospheric oscillations. Gong *et al.* (2002) explored this possibility in relation to snow cover and the AO and NAO. Using large-ensemble atmospheric GCMs, they compared two different surface snow conditions. They concluded that their numerical modeling results agreed with earlier studies that link the AO/ NAO mode with the Siberian high pressure cell and associated variations in snow cover.

4.6 Mid-latitude circulation and teleconnections in the Southern Hemisphere

The SH climate pattern in the mid-latitudes occurs above a significantly different surface structure than that in the NH. Here, the mid-latitudes are located north of the circumpolar trough (CPT) which surrounds Antarctica (Jones *et al.* 1999). The region is dominated by ocean, with the exception of the southern tip of South America, the very southern edge of Southern Africa, Tasmania and the south coast of Australia, and most of New Zealand. Land extends north from Antarctica into the mid-latitudes at 60–70° W longitude, the Palmer Peninsula. This is the area encompassing the "Roaring Forties," the famous SH westerly belt, which extends around the globe virtually unchecked. It is the locational basis of many seafaring stories and novels, such as *Moby Dick* by Herman Melville.

The climate and circulation of the SH mid-latitude westerlies is affected by a range of influences (Trenberth 1987; Tyson *et al.* 1997; Bigg *et al.* 2003 and others). The QBO and ENSO, so important for the tropical climates (see Chapter 2), have overlapping influences in the mid-latitudes. The circumpolar vortex (CPV) and the associated Rossby waves determine the frequency and strength of zonality (wave number 1) or meridionality (wave number 3). In contrast to the NH, there is one center of circulation control, the East Antarctic plateau. Biannual seasonal changes and quasi-decadal changes in circulation patterns and strength determine both spatial and temporal climate variations. Circulation strength and variability are also influenced by the temperature gradients between Antarctica and the tropics, and between ocean, land, and ice. SST variations are also important influences, although not to the extent they are in the tropics. The extent and variations in sea ice also have important effects. Although this section focuses on the mid-latitudes, it is important to remember that links and complex interactions with both the tropics and the polar regions exist.

4.6.1 The role of the oceans

The review by Bigg *et al.* (2003) provides an excellent background to the role of oceans in SH mid-latitude circulation and climate. Aside from the interchange of gases, particulates, heat, and water between the ocean surface and the atmosphere, the ocean redistributes heat across the mid-latitude zone, and creates a thermal lag on climate and climate change. In the SH, two great cold ocean currents, originating from the circumpolar ocean circulation, dominate. These are the Benguela Current, which moves north along the west coast of Southern Africa, and the Humboldt Current, which strongly influences stability, temperature, and moisture along the west coast of South America. By contrast, the warm currents on the east sides of continents tend to have more influence on regional and local scales.

Bigg *et al.* (2003) and Fauchereau *et al.* (2003) define two major differences in the role of the ocean in the mid-latitudes compared to the tropics. First, SST is not as important because its influence tends to be overwhelmed by several other influences on circulation, some of which are listed in the previous section. Divergent circulations in the mid-latitude baroclinic atmosphere are considerably less important than in the tropics, where barotropic circulations are highly influenced by surface conditions. The mid-latitude atmosphere also has considerably greater internal variability than the tropics, associated with shifts in the troughs and ridges of Rossby wave number 3 for example.

Second, the response timing of oscillations in climate in the SH mid-latitudes is controlled mainly by ocean influences. Changes occur over a longer time scale than in the tropics, at the level of inter-decadal rather than inter-annual. For example, the pattern and timing of Rossby wave propagations are strongly influenced by the ocean.

The ocean influence varies spatially, with regions such as the SW Indian Ocean and the central Atlantic Ocean showing stronger coupling and better phasing of climate variability than other areas. In these regions, Fauchereau *et al.* (2003) show that latent heat influences on the atmosphere are most important on the western edge of SST anomalies, with a reasonably close correlation to boundary-layer wind anomalies. SST warming and/or reduced westerly winds lead to reduced latent heat flux, and SST cooling and increased westerly winds create the opposite. However, the significance of ocean influences on mid-latitude atmospheric forcing in this area is still fairly weak. ENSO links explain approximately 20% of the total variance for the Atlantic, and 17% for the SW Indian Ocean, with maximum mid-latitude correlations reaching about 0.3 (>0.4 at about 30° S just south of Madagascar). Correlations with Southern African rainfall, on the slightly more tropical east coast, are on the order of 0.5. Despite its importance as the dominant mode of variability in the hemisphere Karoly *et al.* (1996) and Bigg *et al.* (2003) suggest that the role of ENSO is probably overemphasised.

4.6.2 The influence of jet streams

The jet streams in the middle and upper troposphere in the SH mid-latitudes are not as affected by terrain compared to the NH. They are important steering mechanisms for surface cyclones and anticyclones. Jets are normally strongest in the winter, globally averaging $>40\,\mathrm{m\,s^{-1}}$, compared to about $10\,\mathrm{m\,s^{-1}}$ in summer. However, Trenberth (1987) states that, unlike the NH, zonal mean winds representing jets can be stronger in the summer in some parts of the SH atmosphere. He illustrates this point by comparing vertical wind shears in the 40–50° S latitudinal band between the hemispheres. In winter, the results are comparable, $19.8\,\mathrm{m\,s^{-1}}$ in the SH compared to $20.6\,\mathrm{m\,s^{-1}}$ in the NH. In summer the mean jet is much weaker in the NH, $10.2\,\mathrm{m\,s^{-1}}$ compared to $20.8\,\mathrm{m\,s^{-1}}$ in the SH.

In the southwest Pacific, jets in winter can occur near 40° S (subtropical jet) and 60–65° S (polar front jet) (Sinclair 1996). Trenberth (1987) emphasizes the importance of a split jet occurring during stronger meridional flow, but there is considerable variation depending on the circumstances.

4.6.3 General circulation

Analysis of surface pressure distributions and variations is the main approach taken to describing mid-latitude general circulation variability in the SH, and the importance of zonal flow. One approach was developed in the early 1980s by Pittock (1984), and adopted by Jones *et al.* (1999). This is the Trans-Polar Index (TPI), designed to measure the influence of Rossby wave number 1. The TPI is a similar calculation to the SOI in the SH tropics, in that it compares pressure at two widely spaced locations, Hobart, Tasmania, in Australia, and Stanley in the Falkland Islands. While these locations may not be ideal, the results of TPI calculations for the four seasons, reproduced in Figure 4.7, provide some indication of the relative strength of zonal and meridional circulation.

It is immediately apparent from Figure 4.7 that there is considerable variability in the TPI values between seasons and between years. Positive values indicate higher pressure in Hobart compared to Stanley and a stronger zonal index. Negative values indicate higher pressure in Stanley than Hobart and stronger meridional circulation. The solid line in Figure 4.7 is a 10-year Gaussian filter that provides an indication of trends. In autumn for example, the TPI values are mainly positive between 1900 and 1940, suggesting dominant zonal circulation. Between 1940 and 1980, greater negative years existed,

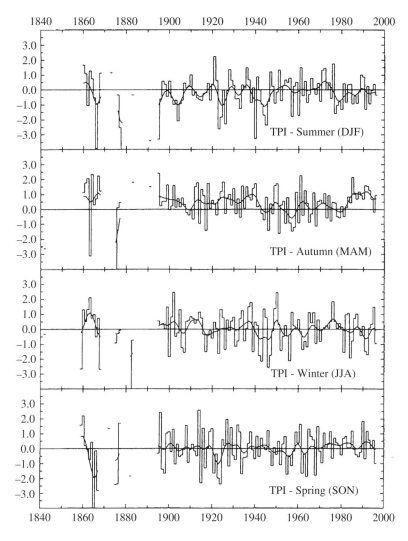

Figure 4.7 Seasonal values of the Transpolar Index (TPI), calculated from MSLP data from Hobart, Australia, and Stanley, Falkland Islands, compared to a base period of 1951–80. (From Jones *et al.* 1999, Figure 6, Copyright John Wiley & Sons Ltd., with permission)

indicating that meridional circulation was more important than in the previous period. A reversion to dominant zonal flow has occurred since 1980.

Karoly *et al.* (1996) measured variance of SH mid-latitude circulation associated with several indices. They discovered that ENSO was still the most important factor (29% of variance explained) with the TPI second (16% variance explained). No other index of measurement was significant. The authors state that the TPI is independent of ENSO, and therefore a reasonable indicator of changes in the mid-latitude zonal pressure gradient linked to the importance of Rossby waves 1 and 3.

The TPI does not have a strong correlation with temperature variations on a regional scale. It does, however, show significant negative correlations (approx -0.3 to -0.5) with temperature in southern South America, in summer and autumn. Opposite correlations also occur with New Zealand temperature in summer. Regional meridional flows tend to be more important, reflecting locational shifts in the ridges and troughs associated with the Rossby waves.

4.6.4 Correlations and spatial distributions

Fauchereau *et al.* (2003) used NCAR/NCEP reanalysis (see Section 4.3) to reproduce sea level pressure distributions, geopotential height variations, and wind flows (at 925 hPa) for the SH mid-latitudes between 1950 and 1999. The authors were able to establish where spatially the strongest in-phase correlations existed. Figure 4.8 presents a summary of the overall pressure results. The strongest correlations occur between pressure fields in the SW Indian and SW Atlantic Oceans, especially in the summer season. The Tasman Sea area between Australia and New Zealand can also be included, but this has a considerably weaker correlation with the other two. Figure 4.8 shows that there are no other areas showing correlations across the mid-latitude zone.

Composite analysis in summer was then used to establish the changes in spatial distribution of the pressure field under anomalous conditions. When either warm anomalies occur in the SW parts of the Atlantic and Indian Oceans, or cool anomalies occur in the NE, there is an associated shift of the subtropical highs to the SE, and they become stronger. Significant anomalies also appear in the Tasman Sea and in the southern Pacific, around 120° W. These positive pressure anomalies can extend to 250 hPa over the Indian Ocean and Tasman Sea, and to 400 hPa over the SE Pacific and the Atlantic. Fauchereau *et al.* (2003) state that, because these centers of composite correlation are distinct from each other, and well separated, as well as being almost barotropic in nature, a wave structure associated with the CPV is directly involved. These mid-latitude variations are out of phase with the climatology of changes in geopotential heights over the subtropical and Antarctic regions (Goodwin *et al.* 2003).

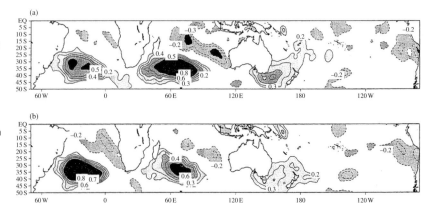

Figure 4.8 Correlations between MSLP pressures in the southwest Indian Ocean (a), southwest Atlantic Ocean (b) and SST anomalies for summer. (After Fauchereau *et al.* 2003, Figure 5, Copyright John Wiley & Sons Ltd., with permission)

Seasonally, the strength of the subtropical highs varies (Leighton 1994). In the Indian Ocean, the high is strongest in winter and weakest in summer. In the Atlantic, the high is strongest in winter, but weakest in both summer and autumn. In the Pacific, the high is strongest in spring, and weakest in autumn.

4.6.5 Characteristics and role of the synoptic-scale pressure systems

At a smaller temporal and spatial scale than the subtropical highs and their anomalies discussed above are the characteristics and role of the synoptic highs and lows that process within the mid-latitude westerly zone (Leighton 1994; Sinclair 1995, 1996; Chen and Yen 1997). These systems transfer heat and moisture between latitude zones and, supported by the jet stream and its variations, help maintain the general circulation.

Anticyclones

In January (SH summer), anticyclonicity dominates the 30–40° S latitudinal band, and especially occurs at the longitudes 10° W, 90° W, south of Australia at 130° E, and around 90° E. These are also areas of higher immobility, suggesting blocking. In July (SH winter), the anticyclonicity zone has moved north, to between 25 and 35° S, with the prevalences in the longitudes listed above reduced by 10–30 percent (Figure 4.9). It is in winter, however, that the frequency of pressure systems with central pressure greater than 1035 hPa increases, especially southeast of Southern Africa, Australia, and New Zealand. The frequency of blocking is the most intense SE of New Zealand, and west of South America (110° W, 55° S), and doubles in winter compared to summer.

Anticyclones tend to form near the SE coasts of continents, especially those with mountain barriers, such as South America. Ridging is enhanced by orography, and stability by low-level cooling (especially the Benguela and Humboldt cold ocean currents). While high mountain chains, such as the Andes, will prevent normal west-to-east movement of anticyclones, Leighton (1994) describes offshoots (or "budding") that form on the lee side of the mountains. Anticyclones will tend to move eastward with the direction of movement affected by the location of the subtropical high ridge axis. Speed of movement is on the order of $10 \, \text{m s}^{-1}$, and is somewhat stronger at higher latitudes. Faster movements are usually linked to the subtropical jet stream.

South of about 45° latitude, there are few anticyclones, since this is where the majority of storm tracks occur, although in the Pacific anticyclones may extend to 55° S. On occasion, if there is a breakdown in westerly circulation, anticyclones may extend below 50° S over any ocean, increasing the mean sea level pressure by up to 30 hPa compared to normal.

Figure 4.9 Mean monthly cyclone and anticyclone track densities (lasting two or more days) for (a) cyclone winter; (b) cyclone summer; (c) anticyclone winter; (d) anticyclone summer. The scale is every one system per 5° latitude radius-circle per month; values >5 and >8 shaded. (From Karoly *et al.* 1998 based on data provided Sinclair 1995, 1996: permission from American Meteorological Society)

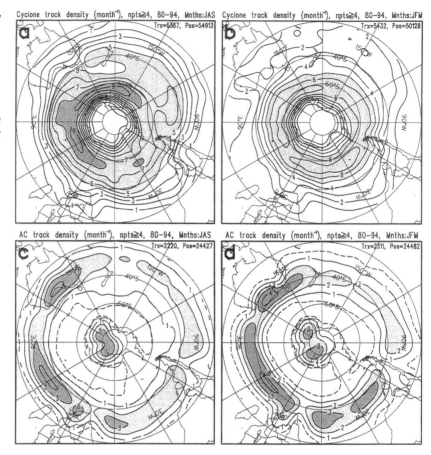

Cyclones

In terms of the SH, cyclonicity is dominated by the CPT around Antarctica (see Chapter 5). Leighton (1994) and Sinclair (1995) used techniques such as geostrophic relative vorticity rather than minimum pressure to remove the CPT bias and allow a focus on cyclones and their movements in the mid-latitudes. Cyclones represent the majority of mid-latitude weather systems, with a maximum in the storm track between 45 and 50° S. In winter, the Pacific dominates the formation area, with frequencies in July averaging about 30. There is a relationship between the strength and location of the SPCZ (see Section 3.2.2) and areas of stronger SST gradient. These areas provide the strongest baroclinic instability and enhanced convection of heat. Maximum frequencies occur in the New Zealand region, the east coast of South America, and SE of Southern Africa extending across the Indian Ocean (see Figure 4.9). Chen and Yen (1997) suggest that cyclones in the last two areas have deepened in strength between

1972 and 1992. In summer (January), the frequency of cyclonicity over the ocean averages about 12. The formation area in the New Zealand region tends to disappear, and the activity is significantly reduced over most of the Indian Ocean.

The jet stream is an important steering mechanism to determine both direction of movement and speed. Cyclones tend to migrate eastward and poleward, but are influenced by standing ridges associated with the CPV (especially wave number 3 in winter). Overall speed of movement is about $15 \, \mathrm{m \, s^{-1}}$ in winter, especially in the Indian Ocean. Summer translation speeds are slower, and the cyclones tend to be weaker overall.

4.6.6 Case study: circulation and teleconnections in the SW Pacific

The discussion above indicates that the SW Pacific (Australia–New Zealand area) is an important region for mid-latitude circulation dynamics in the SH. In this location, correlations between the TPI and local circulation, temperature and moisture are weak (some tendency toward negative) (Jones *et al.* 1999). There are significant interactions between upper tropospheric ridging (90–150° E), and air stream advection toward Antarctica (Goodwin *et al.* 2003). Temperatures and moisture regimes in New Zealand are strongly influenced by regional meridional circulations. For example, increased meridional airflow brings higher precipitation to the north and east of the North Island and to the north of the South Island, but increased zonal airflow increases precipitation in the west and south of the South Island, with cooler summers and warmer winters. Influences such as the föhn effect on the South Island can enhance the connection between zonal circulation and temperature.

Tyson *et al.* (1997) contrasted the advances and retreats of the Franz Josef Glacier on the west coast of the South Island of New Zealand, with changes in area-averaged summer rainfall in Southern Africa. The overall results showed an out-of-phase relationship (glacial advances related to lower rainfall). Goodwin *et al.* (2003, 2004) used NCEP/NCAR reanalysis (see Section 4.3) to assess the relationship between mid-latitude atmospheric circulation and snow accumulation (determined through Na^+ measurements) in Wilkes Land, east Antarctica, in winter. Stronger precipitation was associated with stronger meridional circulation. These studies illustrate the complexity of mid-latitude circulation relationships in the SW Pacific, and are described in more detail below. Both studies emphasize the importance of synchronous adjustments and oscillations associated with wave number 3 in the CPV. A summary of the contrasts under zonal and meridional conditions is presented in Table 4.2.

Under stronger zonal westerly conditions, and fewer waves in the CPV, there are glacial advances in New Zealand, drought in Africa, and lower precipitation amounts in Antarctica. Circumpolar wave number 3, which on average establishes ridges at 40° E (just to the east of South Africa) and at 165° E (just to the

Table 4.2 *Southern Hemisphere mid-latitude circulation variations, glacial changes on the west coast of New Zealand, Southern African rainfall, and snow accumulation in Antarctica*

Dominant zonal circulation	Dominant meridional circulation
Circulation characteristics	
Stronger zonal westerlies	Weaker zonal westerlies
Weak upper westerly waves (shift west)	Stronger upper westerly waves
Westward rotation of wave number 3 by 30–40 degrees	Wave number 3 located more east
Single jet stream	Split jet
Below average cyclone activity east of New Zealand	Above average cyclone activity east of New Zealand
Below average cyclone activity, S. Indian Ocean	Above average cyclone activity, S. Indian Ocean
Increased anticyclonic activity	Decreased anticyclonic activity
Westerlies expand equatorward	Westerlies contract poleward
Link to El Niño events	Link to La Niña events
Weaker trade winds	Stronger trade winds
Subtropical anticyclone moves equatorward	Subtropical anticyclone moves poleward
Weaker poleward transport of heat, moisture, momentum	Stronger poleward transport of heat, moisture, momentum
Positive Antarctic Annual Oscillation	Negative Antarctic Annual Oscillation
Glacial changes in New Zealand	
Glacial advances	Glacial retreats
Strongest impact in accumulation season	Stronger impact in ablation season
Stronger SW wind anomalies	Stronger NE wind anomalies
Increased cloud, lower solar radiation	Decreased cloud, increased solar radiation
Increased precipitation, SW coast	Decreased precipitation, SW coast
Southern Africa	
Drought	Moist
Upper ridge shifts westward	Upper ridge shifts eastward
Negative surface pressure anomalies (south of continent)	Positive surface pressure anomalies
Westerly wind anomalies	Easterly wind anomalies
Antarctica	
Lower Na$^+$ deposits	Higher Na$^+$ deposits
Lower precipitation	Higher precipitation
Contraction of circumpolar trough (CPT)	Expansion of circumpolar trough (CPT)
Ridging limited over CPT S of Australia	Enhanced ridging over CPT S of Australia

Summarized from Tyson *et al.* (1997); Goodwin *et al.* (2003), (2004).

west of New Zealand), shifts westward and decreases in amplitude. The area covered by the circumpolar westerly belt expands, especially toward the equator. El Niño periods tend to dominate, but by themselves to do not control the situation. Tyson *et al.* (1997) state that the Quasi-biennial Oscillation (QBO) has a more important influence. The regional circulation over the South Island of New Zealand is enhanced southwesterly air flows, creating increased cloud, lower temperatures, and increased precipitation on the west coast and slopes of the mountains, but drier conditions in the east and northeast (Jones *et al.* 1999). In the upper troposphere and lower stratosphere, a single jet stream prevails. Transfer of heat, momentum, and moisture toward the poles is reduced.

Over Southern Africa, negative pressure anomalies south of the continent dominate, westerly winds are stronger, and periods of drought occur. The New Zealand glaciers advance after drought occurs in Southern Africa, with a response time of 4 to 5 years lag. Over Antarctica, the combination of negative AAO, a single jet stream, a more compact CPT and stronger zonal flow all combine to minimize the transport of moisture southward past the coast (Fauchereau *et al.* 2003; Goodwin *et al.* 2004). Sodium ion deposition is reduced, indicating lower precipitation amounts.

Virtually the opposite occurs in all three locations when wave number 3 activity in the CPV is enhanced. In the New Zealand region, enhanced meridional circulation is created as the wave ridge shifts eastward. Both the SPCZ (Chapter 3) and the climatological subtropical highs in the Atlantic and Indian Oceans shift south and east. The surface pressure anomalies over the SW Pacific Ocean become positive, and the occurrence of blocking highs increases, both in frequency and strength. NE wind anomalies are created, and the periods and strength of SW airflow are reduced. This leads to increased precipitation to the N and NE, but the west coast has sunnier skies, higher temperatures, and reduced precipitation. Glaciers on the west coast of the South Island retreat. In the upper troposphere, the jet stream is split, enhancing the transfer of heat, momentum, and moisture toward the polar region.

Over Southern Africa, the shift of the wave number 3 ridge eastward encourages more E and NE airflow, enhancing moisture influx into the region, with resultant increased precipitation. In the Antarctic region, the CPT expands, zonal westerlies are weaker, the number of cyclones occurring in the CPT is reduced (see Chapter 5), and enhanced ridging occurs between the Tasman Sea and East Antarctica. The AAO is positive. The stronger meridional flow encourages moisture flux southward across the Antarctic coast, creating increased Na^+ and precipitation deposition.

4.6.7 Are sunspot variations a forcing mechanism?

The reasons for the variations in the CPV and the shifting of standing waves in both hemispheres relates to the forcing mechanisms that drive the atmospheric

Figure 4.10
(a) Relationship between number of days with strong zonal westerly winds (ZWW) over SE Australia (circles) and the sunspot cycle (solid line). (b) Sunspot cycle versus South American (circles), South African (squares) and Australian (diamonds) pressure-difference time series (three-point unweighted running mean). (After Thresher 2002, Figures 1 and 3, Copyright John Wiley & Sons Ltd., with permission)

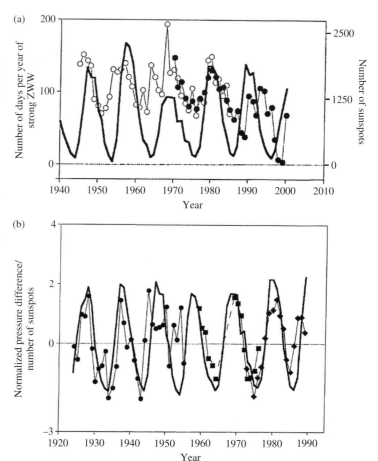

circulation. How these forcing mechanisms operate is not very well known, and there is controversy over interpretations. Despite considerable negative discussion in the past, there is support for the role of sunspots and their changes over time as a mechanism leading to changes in mid-latitude climate variability. For example, Thresher (2002) states that there is a relationship between the sunspot cycle and changes in SH general circulation. Sunspot variations cause changes in shortwave solar radiation, which then affects changes in stratospheric wind speed and zonality. His argument is summarized in Figure 4.10, which shows the sunspot cycle since 1940, the frequency of occurrence of strong zonal westerly winds over SE Australia, and the relationship with pressure time series. Sunspot amounts cycle with a periodicity of approximately 11 years. Strong zonal westerly winds between 1940 and 2000 show a periodicity of 10–13 years.

At least in the Australian region, Thresher (2002) hypothesizes that there is a relationship between the sunspot cycle and the location of the subtropical ridge.

This creates in winter, for example, shifts in the circulation zones. During sunspot maxima conditions, circulation zones are shifted northward, zonal flow enhanced, coastal rain over Australia increases, and winds tend to be calm across the middle of the continent. Comparing this situation to Table 4.2, it is in general equivalent to the zonal conditions in the left-hand column. When sunspots are at minima, the circulation zones shift southward, and the calm area occurs over southern Australia. Table 4.2 suggests that this should represent enhanced meridional circulation.

However, the problems with the argument are illustrated in the correlation analysis between the sunspot cycles and the pressure time series over Australia, South America, and Southern Africa. Only the last shows statistical significance, despite apparent visual relationships (Figure 4.10b). Thus Thresher's results show some evidence of a relationship between sunspots and circulation variations, but there is a lack of consistency, and an apparent dependence on spatial location. The controversy about sunspots as a potential forcing mechanism will continue, with more extensive research leading to a better understanding.

4.7 Chapter summary

Geographically, the mid-latitudes in both hemispheres are located between 35 and 60°. However, the extent of land mass in the NH compared to ocean surface in the SH creates major contrasts in circulation strength and influence. In the NH, the circulation is dominated by troughs and ridges at three locations around the globe. High mountain ranges such as the Rocky Mountains in the United States have an important influence. In the SH, aside from the Andes in South America, there is little influence from terrain. Instead, the ocean, and its contrasts with ice from Antarctica, is the major influence. The circulation anchor for the SH mid-latitude westerlies is the East Antarctic plateau.

The NAO has a strong influence on seasonal and annual temperature and precipitation patterns in the NH. In the SH, the TPI has a much weaker influence, and regional contrasts are a more important determinant of the relationship between zonal and meridional flow.

NCEP/NCAR reanalysis has proven to be an extremely useful tool to provide detailed information about precipitation, temperature, and circulation patterns and trends in both hemispheres. Correlations between circulation areas in the SH, for example, have clearly established relationships between centers in the South Atlantic, south Indian, and Tasman Sea areas, associated with variations in Rossby wave number 3. The latter also determines the extent of zonality in the Australia–New Zealand region, and has been linked to glacial advances on the South Island of New Zealand, the occurrence of drought in Southern Africa, and the level of precipitation in East Antarctica.

The role of precipitation in the global climate system is providing a wealth of research studies. Until fairly recently, the analysis of precipitation was hindered by data quality and availability. Currently, with the availability of models, grid systems, reanalysis and the like, a much greater understanding prevails. But there remain problems, not the least of which relate to current trends in amounts and variability. Much research remains to be completed.

As technology has developed, so have indirect methods for assessing long-term precipitation patterns. For example, using tree-ring analysis, Haston and Michaelsen (1997) developed a 400-year history of precipitation for the area between San Francisco and the US-Mexican border. They concluded that the last 100 years has experienced higher amounts, with less variability, than the previous periods.

The process of mid-latitude circulation is sometimes defined as operating separately from the rest of the global circulation. In reality, influences from both the tropics (SOI for example) and the poles (CPV for example) play important roles, blurring the geographical boundaries within the overall climate system.

4.8 Examples of mid-latitude websites

The NOAA homepage, www.noaa.gov may be used as the entry to a wealth of climatic information for many world areas and topics.

Climate diagnostic information is located at cdc.noaa.gov/index.html. The site provides forecast products and archives for various data series including SSTs.

www.ipcc.ch/ is the home page of the International Panel on Climate Change and provides an abundance of information about organizations and publications.

World Weather Information Service, www.worldweather.org/, provides climatological information and data for many middle-latitude and other world areas.

Southern Hemisphere climate information can be obtained from the Australian Bureau of Meteorology website (www.bom.gov.au). There are similar sites for South Africa and New Zealand.

4.9 References

Alexandersson, H. and Moberg, A., 1997. Homogenization of Swedish temperature data. Part 1: Homogeneity test for linear trend. *International Journal of Climatology*, **17**, 25–34.

Barry, R. G. and Perry, A. H., 1973. *Synoptic Climatology: Methods and Applications*. London: Methuen.

Bigg, G., Jickells, T. and Osborn, T., 2003. The role of the oceans in climate. *International Journal of Climatology*, **23**, 1127–1160.

Chen, T. and Yen, M., 1997. Interdecadal variation of the Southern Hemisphere circulation. *International Journal of Climatology*, **10**, 805–812.

Cullather, R. I. and Lynch, A. H., 2003. The annual cycle and interannual variability of atmospheric pressure in the vicinity of the North Pole. *International Journal of Climatology*, **23**, 1161–1183.

Cusbasch, U. *et al.*, 2001. Projections of future climate change. In J. T. Houghton *et al.* eds., *Climate Change 2001: The Scientific Basis. Contributions of Working Group I to the Third Assessment Report of the IPCC*. Cambridge: Cambridge University Press.

Dai, A., Fung, I. Y. and DelGenio, A. D., 1997. Surface observed global land precipitation variations. *Journal of Climate*, **10**, 2943–2962.

Durst, C. S., 1951. Climate: the synthesis of weather. In T. F. Malone, ed., *Compendium of Meteorology*, Boston: American Meteorological Society, pp. 967–975.

Easterling, D. R. and Peterson, T. C., 1995. A new method for detecting undocumented discontinuities in climatological time series. *International Journal of Climatology*, **15**, 369–377.

Fauchereau, N., Trzaska, S., Richard, Y., Roucou, P. and Camberlin, P., 2003. Sea-surface temperature co-variability in the Southern Atlantic and Indian Oceans and its connections with atmospheric circulation in the Southern Hemisphere. *International Journal of Climatology*, **23**, 663–677.

Fink, A. H., Vincent, D. G., Reiner, P. M. and Speth, P. 2004. Mean state and wave disturbances during phases I, II, and III of GATE based on ERA-40. *Monthly Weather Review*, **132**, 1661–1683.

Garbrecht, J. D. and Rossel, F. E., 2002. Decade-scale precipitation increase in Great Plains at the end of the 20th century. *Journal of Hydrologic Engineering*, **7**, 64–75.

Gates, W. L., Henderson-Sellers, A., Boer, G. J., *et al.*, 1996. Climate models – evaluation. In *Climate Change 1995: The Science of Climate Change, Contribution of Working Group 1 to the Second Assessment Report of the IPCC*, Cambridge: Cambridge University Press, pp. 229–284.

Gong, G., Entekhabi, D. and Cohen, J., 2002. A large-ensemble model study of the wintertime AO-NAO and role of interannual snow perturbations. *Journal of Climate*, **15**, 3488–3499.

Goodwin, I., deAngelis, M., Poole, M. and Young, N., 2003. Snow accumulation variability in Wilkes Land, East Antarctica and relationship to atmospheric ridging in the 130–170° E region since 1950. *Journal of Geophysical Research*, **108**, 4673, 10.1029/2002JD002995.

Goodwin, I., van Ommen, T., Curran, M. and Mayewski, P., 2004. Mid latitude winter climate variability in the south Indian and southwest Pacific regions since 1300 AD. *Climate Dynamics*, **22**, 783–794.

Gulev, S. K., Zolina, O. and Grigoriev, S., 2001. Extratropical cyclone variability in the Northern Hemisphere winter from the NCEP/NCAR reanalysis data. *Climate Dynamics*, **17**, 795–809.

Hanson, C. E., Palutikof, J. P. and Davies, T. D., 2004. Objective cyclone climatologies of the North Atlantic – a comparison between the ECMWF and NCEP Reanalyses. *Climate Dynamics*, **22**, 757–769.

Harman, J. R., 1991. *Synoptic Climatology of the Westerlies: Processes and Patterns*. Resource Publications in Geography No. 11. Washington, DC: Association of American Geographers.

Harrington, J. and Oliver, J. E., 2000. Understanding and portraying the global atmospheric circulation. *Journal of Geography*, **99**, 23–30.

Haston, L. and Michaelsen, J., 1997. Spatial and temporal variability of southern California precipitation over the last 400 yr and relationships to atmospheric circulation patterns. *Journal of Climate*, **10**, 1836–1852.

Houghton, J. T., Meira Filho, L. G., Callandar, B. A. *et al.* (eds.), 1995. *Climate Change 1995. The Science of Climate Change*. Intergovernmental Panel on Climate Change. Cambridge: Cambridge University Press, pp. 229–284.

Jones, D., 1987. Daily Central England temperatures: recently constructed series. *Weather*, **42**, 130–133.

Jones, P. D., 1994. Hemispheric surface air temperature variations: a reanalysis and an update to 1993. *Journal of Climate*, **7**, 1794–1802.

Jones, P. D. and Bradley, R. S., 1995. Climatic variations in the longest instrumental records. In R. S. Bradley and P. D. Jones, eds., *Climate Since A.D.1500*, London and New York: Routledge, pp. 246–268.

Jones, P. D. and Moberg, A., 2003. Hemispheric and large-scale surface air temperature variations: An extensive revision and an update to 2001. *Journal of Climate*, **16**, 206–223.

Jones, P. D., Salinger, M. and Mullan, A., 1999. Extratropical circulation indices in the Southern Hemisphere based on station data. *International Journal of Climatology*, **19**, 1301–1317.

Kalnay, E., Kanamitsu, M., Kistler, R., *et al.*, 1996. The NCEP/NCAR 40-Year reanalysis project. *Bulletin of the American Meteorological Society*, **77**, 437–471.

Kanae, S., Oki, T. and Kashida, A., 2004. Changes in hourly heavy precipitation at Tokyo from 1890 to 1999. *Journal of the Meteorological Society of Japan*, **82**, 241–247.

Karl, T. R. and Knight, R. W., 1998. Secular trends of precipitation amount, frequency and intensity in the United States. *Bulletin of the American Meteorological Society*, **79**, 231–241.

Karoly, D., Hope, P. and Jones, P., 1996. Decadal variations of the Southern Hemisphere circulation. *International Journal of Climatology*, **16**, 723–738.

Karoly, D., Vincent, D. and Schrage, J., 1998. General circulation. In D. Karoly and D. Vincent, eds., *Meteorology of the Southern Hemisphere*, Meteorological Monographs vol. 27, number 49, Boston: American Meteorological Society, pp. 47–85.

Khrgian, A. Kh., 1970. *Meteorology. A Historical Survey*. Jerusalem: Israel Program for Scientific Translations. (Russian edition, Leningrad, 1959.)

Kistler, R., Kalnay, E., Collins, W., *et al.*, 2001. The NCEP-NCAR 50-Year reanalysis: monthly means CD-ROM and documentation. *Bulletin of the American Meteorological Society*, **82**, 247–267.

Klein Tank, A., 2002. Changing temperature and precipitation extremes in Europe's climate. *Change*, **63**, 14–16.

Konnen, G. P., Zaiki, M., Baede, A. P. M., *et al.*, 2003. Pre-1872 extension of the Japanese instrumental meteorological observation series back to 1819. *Journal of Climate*, **16**, 118–131.

Kunkel, K. E., Easterling, D. R., Redman, K. and Hubbard, K., 2003. Temporal variations of extreme precipitation events in the United States: 1895–2000. *Geophysical Research Letters*, **30**, 1029.

Lamb, H. H., 1972. *Climate: Past, Present and Future, Vol. 1*. London: Methuen.

Landsberg, H. E., 1987. Climatology. In J. E. Oliver and R. W. Fairbridge, eds., *The Encyclopedia of Climatology*, New York: Van Nostrand Reinhold, pp. 327–338.

Lanzante, J., Klein, S. and Seidel, D., 2003. Temporal homogenization of monthly radiosonde temperature data. Part I: Methodology. *Journal of Climate*, **16**, 224–240.

Leighton, R., 1994. Monthly anticyclonicity in the Southern Hemisphere, averages for January, April, July, and October. *International Journal of Climatology*, **14**, 33–46.

Manley, G., 1953. The mean temperature of Central England. *Quarterly Journal of the Royal Meteorological Society*, **79**, 242–261.

McGuffie, K. and Henderson-Sellers, A., 2005. *A Climate Modeling Primer*, 3rd edn. Chichester: John Wiley.

Michaels, P. J., Knappenberger, P. C., Frauenfeld, O. W. and Davis, R. E., 2004. Trends in precipitation on the wettest days of the year across the contiguous USA. *International Journal of Climatology*, **24**, 1873–1882.

Moberg, A. and Alexandersson, H., 1997. Homogenization of Swedish temperature data. Part II: Homogenized gridded air temperature compared with a subset of global gridded air temperature since 1861. *International Journal of Climatology*, **17**, 35–54.

Moberg, A. and Bergström, H., 1997. Homogenization of Swedish temperature data. Part III: The long temperature records from Uppsala and Stockholm. *International Journal of Climatology*, **17**, 667–699.

New, M., Todd, M., Hulme, M. and Jones, P., 2001. Precipitation measurements and trends in the twentieth century. *International Journal of Climatology*, **21**, 1899–1922.

Palmen, E. and Newton, C. W., 1969. *Atmospheric Circulation Systems*. New York: Academic Press.

Parker, D. E., Legg, T. P. and Folland, C. K., 1992. A new daily Central England temperature series, 1772–1991. *International Journal of Climatology*, **12**, 317–342.

Petterssen, S., 1956. *Weather Analysis and Forecasting. Volume II: Weather and Weather Systems*, 2nd edn. New York: McGraw-Hill.

Pittock, A. B., 1984. On the reality, stability and usefulness of Southern Hemisphere teleconnections. *Australian Meteorological Magazine*, **32**, 75–82.

Rossby, C. G. *et al.*, 1939. Relations between variations in the intensity of the zonal circulation and the displacement of the semi-permanent centers of action. *Journal of Marine Research*, **2**, 38–54.

Saucier, W. J., 1955. *Principles of Meteorological Analysis*. Chicago: University of Chicago Press.

Schoof, J. T. and Pryor, S.C., 2003. Evaluation of the NCEP-NCAR reanalysis in terms of synoptic-scale phenomena: a case study from the Midwestern USA. *International Journal of Climatology*, **23**, 1725–1741.

Schubert, S. D., Rood, R. B. and Pfaendtner, J., 1993. An assimilated dataset for earth science applications. *Bulletin of the American Meteorological Society*, **74**, 2331–2342.

Simmons, A. J. and Gibson, J. K., 2000. *The ERA-40 Project Plan*. The ERA-40 Project Report Series No.1, ECMWF (www.ecmwf.int/research/era/Products/Report_Series/ERA40PRS_1).

Simpson, M. S., 2005. Climate data centers. In J. E. Oliver, ed., *The Encyclopedia of World Climatology*. Dordrecht: Springer.

Sinclair, M., 1995. A climatology of cyclogenesis for the Southern Hemisphere. *Monthly Weather Review*, **123**, 1601–1619.

Sinclair, M., 1996. A climatology of anticyclones and blocking for the Southern Hemisphere. *Monthly Weather Review*, **124**, 245–263.

Sorre, M., 1934. Climatophysique et climatochimie. Introduction. In M. Piery, ed., *Traité de climatologie. Biologique et medicale*. Paris: Masson, pp. 1–9.

Sturaro, G., 2003. Patterns of variability in the satellite microwave sounding unit temperature record: comparison with surface and reanalysis data. *International Journal of Climatology*, **23**, 1799–1820.

Thresher, R., 2002. Solar correlates of Southern Hemisphere mid-latitude climate variability. *International Journal of Climatology*, **22**, 901–915.

Trenberth, K., 1987. The zonal mean westerlies in the Southern Hemisphere. *Monthly Weather Review*, **115**, 1528–1533.

Trenberth, K. A. and Paolino, D. A., 1980. The Northern Hemisphere sea-level pressure data set: trends, errors and discontinuities. *Monthly Weather Review*, **108**, 855–872.

US Department of Agriculture, 1941. *Climate and Man*. Washington, DC.

Tyson, P., Sturman, A., Fitzharris, B., Moon, S. and Owens, I., 1997. Ciculation changes and teleconnections between glacial advances on the west coast of New Zealand and extended spells of drought years in South Africa. *International Journal of Climatology*, **17**, 1499–1512.

Washington, R., Todd, M., Middleton, N. J., and Goudie, A. S., 2003. Dust-storm source areas determined by the Total Ozone Monitoring Spectrometer and surface observations. *Annals of the Association of American Geographers*, **93**, 297–313.

WCRP, 1998. *Proceedings of the First WCRP International Conference on Reanalysis (Silver Spring, MD, USA, 27–31 October 1997)*. WCRP-104, WMO/TD-N 876, World Climate Research Program, Geneva.

WCRP, 2000. *Proceedings of the Second WCRP International Conference on Reanalysis (Wokefield Park, Reading, UK, 23–27 August 1999)*. WCRP-109, WMO/TD-N 985, World Climate Research Program, Geneva.

Woodruffe, S. D., Slutz, R. J., Jenne, R. I. and Steurer, P. M., 1987. A comprehensive ocean-atmosphere data set. *Bulletin of the American Meteorological Society*, **68**, 1239–1250.

Yan, Y. Y., 2002. Temporal and spatial patterns of seasonal precipitation variability in China. *Physical Geography*, **23**, 281–301.

Zhou, J. Y. and Lau K. M., 2002. Principal modes of interannual and decadal variability of summer rainfall over South America. *International Journal of Climatology*, **22**, 1623–1644.

Chapter 5
Climate of the polar realms

5.1 Introduction

Gerd Wendler, *University of Alaska*

The polar regions are defined as the areas north of the Arctic Circle and south of the Antarctic Circle. These regions represent the sinks in the global energy system of planet Earth. They have a number of common characteristics such as cold temperatures, ice covered oceans, glaciers and ice sheets. Both areas display an extreme seasonal variation in day length, with continuous daylight and darkness of up to six months. However, there are also large differences between the two regions, which are caused mainly by their geographic settings. In the Arctic there is a centrally located Arctic Ocean, which is surrounded by continents, while in Antarctica there is a centrally located continent, which is surrounded by oceans. Consequences of this are that Antarctica is covered by an ice sheet, and only some 2% of its area is ice free, while the only extensive ice sheet in the north can be found in Greenland. Owing to the high altitude of Antarctica – it has the highest mean elevation of all continents due to the snow and ice accumulation – the temperatures drop to very low values and in Vostok, a Russian inland station, a minimum temperature of $-89.5\,^{\circ}\text{C}$ has been observed, the coldest temperature measured on Earth. This value is much lower than anything measured in the Arctic, where the coldest temperatures are found in Siberia. Sea ice is, on average, much thinner in the Southern Ocean, which at first glance might be surprising. However, here the sea drifts steadily north into the South Pacific, Indian and South Atlantic Oceans where it melts. In contrast to this, the Arctic Ocean has a much thicker ice cover and multi-year sea ice is common, as the continents prevent a southward drift with the exception of the Bering Strait and the Northern Atlantic Ocean.

The temperature differences between the tropics (see Chapter 3) and the polar regions are, of course, the driving force of the general circulation systems. In Figure 5.1 the radiative budget on top of the atmosphere is shown, the tropics and the subtropics being the areas with an energy surplus. The energy is lost in mid-latitudes and especially in polar regions. The resultant poleward energy transfer for equilibrium, both by the atmosphere

Figure 5.1 Meridional cross section of the radiation budget and the resulting energy transport for equilibrium from the tropics to the polar regions.

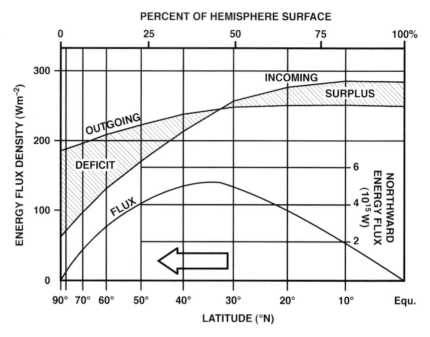

and the ocean, is also shown. Differences and similarities between the two hemispheres will be discussed in greater detail separately for the two regions. Furthermore, these regions play a major role in global change, as here the climate change is enhanced due to the strong polar albedo feedback mechanism (IPCC 2001).

5.2 ESSAY: Antarctic Climate

5.2.1 Introduction

It is thought that Antarctica was first sighted by Captain James Cook on January 26, 1774, when he was reported to have seen "an appearance of land." The first systematic meteorological observations were obtained more than a century later in connection with the exploration of the continent and the race to the South Pole, the so-called heroic period in Antarctic exploration. Expeditions led by Shackleton, Scott, Amundsen, and Mawson had to overwinter in Antarctica so that they had an early start the following spring, at which time the sea ice surrounding Antarctica made the access to the continent proper impossible. Hence, the first year-round meteorological observations on the continent were obtained early in the twentieth century. After the South Pole had been conquered in 1912, the activity in Antarctica decreased. An early summary of our knowledge is given by Meinardus (1938). The next great push in scientific

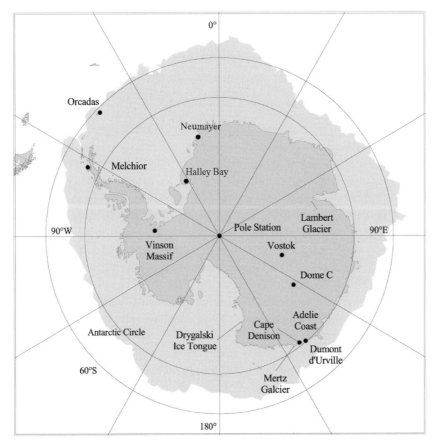

Figure 5.2 Area map, Antarctica. The light grey area is the maximum sea ice extent in early spring.

exploration came with the International Geophysical Year (IGY) in 1957, and ever since, manned year-round stations have been maintained by several nations.

Antarctica (Figure 5.2) has an area of 14 million km^2, about 10% of the land surface on Earth. In late winter/early spring the actual size roughly doubles due to the surrounding sea ice. The high plateau of East Antarctica extends to a height in excess of 4000 m and the highest point is the Vinson Massif (5440 m) in the Ellsworth Mountains. The average height of the continent is 2200 m and it is thereby the highest of all continents on Earth. There are two major ice sheets in Antarctica, the larger eastern one with more than 10 million km^2 and the smaller western one, which has a much lower elevation (850 m) and a size of about 2 million km^2. Antarctica boasts about 90% of all freshwater stored, with an average ice thickness of 2400 m; if all the ice should ever melt, all oceans would rise by some 65 m. However, there is no indication that this could happen in the near future, and it is not even clear in what direction the mass balance of Antarctica has gone in the last century, in which time a climatic warming has been observed (Giovinetto and Bentley 1985).

The ice that has been accumulating over hundreds of thousands of years in interior Antarctica flows slowly towards the coastline. In so-called "ice streams," speeds of up to 500 m/year have been measured. When it reaches the ocean, it can form ice shelves, the Ross and the Filchner Ice Shelves being the largest, each with an area in excess of $500\,000\,\text{km}^2$; these ice shelves make up 44% of the coastline of Antarctica. In other areas glacier tongues penetrate into the ocean, as for example the Lambert and Mertz Glacier Tongues in East Antarctica, or the Drygalski Ice Tongue in the Ross Sea. These can extend into the sea for tens of kilometers. The ice that breaks off from these glaciers forms the spectacular and fairly common icebergs (Figure 5.3a), and large tabular icebergs are formed when parts of shelves break off, which appears to have occurred more frequently in the last decade. These latter ones can last for long time periods. For example, a large tabular iceberg (B9) was formed in 1989 by the Ross Ice Shelf; later, it broke in half. One of the parts drifted north (B9A), where it eventually melted. The other half is presently still well preserved (B9B), after having drifted eastwards for 1200 km, and getting grounded off the Adélie Coast. The remainder of the coastline is often rocky, and this is the area where ablation takes place in summer. These areas together with the nunataks, or mountaintops, which penetrate the ice sheet, represent the relatively limited area not covered by snow and ice permanently.

It should also be pointed out that modern technology has increased our knowledge of Antarctic climatology substantially. Some 20 years ago, automatic weather stations (AWS), reporting via satellites, were originally installed at numerous locations by the USA and Australia, but now are operated by many nations. Antarctica was an ideal location for these, as it is so expensive to run year-round manned stations. A book reporting on some of the results from US stations was edited by Bromwich and Stearns (1993), while Allison *et al.* (1993) reported results for East Antarctica.

5.2.2 Climatology

Surface observations

Antarctica has a very low surface temperature and represents one of the two heat sinks on Earth. It is part of the global energy cycle, and closely tied to global climate and its variations. Teleconnections between Antarctica and the subtropics have been found with various time lags. For example, the Southern Oscillation Index (SOI, Section 2.8) is correlated to various Antarctic climate parameters (Simmonds and Jacka 1995; Turner 2004), as shown in Section 5.6.

(a)

(b)

Figure 5.3 (a) Icebergs as seen off the shore of East Antarctica. (b) Hercules aircraft (LC 130) in East Antarctica, nearly completely covered by drifting snow.

Antarctica's low temperature regime is the combined result of the high latitude and high elevation. In midsummer the solar radiation on top of the atmosphere at the Pole is substantial, actually more than that which is received at the equator. However, the long and slanted pass through the atmosphere reduces the amount of radiation observed at the surface, and even more importantly, the high albedo of the snow cover reflects a large amount of it back to space. Typical snow albedos in Antarctica are on the order of 80%, the exact amount being a function of several factors including solar elevation, crystal size, water content, and others. This means that 80% of the radiation received at the surface is reflected back to space, which compares to water that reflects only some 10%, or land surfaces with vegetation that reflects 10 to 20%.

The coldest area is the high plateau of East Antarctica, where the Russian station Vostok is located at an altitude of 3488 m. Records since the IGY are available, and aside from Pole Station this is the longest lasting data record for the interior of Antarctica. As mentioned in the introduction, the coldest temperature on Earth has been measured here, but probably even more impressive is a mean annual temperature of around −55 °C, with five months of mean temperatures below −65 °C. For two summer months it "warms up" to mean monthly temperatures just below −30 °C. Another typical characteristic of the temperature in high latitudes is the flat winter temperature, referred to as the "coreless" winter. Any one of several winter months such as May, June, July, August, or September might display the coldest monthly mean temperature for a specific year. This is in contrast to the mid-latitudes, where the annual course of the temperature is more sinusoidal. After sunset in winter all heat has to be supplied by advection, which does not show a systematic variation during the winter months. Finally, it might be of interest to note that recently the Japanese established a new, somewhat higher station at Dome Fuji in East Antarctica, and under the assumption that this station will be operating long enough, a new absolute temperature minimum for the Earth might be expected.

In contrast to the interior, the greater number of coastal stations experience more pleasant temperatures. Typical winter temperatures are between −20 and −25 °C, while the warmest month is close to the freezing point. In these areas some snow normally melts in summer. Mean annual temperatures are typically around −15 °C. Normally, the summer is short, typically December and January, the two intermediate seasons are short too, while the winter lasts for the rest of the year. Hence, the mean annual temperature is below the average of the warmest summer month and coldest winter month. This phenomenon is especially well pronounced in the interior.

The temperature in Antarctica has increased. The longest record is available for the South Orkney Islands, where the station Orcadas (60.7° S,

45.7° W) is located, at which systematic measurements have been carried out since 1905. It is, of course, located north of the Antarctic Circle, hence not a proper polar station. The temperature record displayed an increase of 1.8 °C for the last 98 years. The accepted worldwide value is 0.6 °C. This higher value is in good agreement with models, which predict an enhancement in temperature change for polar regions. The Peninsula showed even higher values, and for the coastal stations of Antarctica proper, warming was observed, the magnitude of which varies from station to station. Further, it is, of course, difficult to deduce long-term temperature trends from shorter and non-identical time periods, as temperature change with time is seldom linear. However, a mean of about 2 °C for the last century could be obtained.

There are only two long-term interior stations: Vostok (3488 m) and Pole Station (2835 m). The distance between the stations is 1260 km and both are located on the high plateau of East Antarctica. While Vostok shows an increase in temperature with time, similar to those observed for the coastal stations, Pole Station recorded a decrease in temperature. When analyzing the data, obvious errors in the measurements could not be detected. It is possible to explain the opposing trend between the coastal and interior stations as due to changed advection of warm air from the north, in which the lapse rate could be modified. However, the observed opposing trends of the two inland stations are difficult to explain; both stations are located at similar altitudes over uniform surfaces of snow, hence microclimatology cannot be the culprit.

The air above the surface of Antarctica cools due to the negative radiation balance. As the solar radiation is mostly reflected, it is unable to balance the continuous losses in the infrared region. This process develops an inversion, which means that the air at the surface is colder than the air above. These inversions are common all over Antarctica, but especially well developed in the interior in winter, in the absence of solar radiation. Phillpot and Zillman (1970) were the first to carry out a systematic study of the inversion strengths for the Antarctic continent, which are a unique feature of Antarctica insofar as frequency and strength are concerned. At lower latitudes inversions are less steep and mostly limited to nighttime and winter.

Katabatic winds

If terrain is sloped, cold air close to the surface will start flowing downhill due to its higher density, and the so-called katabatic wind is formed. Eventually, an equilibrium between slope angle, inversion strength, and Coriolis force will be established, which was first modeled by Ball (1957). Katabatic winds are very common in Antarctica, and can reach very high wind speeds due to the low roughness of the snow surface, the large fetch often in excess of 1000 km, a persistent strong inversion, and funneling due to topographic features. They normally have their maximum speed during winter, when

Figure 5.4 Spatial distribution of surface wind streamlines over Antarctica. (From King, J. and Turner, J., 1997, with permission Cambridge University Press)

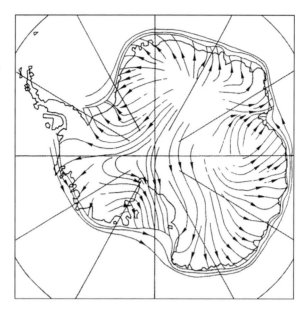

the inversion is well developed. While in winter there is no systematic diurnal cycle due to the absence of sunlight, the winds in summer are normally stronger during nighttime. These winds blow from a prevailing direction, normally somewhat to the left of the gravity fall line, and are represented by streamlines in Figure 5.4. The directional constancy, defined as the vectorial wind divided by the sum of all winds, is frequently in excess of 0.9, surpassing even the trade winds in their unchanging direction. The winds start out at the high plateau with low speeds, accelerating on their way to the coast, and can reach extremely high wind velocities at the coast, especially in confluence zones. Such a zone is Cape Denison in Commonwealth Bay, where Douglas Mawson over-wintered between 1911 and 1913. He named his book appropriately *The Home of the Blizzard* (Mawson 1915). Here are some statistics from his observations:

Mean annual wind speed	$19.4\,\mathrm{m\,s^{-1}}$
Mean of quietest month	$11.7\,\mathrm{m\,s^{-1}}$
Mean of stormiest month	$25.8\,\mathrm{m\,s^{-1}}$
Mean of stormiest day	$36.0\,\mathrm{m\,s^{-1}}$
Mean of stormiest hour	$42.9\,\mathrm{m\,s^{-1}}$

After his return to Australia, Mawson's measurements were doubted, the instruments were recalibrated, and the values were adjusted downwards. Recent observations with automatic weather stations (Wendler *et al.* 1997) suggest that the downwards adjustment might have been overdone. To live under such conditions is extremely difficult. Madigan, the meteorologist of Mawson's expedition, wrote the following: "For nine months of the year an almost continuous blizzard rages, and for weeks on end one can only crawl about outside the shelter of the hut, unable to see the arm's length owing to the blinding drift snow. After some practice, the members of the expedition were able to abandon crawling, and walked on their feet in these torrents of air, 'leaning in the wind'". That such conditions are dangerous goes without saying. A. Prudhomme, a meteorologist at Dumont d'Urville, a French station 120 km to the east of Cape Denison, never returned after a meteorological observation, and it was assumed that he was carried by the wind out to sea.

Katabatic winds are quite shallow, having a maximum in speed a few hundred meters above ground, and the onset and secession can be quite abrupt. Parish (1982) was the first to model these. They can transport large amounts of snow (Radok 1970), which can bury any exposed object (Figure 5.3b). The mass, which is transported from the continent to the ocean, is also of importance for the mass balance in windy areas. While extreme winds, like those reported here, occur in confluence zones, more moderate winds are very common in Antarctica. They are indeed so common that it can be stated that on no other continent does a single meteorological parameter have such an influence on the surface climate of the whole continent as the katabatic wind for Antarctica.

Upper air data

More difficult to analyze than the surface data is the meteorology aloft due to the paucity of radiosonde data. In addition, different radiosonde types are being used, as stations of different nationalities use the radiosondes of their home countries. Different radiosondes have different characteristics and are not necessarily comparable. The coverage for the Peninsula and coastal areas is representative; however, there are only two long-term inland stations. Further, profile measurements from satellite data also have their problems due to the strong surface inversion and the snow-covered surface. The reanalyzed NCEP/NCAR data sets (Section 4.3), which have been carried out for more than a decade, are a reliable data source.

Antarctica is surrounded by a pressure minimum, the so-called circumpolar trough (Section 5.4). Cyclones are frequent (Section 5.5) and these areas are well known for their fierce storms. Before the Suez and Panama Canals were in existence, the rounding of Cape Horn and the Cape of Good Hope by sailing ships was a dangerous and challenging affair, and a great number of ships were lost. Over the continent, an anticyclone can be observed, the reality of which has been debated, due to the problems of reducing data from high altitude stations to sea level. This anticyclone is substantially stronger in winter than in summer, but exists throughout the year. The coldest temperature over the continent in summer is around the 250 hPa level at $-50\,°C$; in winter the atmosphere is much less stratified, and the coldest temperatures with values between -80 and $-90\,°C$ are found at the 30 hPa level. The zonally averaged eastward wind component is, for all altitudes and seasons, weak over Antarctica. The surface winds are, to a great extent, decoupled from the wind aloft due to the strong surface inversion. Dome C (now called Place Concordia), located on the top of a dome at an altitude of 3280 m, measured a mean annual wind speed of less than $3\ \mathrm{m\ s}^{-1}$; such a low value is found nowhere else on Earth for a freely exposed station at that altitude.

Sea ice

Sea ice surrounding Antarctica displays a large annual variation in coverage. The maximum is normally found in late winter/early spring with $19 \times 10^6\ \mathrm{km}^2$, while the minimum occurs in late summer with only $3.5 \times 10^6\ \mathrm{km}^2$. The presence or absence of sea ice strongly influences the energy transfer between the ocean and atmosphere and is of great importance for understanding the climate. Systematic observations of sea ice go back some 30 years, and became possible only by remote sensing from satellite. Passive microwave sensors, not affected by darkness or clouds, are especially useful. As the Antarctic continent is surrounded by oceans, the ice drifts steadily north where it melts. The ice pack consists mostly of first-year ice which is thin (<1 m), whereas in the Arctic a large amount of multi-year ice is present. These differences have been known for a long time owing to whaling activities. For example, in March 1840, Charles Wilkes wrote in a letter to Sir James Clark Ross: "The ice of the Antarctic is of totally different character from that of the Arctic." Most long-term trend studies have found a decrease in ice extent (Ackerley et al. 2003). However, new studies using satellite data and in part new algorithms to determine the ice amount came to the opposite conclusion, that is that the amount of sea ice is increasing (Parkinson 2002). This increase has been verified for the last two decades and cannot be doubted (Zwally et al. 2002).

The very strong katabatic winds can drive the ice away from the coastline, and coastal polynyas (open water areas) can occur, even in midwinter. The energy exchange over these polynyas can be two orders of magnitude higher than the heat conduction through sea ice, because cold air, originating from the interior of Antarctica, is moved over this "warm" water with a high velocity. These are the areas where large amounts of new ice are formed. During this process, brine is released, which increases the density of the water. This water sinks, forming the Antarctic bottom water. In fact, 75% of the bottom water in all oceans originates from Antarctica. The production areas along the Antarctic coast are limited to confluence zones, such as Adélie Land (Mertz polynya) and Terra Nova Bay (polynya with the same name). Hence, climate change in Antarctica will affect the production of such high-density water, and with it the climate of Earth on a century time scale.

Other meteorological parameters

As the water vapor saturation pressure is a logarithmic function of temperature and decreases to about half its value for each 5 °C decrease, precipitation is, in general, light at high latitudes. The Antarctic Peninsula is an exception, however. Melchior (65.3° S, 63.0° W), located close to sea level, recorded a mean annual value of 1189 mm. Otherwise, coastal stations in Antarctica record values around 500 mm. In windy areas it is not possible to measure precipitation due to blowing snow, and, for example at Dumont d'Urville, no amount but only the frequency of precipitation events is recorded. The precipitation amount decreases steeply when going inland; less than 100 km from the coast, values under 200 mm are observed. Further inland and on the high Antarctic Plateau values of less than 30 mm annually are observed. Here all precipitation falls in solid form; frequently clear sky precipitation is observed. No clouds are present, but in the cooling air mass, ice crystals are formed, which float slowly to the surface and can result in impressive optical phenomena such as halos and parhelia (sun dogs). The snow accumulates over time and by analyzing ice cores, the past climate can be reconstructed (see Chapter 6). The most famous ice core is from Vostok and dates back more than 300 000 years. It was the analysis of such ice cores that supported the Milankovitch theory being accepted in science. The theory says that the climate on Earth varies with three parameters of the Earth's rotation (changes in eccentricity, in axial tilt and in axial precession), having different time periods of tens of thousand of years, and affecting the solar radiation and with it the temperature.

In the coastal region of Antarctica, the total mean annual cloud cover is about 50%; it decreases slightly in winter. When going inland, the amount of total cloud cover decreases strongly, which is in agreement with the decrease in precipitation and the frequency of occurrence of active weather systems. Further, the annual course in cloudiness is more pronounced inland.

The global sea level has been rising annually by about 2 mm, and for a long time the question has been asked if Antarctica is contributing to this sea level rise. The question is not trivial, as warming will result in more ablation in the coastal areas of Antarctica, and also most likely in an increase of precipitation. If the snowfall occurs somewhat inland, where no ablation takes place, more snow might be stored in these areas. The sum could be either positive or negative. While studies using both direct glaciological methods and atmospheric methods to calculate the moisture fluxes across the Antarctic Circle have been carried out, the results are still inconclusive. However, satellite-derived altimetry should solve this problem. In 1997, the Canadian satellite

RADARSAT was turned around to look to the south so that it could obtain accurate altitudes of all of Antarctica. In 2000, the mission was repeated, this time without turning the satellite around, hence not obtaining data for the center of Antarctica. A new satellite with laser altimetry has recently been launched. Highly accurate data are expected to be obtained, and when repeated, they will give elevation changes from which the mass balance of Antarctica can be accurately calculated.

5.3 Upper air circulation and wind

The interaction between the high-altitude ice mass that makes up Antarctica, the surrounding ice shelf and ocean, and the Southern Hemisphere general circulation creates a complex climatology of airflows in the region. Detailed and accurate information about the upper atmosphere and its processes over the Antarctic region is limited due to lack of long-term data, lack of radiosonde data, and difficulties in interpreting information from satellites that are normally oriented toward the mid-latitudes (Simmonds 1998).

Figure 5.5a illustrates a simple model of the overall circulation structure over the continent (King and Turner 1997). Inflows from the mid-latitudes occur in the upper troposphere and lower stratosphere, creating confluence and convergence, and bringing warmer air to the polar region. This air subsides very slowly over the land mass. The subsidence is centered over the East Antarctic Plateau, the region of highest altitude. Cyclonic vorticity associated with the subsiding air creates a westerly vortex in the atmosphere above 500 hPa (Figure 5.5b). The climatological center of the vortex tilts poleward with height, being located over the Ross Sea in winter at 500 hPa and over the South Pole at 300 hPa. On a day-to-day basis, cyclonic weather systems can disrupt the circulation pattern, allowing irregular meridional transport of moisture and energy from the north (König-Langlo et al. 1998). This upper air circulation structure acts as a compensation for, and a controller for, surface airflow, especially the katabatic winds. At the surface, below about 2 km, the vortex circulation is easterly (Figure 5.5b).

In late winter, when the Sun is in the Northern Hemisphere, and the polar–equatorial and land–ocean temperature gradients the strongest, upper air circulation around Antarctica is the strongest for the year. The polar night jet (PNJ) (>75 m s^{-1} at 10 hPa) dominates the circulation above the tropopause between 50 and 60° S (see Section 5.7). The long polar night, and the low temperature associated with the "coreless winter" strengthen both the upper air and near-surface circulation (Bromwich and Parish 1998).

In summer, the westerly circulation is weaker, especially above 300 hPa, and becomes stratospheric easterlies above 50 hPa. In this season, the tropopause drops in height from about 100 hPa at 50° S to 300 hPa over the continent. The

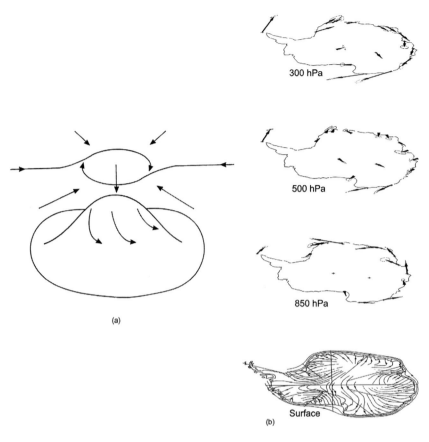

Figure 5.5 (a) Simple upper-tropospheric/lower-stratospheric circulation model over Antarctica. (b) Representative wind directions at four altitudes over Antarctica. (After King, J. and Turner, J. 1997, with permission Cambridge University Press)

meridional temperature gradient reverses between 200 and 300 hPa, defining the location of the seasonally weaker polar front jet stream (28 m s^{-1}) near 55° S.

The East Antarctic Plateau anchors the upper tropospheric and lower stratospheric circumpolar circulation, and strongly influences the standing and transient waves in the circumpolar westerlies; an influence that extends well into the mid-latitudes (see Section 4.6). King and Turner (1997) state that, in comparison to the Arctic, the standing waves in the Southern Hemisphere are more barotropic in structure, and transport less moisture and heat. Therefore the transient waves and eddies become very important to the energy and moisture transport process. Simmonds (1998) emphasizes that knowledge about how this influence operates is incomplete and unclear, and a matter of some debate.

5.4 Surface pressure variations

Figure 5.6, based on NCEP/NCAR reanalysis (see Section 4.3) over a 12-year period by Simmonds (1998), establishes a clear pattern of low pressure (below 990 hPa except over the Antarctic Peninsula) around the continent edges, the circumpolar trough (CPT). Figure 5.6 shows that there are three centers of lower

Figure 5.6 Mean sea level pressure for (a) summer and (b) winter in the Antarctic region. (Courtesy of Ian Simmonds, University of Melbourne)

(a)

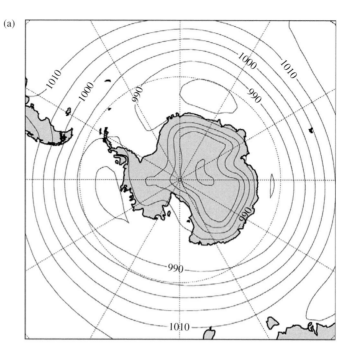

(b)

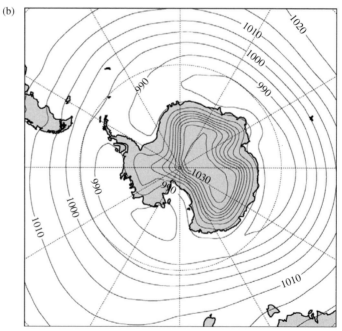

pressure in the CPT (Bromwich and Parish 1998). These centers, reaching around 980 hPa, exist between 0 and 40° E longitude (south of Southern Africa), 80 to 100° E longitude (Southern Indian Ocean), and 160–170° W longitude (southeast of New Zealand). The physical barriers created by the continent to the south, the strength of the pressure gradients, plus the open ocean to the north and east, encourage the strong zonal flow that defines the mid-latitude westerlies in the Southern Hemisphere.

The climatic picture shown in Figure 5.6 obscures the strong pressure variations that occur within the CPT (Simmonds 1998). Standard deviations in pressure can vary by more than 10 hPa over periodicities that exceed 10 days. Over shorter time periods, pressure may change by up to 25 hPa. One major result is the creation of a large number of intense synoptic-scale low pressure systems (see section 5.5). The seasonal variations in pressure in the CPT are controlled by the Antarctic Semi-Annual Oscillation (AAO, see Section 2.7) in pressure. Spatial variations in the CPT are also linked to sea ice extent. The coastal locations closest to the CPT, such as Dumont d'Urville, thus have the greatest pressure variations associated with the AAO (König-Langlo *et al.* 1998). A pressure minimum occurs in October when the CPT is also the furthest south in location, and sea ice extent is at maximum. A secondary minimum in pressure, and a secondary maximum in latitude, occurs in March and April. North of 60° latitude, the influence of AAO on pressure variations diminishes considerably, and the phase reverses. Table 5.1 lists some reasons for the development and structure of the AAO, and its major influences.

Over the continent, seasonal surface pressure variations are greatest near 90° S (Bromwich and Parish 1998). Here, in August and September, average pressure is 15–18 hPa lower than in January. Seasonal variations become much less close to the coast. The terrain influences the spatial distribution of surface pressure

Table 5.1   *Reasons for the development of the AAO around Antarctica, and its major influences (after Simmonds 1998; Turner 2004)*

Reasons for development of AAO	Major influences on climate by the AAO
Differing annual temperature variations between land and water	Semi-annual cycle of pressure
Differences in thermal inertia between land and water	Variations in the circumpolar trough
Oceans cool in autumn faster than warm in spring	Mean wind strength and direction
Antarctic coast warms faster in spring than cools in autumn	Mean distribution of sea ice
Phase of annual temperature cycle differs between land and water	Expansion and contraction of circumpolar vortex
	Meridional shifts of the Polar Front jet stream

changes, with diabatic cooling rates creating the strongest pressure differences on steeper coastal slopes.

5.5 Cyclogenesis and cyclonicity

The CPT around Antarctica is one of the most active cyclonic regimes on Earth (Bromwich and Parish 1998; King and Turner 1997; Simmonds 1998). Not only does the CPT spawn storm systems through cyclogenesis, there are also a large number of low pressure systems circulating within the CPT. More than half of the extra-tropical cyclonic systems in the Southern Hemisphere form in the CPT. Strong baroclinic influences associated with wind and temperature are a major factor.

Due to the barrier created by the land mass, cyclogenesis is restricted to the offshore area. There are significant spatial variations, but a similar seasonal pattern. Cyclogenesis occurs mainly in the West Antarctic section of the region, on either side of the Antarctic Penisula. The East Antarctic is a relatively quiet zone. Three nodes dominate the spatial distribution. In both seasons, cyclogenesis is strong in the Ross Sea area, where the combination of cyclonic turning of katabatic flows, the instability associated with the ice–ocean temperature contrast, and confinement from the surrounding coastal elevations minimize horizontal movement. A second area is the lee of the Antarctic Peninsula, around 65° S, where the higher elevations of the Peninsula interrupt the westerly airflow, and create a cyclonic circulation area. The third area is in the Weddell Sea. A considerably greater number of cyclones form in winter, when the described contrasts are the greatest. Particularly in the Ross and Weddell Sea areas, the density of cyclogenesis doubles in winter compared to summer.

The spatial differences in cyclone system density reflect their development and mobility within the CPT circulation, and thus differ from the major centers of cyclogenesis (Simmonds 1998). Figure 5.7a and b compare summer and winter cyclone system density, and again indicate three areas of major activity: off the East Antarctic coast south of Southern Africa; off the East Antarctic coast south of Australia; and off the West Antarctic coast west of the Antarctic Peninsula. These locations relate strongly to the AAO, and although the frontal systems may not be as well developed as in the mid-latitudes, they are the main source for the poleward flux of energy and moisture.

The offshore flow of katabatic winds, the influence of Coriolis force, and the channeling effect of the Antarctic coast itself encourage further easterly flow and cyclonic circulation (Bromwich and Parish 1998). Depending on location, coastal locations show variations in wind (and temperature) according to katabatic or cyclonic dominance (König-Langlo *et al.* 1998). At Dumont d'Urville, katabatic flows dominate, bringing cold dry air from the continent (wind direction 125–185°). Winds occasionally exceed $50\,\mathrm{m\,s^{-1}}$. On the opposite (NW) side of the continent, winds at both Halley Bay and Neumayer are dominated by directions from 80 to 90°. Katabatic influences and cyclonic disturbances control the

(a)
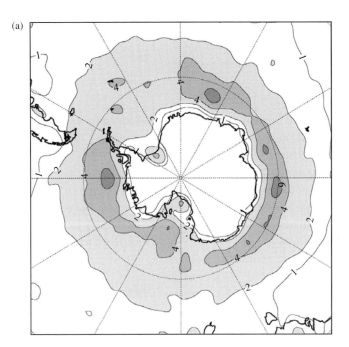

(b)
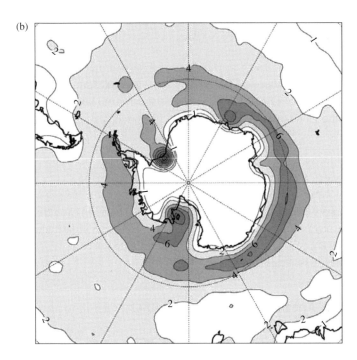

Figure 5.7 Cyclone system density in (a) summer and (b) winter around Antarctica. (Courtesy of Ian Simmonds, University of Melbourne)

Table 5.2   *Some potential teleconnections between El Niño events and Antarctic climate (after Bromwich and Parrish 1998; Turner 2004). Links are highly variable and inconsistent*

Antarctic climate parameter	Comments
Increased surface winds and colder surface temperatures	Depends on year grouping considered; irregular impacts; time lags
Polar Front Jet weaker	PFJ stronger during La Niña
Geopotential heights higher	Geopotential heights lower during La Niña
Weaker circumpolar westerlies	Modulations not the same between events
Positive height anomalies over Amundsen-Bellingshousen Sea area	Strongest signal, but considerable variability in occurrence (opposite during La Niña)
Diminished number of cyclones in CPT	
Amplified seasonal cycle creates more meso-cyclones	Shifts in locations
MSLP anomalies change from positive to negative across SOI minimum	Occurs generally across the continent
Variations in the circumpolar wave pattern	Dynamics not well understood
Precipitation lower	Irregular/inconsistent impacts; strongest in West Antarctica
Higher Ross Sea SST	2–4 month lag
Reduced Ross Sea sea ice area	Limited data

wind regime. Wind speeds at Neumayer can exceed $50\,\mathrm{m\,s^{-1}}$, but at Halley Bay, which is more sheltered, winds rarely exceed $40\,\mathrm{m\,s^{-1}}$.

King and Turner (1997) provide an extensive description of the range of mesoscale cyclones that can develop and die within the CPT, at a smaller spatial scale than the synoptic lows. These systems can create a variety of short-term severe weather conditions, adding to the extreme complexity of storms and their interactions along the Antarctic coast.

5.6 Antarctic climate and ENSO

An excellent review by Turner (2004) provides a detailed evaluation of the potential impacts of ENSO (see Section 2.8) on Antarctic climate patterns. Table 5.2 summarizes the major relationships, but also emphasizes that there is a strong lack of consistency and stability over time. Other reasons for climatic variations, such as the AAO, are therefore more important. ENSO signals are most likely transferred through the Rossby wave train, associated with topical forcing of the equatorial to polar circulation processes.

Turner (2004) found that the strongest relationship was between height anomalies over the Amundsen–Bellingshousen Sea and the SOI. During El Niño, height anomalies are positive; during La Niña, they are negative. Pressure anomalies change from positive to negative across the SOI minimum. During El Niño there tend to be fewer cyclones in the Bellingshousen Sea, winds exhibit a greater southerly direction, and the polar front jet tends to be weaker. However, there are no linear relationships, and considerable variability from event to event, with some El Niño years or periods having no effect. Considerable further research is needed to define and resolve ENSO/Antarctic climate relationships.

5.7 Polar night jet and stratospheric ozone depletion

Toward the end of winter and into early spring (August to mid-October), the combination of a strong polar–equatorial temperature gradient, strong land–ocean temperature contrasts, and the coreless winter, create the strongest circumpolar vortex during the year. The PNJ, the leading edge of the circumpolar vortex, shifts poleward, and downward in altitude to the lower stratosphere and upper troposphere (Kuroda and Kodera 2001), and intensifies to over $75\,\mathrm{m\,s^{-1}}$. Transport of warmer, more humid air from the mid-latitudes toward the poles, usually through planetary waves, is blocked, and the atmosphere over the poles becomes isolated. The clear skies and high altitude of the Antarctic ice plateau encourage longwave radiation loss to space from the surface, enhancing the development of very cold temperatures. Anomalously cold stratospheric air descends into the upper troposphere (see section 5.3). Air temperatures fall to $190\,\mathrm{K}$ in the upper troposphere and lower stratosphere over much of the continent (Simmonds 1998).

The lack of interaction with lower-latitude air, and the extremely cold temperatures have a major influence on stratospheric ozone levels over Antarctica, and are the major controllers of the chemistry creating the Antarctic "ozone hole." Stratospheric ozone absorbs ultraviolet radiation in the 280 to 320 nm wavelength band (UV-B), which would cause skin cancers on human beings, and be harmful to other life at the Earth's surface (GEO 2003). Damage to the ozone layer is of global concern because of this threat.

Background Box 5.1 summarizes the chemistry process and the elements involved. Since the 1920s, a wide range of chlorine (Cl, chlorofluorocarbons) and bromine (Br) compounds, have been released into the troposphere through human activities (Turco 2002). After several decades (depending on the compound), these molecules reach the stratosphere, where they are dissociated by ultraviolet radiation. The now-free Cl and Br atoms react with ozone, creating the more stable molecule oxygen, and ClO or BrO. The latter molecules are dissociated by ultraviolet radiation, or through reaction with single oxygen atoms, and the cycle begins again. It is estimated that one Cl atom can destroy 100 000 ozone molecules, until the cycle is broken and Cl is "stored" in a reservoir molecule.

Background Box 5.1

The chemistry of stratospheric ozone depletion (after Bridgman 1997; Turco 2002)

The anthropogenic destruction of stratospheric ozone is caused mainly by chlorine (Cl) and bromine (Br) atoms. This summary uses chlorine chemistry to illustrate the process: Chlorofluorocarbons (example CFC-11) enter the stratosphere from the troposphere where they react with sunlight

$$CFCl_3 + hv \rightarrow CFCl_2 + Cl$$

The Cl atom then reacts with ozone (O_3)

$$Cl + O_3 \rightarrow ClO + O_2$$

ClO then reacts with single oxygen atoms

$$ClO + O \rightarrow Cl + O_2$$

and the cycle starts again.

Eventually, Cl is stored in reservoir species, when it reacts with methane

$$Cl + CH_4 \rightarrow CH_3 + HCl \text{ (reservoir, hydrogen chloride)}$$

or with nitrogen dioxide

$$ClO + NO_2 \rightarrow ClONO_2 \text{ (reservoir, chlorine nitrate)}$$

The reaction chain is illustrated in Figure 5.8a.

On the surface of a polar stratospheric cloud, Cl is released from reservoir storage

$$HCl + ClONO_2 \rightarrow \rightarrow \rightarrow Cl_2 + HNO_3 \text{ (summary reaction)}$$

and

$$Cl_2 + hv \rightarrow Cl + Cl$$

The strength and consistency of the PNJ and the CPV in the late winter and early spring create the conditions that lead to major destruction of stratospheric ozone in the Antarctic atmosphere (Turco 2002). The vortex prevents ozone replenishment from the mid-latitudes. The extremely cold temperatures within the vortex allow the creation of ice and nitric acid polar stratospheric clouds (PSCs), especially above the coast. The PSCs act as a reaction surface, allowing the reservoir species to break apart, freeing Cl atoms once again. In the weak spring sunlight, Cl destruction of ozone continues unabated.

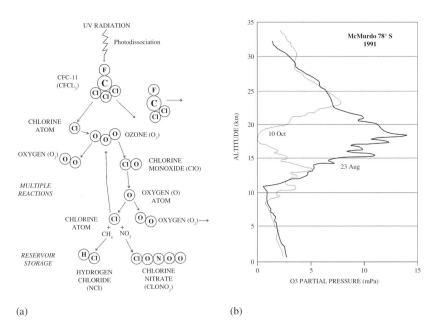

(a)

(b)

Figure 5.8 (a) Chlorine chemistry associated with stratospheric ozone depletion. (b) Loss of ozone over Antarctica in early spring compared to early winter. (After Bridgman, H., 1997, Figures 22.3 and 22.4 with permission from Routledge)

Most ozone loss occurs at altitudes between 13 and 23 km, with almost 100% loss at the altitudes around 18 km. Aircraft and satellite data from several studies have confirmed that the CPV, with the PNJ at its leading edge, is a clear barrier between ozone destruction and ozone survival (Figure 5.8). Finally, some time between about mid-October and early November, the sunlight becomes strong enough, and the polar–equatorial temperature gradient has become weak enough, to diminish the circumpolar barrier. Meridional transport of mid-latitude air resumes, and ozone levels recover in the Antarctic stratosphere.

The size and location of the circumpolar vortex and the area of major ozone loss varies from year to year and day to day. It is strongly linked to the AAO. Coastal research stations such as Neumayer and Halley Bay are far enough south to generally remain within the vortex each year (König-Langlo *et al.* 1998), and ozone depletion in spring is regularly measured. Stations like Dumont d'Urville, however, are under the leading edge of the vortex. Ozone loss here depends on the spatial fluctuations of the vortex.

5.8 ESSAY: Arctic Climate

Gerd Wendler, *University of Alaska*

5.8.1 General observations

The definition of the Arctic is not uniform in the literature. We use here the astronomical definition, the Arctic Circle. Other authors have used a

Figure 5.9 Area map, Arctic. The light grey area is the maximum sea ice extent in early spring.

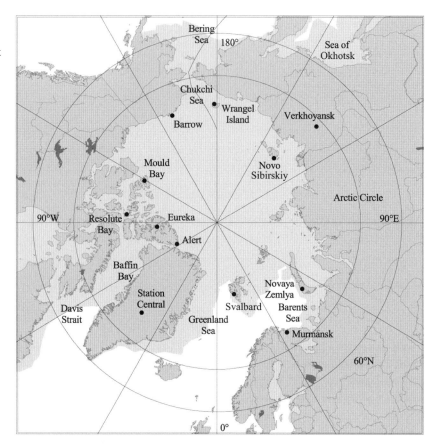

climatological definition, that is the position of the 10 °C isotherm of the warmest month (Köppen 1936). The position of this isotherm closely follows the line that differentiates the treeless tundra from the taiga. The Arctic Ocean is fairly centrally located around the North Pole (Figure 5.9). The largest land mass within the Arctic Circle is found in Siberia, followed by Canada. Greenland lies within the Arctic Circle, with the exception of its southern tip. Here we find the only large ice sheet in the Northern Hemisphere, with an elevation of about 3000 m; this ice sheet contains 9% of all fresh water on Earth. Further, parts of Finland, Sweden, Norway, and Alaska lie north of the Arctic Circle. These areas have been inhabited for millennia, originally by Inuits in Greenland, Canada, and Alaska, Chukchis in Eastern Siberia, and Scandinavians and Russians in their respective countries. As agriculture is limited at these high latitudes, the people have lived from hunting and fishing and the population density is very low.

The first systematic meteorological observations of the Arctic Ocean were obtained in the nineteenth century, when great interest was given to the potential navigational routes through the Arctic Ocean. Rounding Africa or South America for commerce with Pacific nations was not only time consuming, but also dangerous in the days of sailing ships. Maybe the most famous trip North was the tragic Franklin expedition of 1845 in which 128 people on two ships disappeared. A total of 32 rescue attempts were carried out. In 1878/9 Nordenskjöld succeeded in exploring the North-East Passage, and in the early twentieth century, Amundsen was the first to sail the North-West Passage (1903/4). Finally, coming from Ellesmere Island, Peary stated that he had reached the North Pole in 1909, a claim that is now doubted. My personal favorite story is Fridtjof Nansen's drift with the *Fram* starting in 1896. His ship was built specifically for Arctic exploration with a hull made out of oak, so that it could not easily be crushed by sea ice. He sailed along the Russian and Siberian coasts, and at about $150°$ E close to the New Siberian Islands he turned north. With approaching winter, his ship froze in at close to $79°$ N and drifted with the sea ice. It had been known that the ocean currents were generally to the west, as wood originating in Siberia had been found off the coast of northern Scandinavia, however, not much was known about the speed of the current and if such a trip would take a few months or many years. In the event, three years later the *Fram* returned to Norway to a triumphant welcome, having collected a great amount of climatological, sea ice and oceanographic observations from the central Arctic Ocean.

In contrast to Antarctica, the surface characteristics are much more varied in the Arctic. Here, besides snow cover, sea ice, and water, we also find tundra, permafrost, taiga, and boreal forest. Different surface characteristics affect the surface energy budget and hence result in a larger number of different micro-climates.

5.8.2 Climatology

The Arctic region is the second energy sink on Earth. The radiative input from the Sun is similar to Antarctica. However, during the northern summer, the Earth–Sun distance is at its maximum, reducing the solar radiation on top of the atmosphere by some 3%. On the other hand, owing to the increased distance, the Earth's rotational speed around the Sun decreases, increasing the length of the summer. In general, winters are long, typically between 6 and 9 months, and can be very cold. Summers are short but have an abundance of sunlight, as day length is very long, highly important for the growth of the vegetation.

Temperature
The coldest temperatures in the Northern Hemisphere are not found in the central Arctic Basin, but in Eastern Siberia. In the central Arctic Ocean, sea

ice is present year-round, consisting typically of thick ($\sim$3 m), multi-year ice. This sea ice is in steady movement, and through dynamical actions, pressure ridges and leads (polynyas) are formed. Open water or thin ice that might be present in winter is much, much warmer than the atmosphere, and the heat transfer from the ocean to the atmosphere can be large, warming the atmosphere somewhat. In contrast to this, the energy transfer from snow-covered soil is small. Hence, it is not surprising that the lowest recorded temperature of $-67.8\,^{\circ}$C was measured in Verkhoyansk, Eastern Siberia, where we find a very continental climate, with very low winter temperatures and relatively warm summer temperatures, resulting in a large annual variation in temperature. The minimum in winter is fairly flat, and any of several winter months for a specific year might record the coldest temperature, a so-called "coreless" winter, which is however less pronounced than in Antarctica. Conrad (Köppen 1936) used Verkhoyansk for defining his continentality index, giving it the value of 100%, while giving Tromsø, on the subarctic Norwegian coast, where the annual variation in temperature is very small, the value of 0% (maritime climate). There are large variations in mean annual temperatures of the Arctic region (Figure 5.10a). Murmansk (69° N) in Russia is a year-round ice-free port with a mean temperature of 0 °C, very warm as it is still influenced by the Gulf Stream. The mean annual temperatures are between -10 and $-20\,^{\circ}$C for other coastal stations not influenced by the Gulf Stream as well as for the Arctic Ocean, while the crest of Greenland records values around $-27\,^{\circ}$C; this cold temperature is the additional effect of altitude. In Table 5.3 temperature values for selected stations in the Arctic are presented.

Further, surface temperature inversions are frequent in the Arctic, especially in winter, at which time they are semi-permanent. They are especially strong in sheltered valleys, as the wind aloft cannot penetrate to the surface. In contrast to this, they are less strong over the central Arctic, where the wind aloft may destroy them through forced mixing, especially as the surface roughness is low and not hindered by any topography. This might be an additional reason for the relatively benign winter temperatures at the Pole.

In general, temperatures have increased in the Arctic, and the increase has been larger than the worldwide average, which is about 0.6 °C for the last century (Jones 1995). This enhancement in temperature increase for the polar region was expected and has also been observed for Antarctica. However, the rate of increase is not at all uniform over the past century, a result that would have been expected if it were solely caused by the increasing amount of CO_2 and other trace gases in the atmosphere. Further, the observed temperature increases vary strongly not only in time, but also from place to place (Chapman and Walsh 1993; Frey and Smith 2003; Whitfield et al. 2004). Very large changes in temperature sometimes reported in the literature hold true only for a specific place and time period, and are not valid for the Arctic

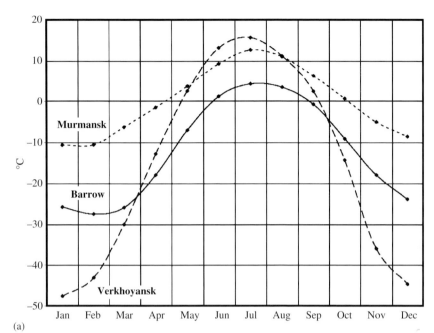

(a)

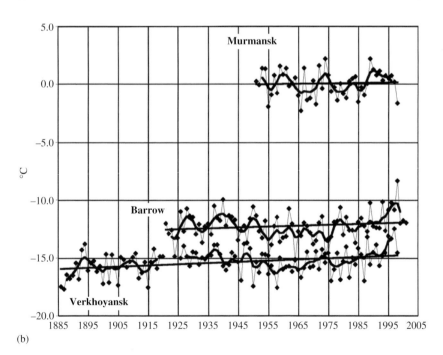

(b)

Figure 5.10 (a) Mean annual temperature variation and (b) mean annual temperatures, 5-point weighted mean and linear trend of temperature, for Murmansk (maritime), Barrow (intermediate) and Verkhoyansk (continental).

Table 5.3   *Location and mean temperatures for the coldest month, warmest month, and the mean annual value for selected stations in the Arctic*

	Latitude °N	Longitude °	Altitude (m)	Coldest month (°C)	Warmest month (°C)	Mean annual (°C)
Coastal stations						
Svalbard	78°04′	13°38′ E	9	−12.1	5.5	−5.4
Murmansk	69°00′	33°06′ E	46	−10.6	12.6	0.0
Novaya Zemlya	72°23′	52°44′ E	16	−15.0	6.7	−5.2
Novo Sibirskie	76°00′	137°54′ E	10	−29.6	2.8	−15.2
Wrangell Island	70°58′	178°32′ W	3	−25.6	2.8	−11.8
Barrow	71°18′	156°47′ W	9	−27.7	5.3	−12.2
Mould Bay	76°14′	119′20′ W	15	−35.4	5.0	−17.4
Eureka	80°00′	85°56′ W	2	−37.6	5.7	−19.1
Alert	82°30′	62°20′ W	63	−33.0	3.9	−17.8
Other stations						
Station Central	70°55′	40°38′ W	2993	−39.9	−10.4	−26.8
N. Pole Drifting	75–90°	147–171° E	0	−36.5	−0.3	−18.8
Verkhoyansk	67°36′	133°24′ E	127	−47.7	15.6	−15.3

as a whole over longer time periods. Models used to support such claims may not always correctly represent the polar atmosphere and the complex feed-back mechanisms. There are many variations in surface types, each develop-ing their own micro-climate, and circulation pattern changes do not affect all surface types uniformly. Figure 5.10b displays the temperature increases for Murmansk, Barrow, and Verkhoyansk. It can be seen that the temperature increase is strong early in the century, then from 1955 to 1975 a cold period is observed, and thereafter the temperature continued to increase again. In the last years this temperature increase has stopped, however, it is too early to judge if it is a real reversal in the trend. In general it can be seen that the temperature increase for Verkhoyansk is substantial, Barrow displayed a value closer to the mean value for the Northern Hemisphere, while the shorter time series for Murmansk is flat. It should be further noted that the tempera-ture increase displays a strong seasonal variation; it is large in winter and spring, less so in summer, and in autumn even a small cooling has been observed (Curtis *et al.* 1998).

Low-frequency variability

Looking at the temperature record one observes cyclic or quasi-cyclic beha-vior. One of the climate forcing cycles is the North Atlantic Oscillation

(NAO, see Section 2.2), which has a major impact on the weather in Northern Europe. Differences in the sea surface temperature change the atmospheric pressure distribution. There are two semi-permanent low pressure systems that strongly influence the circulation pattern in the Arctic, namely the Icelandic and the Aleutian Lows. When the gradient is above average (positive), strong westerly winds blowing over the relatively warm ocean (Gulf Stream) advect warm air to Northern Europe but cold air to Alaska, Canada, and Greenland. Such a situation occurred from 1986 to 1995. In 1995 the NAO reversed phase, reducing the low pressure of the Icelandic Low, which in turn reduced the advection, resulting in a far below normal winter temperature (1995/6) in Northern Europe.

The Arctic Oscillation (AO, see Section 2.7) effect on weather patterns is far reaching. In its negative phase, the AO's ring of air spins more slowly and is more easily disturbed, allowing warm air from lower latitudes to penetrate the Arctic. In the positive phase, the ring of air spins faster and impedes the frigid Arctic air from moving south. Normally the index flops back and forth within a winter. There has been a tendency in the last two decades for this index to spend more time in the positive phase, reducing the outbreak of cold air from the Arctic and, for example, the frequency of snowfall events in Seattle, Washington.

The Pacific Decadal Oscillation (PDO, see Section 2.3) has a substantial effect on the temperature in the western Arctic. In 1975/6 the index changed sign from negative to positive and a sudden temperature increase was observed for Alaska. In Barrow (Figure 5.10b) there is a 1 °C temperature difference for the mean of the 10 years before and after the change in the index. These differences are even more pronounced in winter, especially in January. This sudden change is of the same order as the total change for the 81-year time period, assuming linear regression analysis (Hartmann and Wendler 2005).

As mentioned previously, the Icelandic and Aleutian areas are preferred areas for cyclonic activities. Anticyclones are located in Eastern Siberia, Alaska/Yukon, and to a lesser extent over the central Arctic and Greenland. These anticyclones are well pronounced in winter, while in summer the cyclonic activity increases, as the cold core circumpolar westerly circulation decreases in intensity due to the decreased latitudinal temperature gradient (see section 5.11).

Precipitation

The precipitation is in general light with typical values of 200–300 mm annually. A great percentage of this amount falls as snow and is redistributed by wind as drifting snow. The amount of snow is normally under-reported. Since frequently snowfall amounts are small and accompanied by wind, it is not easy to obtain an accurate measurement. Northern Alaska (e.g. Barrow

104 mm/year), Northern Canada (e.g. Resolute Bay 130 mm/year) and the central Arctic Ocean (Russian drifting stations ~100 mm/year) report even less. In other parts of the world, areas with such small amounts of precipitation would be considered deserts.

Snow covers the tundra for 8–9 months of the year with a relatively shallow snow cover, typically of 30 cm thickness. The transition in spring (June) is fast and impressive. While snow is present, most of the solar radiation is reflected back to space, while in contrast to this, the bare tundra absorbs some 80% of the radiation. This leads to a sudden temperature increase (Weller and Wendler 1990), and the tundra becomes green and starts flowering. This transition period is very short, but the "summer" last only 3 months. Further, as can be seen from Table 5.3, the highest monthly mean temperature is less than 10 °C. In recent decades, the spring melt has occurred somewhat earlier in the western sector of the Arctic. This might have been caused by the observed temperature increase, which was well pronounced for the spring. On the other hand, the annual and winter precipitation have decreased for this area. This would also lead to an earlier break-up, more positive surface energy budget, and warmer temperatures. So it is difficult to judge if the warmer temperatures are the sole cause of the earlier break-up.

Other atmospheric influences

In general, the air in the Arctic is clean and visibility in winter frequently exceeds 100 km. However, in spring "Arctic haze" can often be observed (see Bokoye et al. 2003 for a useful summary and references). The existence of Arctic haze has long been known, and some 50 years ago the US Air Force investigated it with the so-called Ptarmigan flights, which originated in Central Alaska and went over the Arctic Ocean. In the meantime, chemical analyses showed that these are products consisting of mid-latitude emissions from fossil fuel consumption, smelting, and industrial processes. Large source regions are Europe and Russia, and the pollutants are transported to the Arctic in spring, when the washout due to precipitation is minor and the meridional temperature gradients and with it the circulation intensity are strong. Further, sand particles from the Gobi desert have been observed in Alaska, again indicating that the air over the Arctic is well connected with the lower latitudes.

Permafrost, in some areas hundreds of meters in thickness, underlies the tundra. The active layer, the layer that annually thaws, is only 30–50 cm deep. This underlying permafrost prevents drainage of surface water, and the tundra is spotted with lakes and the surface is wet. Hence, the term "polar desert," which is sometimes used, is not appropriate, as desert implies the lack of water (Köppen 1936), and no river can originate in such areas. Permafrost temperatures have in general increased, and in areas where permafrost is shallow and discontinuous, permafrost is melting. However, these occurrences are a function not solely of observed temperature increase, but also of the snow cover. Snow cover insulates the soil from the air, hence the amount as well as the distribution over the season are important. A heavy snowfall in autumn will insulate the soil from the cold winter air temperature. A similar snowfall in spring will have much less of an effect on the soil temperature.

The katabatic wind, which dominates the surface climate of Antarctica, is widespread in Greenland. However, neither the persistence nor the intensity is as high as in Antarctica. Greenland's surface radiation budget is less negative, and further, the fetch area does not have the same dimension as in Antarctica. A smaller cousin of this wind is the glacier wind, which is often observed on mountain glaciers. The air directly over the glacier surface is cooled by the snow and ice, while the air in the surrounding valley warms up during a summer day. The air over the glacier being colder (heavier) starts moving down the glacier, a phenomenon well known to mountaineers.

5.8.3 Sea ice

The annual variation in sea ice extent is much less pronounced in the Arctic when compared to Antarctica. This is, of course, an effect of the geography, as the Arctic Ocean is surrounded to a great extent by land masses. The greatest variation is found in the peripheral seas, namely the Sea of Okhotsk, the Bering Sea, Chukchi Sea, Davis Strait, Baffin Bay, Greenland Sea, Norwegian Sea and Barents Sea. Some of these areas are located south of the Arctic Circle, and do not belong to the Arctic proper. In the Arctic Ocean the sea ice is mostly multi-year ice with a thickness of about 3 m. In summer, it melts at the surface, and in winter new ice is formed at the bottom; Russian scientist have been especially active in this field (Makshtas 1991). Observations from land and from ships go back to the nineteenth century, but systematic high quality satellite observations have been obtained for about three decades. The annual maximum in ice extent for the Arctic Ocean and the adjacent seas is normally reached in February with 14 million km^2, the minimum in late August with about half this value.

A number of sea ice studies have been carried out, and the majority of these show a decrease in ice concentration, both for the Arctic Ocean and for the surrounding seas; however, the variations in ice extent are large from year to year. Further, the ice thickness has decreased, as shown by submarine cruises below the sea ice and anchored buoys, both being equipped with upward looking sonar (Rothrock *et al.* 1999). On the other hand, a more recent study doubts the magnitude of the frequently cited rapid thinning of sea ice (Holloway and Sou 2002) and states that the volume of sea ice in 2000 was similar to that in 1950. They further found that the sea ice has cyclic behavior. A decrease in ice mass is in agreement with observations of mountain glaciers in the Arctic, the majority of which showed shrinkage. Further, the Greenland ice sheet, the greatest ice sheet after Antarctica, has lost mass during the last five decades, during which time systematic observations have been carried out, and has contributed to the observed sea level increase.

5.8.4 Climate change

The climate of the Arctic has undergone large changes in the past, at times when human activity could not have influenced the climate. Actual measurements in the Arctic go back some 200 years, but proxy data such as tree-rings go back for a millennium, and ice core data from Greenland to 100 000 years. The European Arctic was warm during the Medieval Warm Period (MWP), the tenth century. It was at this time that the Vikings settled Greenland (see Section 8.2), and from the name one can deduce that vegetation was abundant. Around AD 1300 the Little Ice Age (LIA), lasting some five hundred years, commenced; the Viking colonies died out in Greenland by around AD 1500, two hundred years into this cold spell. Going further back, the ice core analyses of Greenland give detailed information. One astounding result was the abrupt changes in temperature that occurred, also called Dansgaard–Oeschger cycles, lasting from centuries to millennia. The last ice age ended some 13 000 years back, with one 1300-year reversal to cold temperatures, the Younger Dryas. While long-term temperature cycles are related to the Milankovitch theory (Section 5.2), a full understanding of shorter cycles and abrupt changes is still missing.

Climate change occurred in the Arctic during the past century; however, the processes have been anything but linear, as might have been expected from the fairly steady increase of CO_2 and other greenhouse gases in the atmosphere. This shows clearly that climate change, along with anthropogenic influences, is also affected by natural cycles. The Arctic climate system appears very complex and is well connected to events at lower latitudes, a result not surprising considering that the polar regions present the heat sinks of the global energy budget. The behavior of different climatic parameters appears rather cyclic, with different time periods. Furthermore, temporal changes vary widely spatially. Nevertheless, the Arctic is today warmer than it was 100 years ago, and this warming is more pronounced than that for lower latitudes. Further, there is less and thinner sea ice, permafrost temperatures have increased, most of the mountain glaciers are in recession, and the Greenland Ice sheet has lost mass. A better understanding of these processes will be a major undertaking for the next decades, combining ground-based data with remotely sensed observations and models.

5.9 Arctic general circulation

As in the Antarctic, the interactions between sea, ice, and land have an essential role in determining the strength and variability of the general circulation in the Arctic. As mentioned earlier, the major physical difference, however, is ice over water in the Arctic polar region, the surrounding land masses, and the scattered high elevation areas such as Greenland. Maxwell (1992) emphasizes the importance of open water (leads) within the ice pack, and the coastal impacts of land, islands, and the sea channels in between. The atmospheric circulation provides 95% of the heat advection from lower latitudes into the Arctic Basin, with the rest coming through the ocean (Bobylev *et al.* 2003).

The circulation patterns over the Arctic are controlled largely through the strength and variations in the CPV. A deep, cold low extending through the mid-troposphere to the lower stratosphere is a climatologically permanent fixture. Geopotential heights at 500 hPa drop from 5800 m in the mid-latitudes to 5100 m over the pole (Maxwell 1992). Geographically, within the circumpolar vortex there are three core troughs: over northern Canada and the Baffin Bay area (the strongest); over the island of Novaya Zemlaya; and over the east coast of Siberia. Vortex fluctuations are normally strongest in the Eurasian sector, and weakest over North America (Frauenfeld and Davis 2003).

The AO/NAO variations explain a significant portion of the CPV fluctuation pattern. Bobylev *et al.* (2003) and Frauenfeld and Davis (2003) state that, as the strength of the AO/NAO has increased, a more contracted and stronger CPV creates stronger westerlies, and enhances warming in the Arctic. Other impacts include stronger heat and moisture transport into the Basin, enhanced convective activity leading to more frequent low pressure systems, and changes in sea ice extent and thickness. Vortex circulation is strongest in winter. This season also produces the greatest variations from mean temperatures, largely due to warm air advection, but also caused by variations in longwave radiation loss, and in the land–ocean–atmosphere interactions (Bobylev *et al.* 2003). The intensity of the circumpolar vortex is enhanced by winter cooling of the polar stratosphere. However, there is considerable meandering of the vortex, and irregular breakdowns allow incursions from lower latitudes.

5.10 Surface pressure and wind

Figure 5.11 provides a summary of surface pressure variations across the Arctic Basin in summer and winter. The isobaric pattern shows higher pressure (>1021 hPa) in the eastern Arctic in winter. However the climate average shown obscures a highly variable day-to-day situation. The Atlantic sector to the North Pole is dominated by lower pressure, and, on average, strong pressure gradients, which enhance the warm air incursions from this source. The control

Figure 5.11 Surface
pressure distribution and
representative wind
streamlines in the Arctic
Basin for (a) winter and
(b) summer. (Adapted from
Przybylak 2003)

(a)

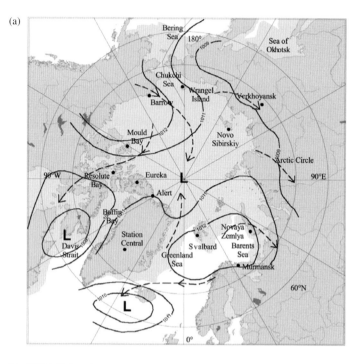

(b)

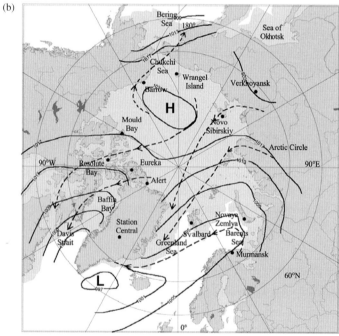

is the Icelandic Low. Incursions of air from the Pacific, through the Bering Strait, are much less frequent.

The summer surface pressure distribution (Figure 5.11b), is much simpler than for winter, and more symmetric around the pole. A weak low (<1010 hPa) exists at the pole itself, attracting weather systems from several different sources. Two centers of weak high pressure (>1012 hPa) exist in the Atlantic sector, and over the Beaufort Sea area.

Wind in either season is highly variable, with only very broad coherence across the pole (see streamline indicators, Figure 5.11). In winter, there are more frequent easterlies in the Atlantic sector, while over the Canadian Arctic, the streamlines trend toward northerly. In the summer, weak airflows move toward the pole, but also in a general southerly direction along the continental edges. In both seasons, wind is strongly influenced by the presence of cyclones, and by local land–ice–sea contrasts.

5.11 Extra-tropical cyclones

The development, life, and decay of Arctic cyclones exhibits a complex relationship between general circulation, latitudinal temperature gradients, and ocean–ice–land distributions. Extra-tropical cyclones are the primary mechanism for the distribution of heat and moisture into and within the Arctic Basin. With no circumpolar trough similar to Antarctica, cyclones may develop within the polar area during any season, or may develop in the ocean areas to the south and then transport into the basin. Although they are generally weaker in strength than those in the Antarctic, mainly due to a weaker polar–equatorial temperature gradient (King and Turner 1997), on a day-to-day basis, they can create very rapid and unpredictable weather changes. Figure 5.12 shows how spatial distributions and frequency vary by season. Zhang et al. (2004) state that, on average, 50 cyclones are created within the Arctic region each year, and about 17 originate from outside the region.

Several factors have important influences on the development and distribution of cyclones in the Arctic. They are very sensitive to the variability in atmospheric circulation (Przybylak 2003). Surface and air temperatures assist or retard development. The diabatic process, identified as important on the slopes of the Antarctic coastal area, is also essential in the Arctic (Bobylev et al. 2003). Over warmer surfaces, such as open water, cyclogenesis is enhanced. Over colder surfaces, such as the ice pack, anticyclogenesis is encouraged. There is a significant negative correlation with cyclonic activity in the mid-latitudes (Zhang et al. 2004), and a positive correlation with precipitation (Holland 2003). This is related to the strength of the NAO/AO. When the NAO is positive, circulation is more meridional, and cyclones will migrate into the basin. On average, between 1952 and 1989, there were increases in Arctic cyclone activity in all seasons, with significant results in winter, spring, and summer. When the

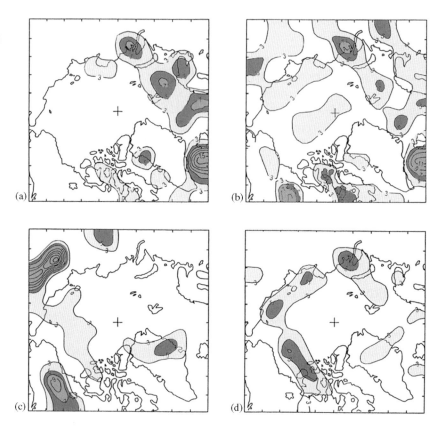

Figure 5.12 Mean distribution of extra-tropical cyclones ((a) winter, (b) summer) and anticyclones ((c) winter, (d) summer) in the Arctic Basin. (From Przybylak 2003)

NAO/AO is negative, the circulation is more zonal, and cyclones tend to remain on a more latitudinal track.

Figure 5.12a and b shows the spatial distribution of cyclone frequencies in summer and winter. There is a strong relationship to the pressure distributions shown in Figure 5.11. Within the Arctic, in both seasons, cyclones are most frequent in the Atlantic sector, especially Baffin Bay, and extending from the Scandinavian coast across to the Novaya Zemlaya area. Cyclones are noticeably absent in the eastern half of the basin, especially in winter. In summer, the weak polar low attracts occasional development along the east Siberian coast.

In winter, cyclones originating from outside the Arctic enter through three major pathways (Przybylak 2003; Zhang *et al.* 2004). The most active is the Atlantic corridor, originating out of the Icelandic Low. Fewer cyclones originate through the NW Pacific Ocean/Barents Sea route, and from Northern Europe to the coast shelf off Eurasia. Cyclones are strongest, with lowest centers of pressure, in July. Transport can be interrupted or channeled by areas of higher elevation, such as Greenland. According to Bobylev *et al.* (2003), cyclones are often separated from the surface due to the inversion, but cyclones can destroy the surface inversion layer over the exposed ice pack.

Summer sources and transport patterns are more diffuse, reflecting the attraction of the weak polar low. Summer cyclones are weaker (averaging 1.5 hPa higher in central pressure than in winter), but have longer lifetimes (Zhang *et al.* 2004). Speed of movement is highest in the winter half of the year (especially March), and lowest in August and September (Przybylak 2003).

Figure 5.12c and d shows the spatial distribution of anticyclone frequency over the Arctic Basin. Anticyclones are weak and slow moving with sluggish circulation. They are not particularly well defined. Stronger in winter, they are most frequent over the Beaufort Sea and East Siberian Arctic, and inland Canada and Siberia. The contours over Greenland in Figure 5.12 may reflect altitude more than the existence of high pressure. In summer, Arctic anticyclonic frequency is similar to winter, with an additional center over the Novaya Zemlaya area.

As in Antarctica, meso-scale lows can develop and decay quite rapidly (Przybylak 2003). Often independent of the larger-scale circulation, these can cause significant local changes to weather, and bring considerable precipitation.

5.12 Polar night jet and stratospheric ozone depletion

The late winter and early spring general circulation over the Arctic has some broad similarities to Antarctica. The strength and variability of the PNJ, as the leading edge of the circumpolar vortex, are strongly linked to the semi-annual oscillations (AO, AAO) in both hemispheres (Kuroda and Kodera 2001). The Arctic stratospheric circulation (zonal-mean zonal winds) is poleward, and subsiding, bringing colder air toward the surface, in a similar manner to Antarctica (Kuroda and Kodera 2004). Changes in Rossby wave number 1 mainly determine variability in the PNJ/circumpolar circulation.

However, major differences occur between the hemispheres, which allow stratospheric warming to occur over the Arctic, but not the Antarctic, at this time of year. There is no high-altitude continental "anchor" in the Arctic, and the circulation meanders over wide areas. The circulation can break down on an irregular basis, allowing major incursions of warm air to offset the winter cooling associated with stratospheric subsidence. The overall wind speeds in the vortex are lower than over Antarctica, and more variable.

This type of circulation pattern does not allow polar-wide stratospheric ozone depletion in the Arctic. While cold, protracted winters can occur, the overall decrease in Arctic stratospheric ozone in six worst-case years in the 1990s averaged less than half the decrease over Antarctica (Bobylev *et al.* 2003). Overall average ozone depletion is less than 15%. There are periods during the season when temperatures in parts of the Arctic drop low enough to form polar stratospheric clouds, and therefore significant ozone depletion can occur, but these periods are not very consistent (Turco 2002). Over the Arctic, stratospheric aerosols can be important as a substitute for PSCs, activating heterogeneous chlorine chemistry processes if the temperatures fall below 200 K. As a result, "pockets" of ozone depletion occur for relatively short periods of time, due to chlorine chemistry. These are highly variable in location. Pockets can occur outside the vortex on occasion (Bobylev *et al.* 2003), but these are created by a decrease in single oxygen production, and not chlorine chemistry. Since people live in the higher latitudes of the Northern Hemisphere, such pockets are of concern to human health, due to risks from increased ultraviolet radiation.

5.13 Concerns about future warming

The Arctic atmosphere is far less isolated from the interactions with the rest of the atmosphere than the Antarctic. While there is evidence of recent warming in both locations, sometimes attributed to the greenhouse effect, the potential impacts in the Arctic are currently of greater concern (Whitfield *et al.* 2004 and others). Bobylev *et al.* (2003) review the summaries from IPCC (2001) regarding near-future climate change in the Arctic, associated with greenhouse warming. Results from combined model scenarios indicate that, by the end of the twenty-first century, warming in the Arctic will be 40% greater than the global average. The Arctic Ocean is likely to become nearly ice-free, creating large reductions in albedo and major changes in the structure of the energy budget. Open water rather than ice will result in increased carbon dioxide absorption and changes in the marine ecosystem. Further changes include ocean circulation, meltwater pathways, reductions in deepwater formation and so forth. The interaction between the all-ocean surface and the atmosphere will create a different feedback system than that occurring now with the ice pack, resulting in changes in precipitation distribution, surface temperatures, and atmospheric circulation. The complexities of what is likely to occur are not well understood.

5.14 Chapter summary

Table 5.4 summarizes the major geographical and climate characteristics of the Arctic and Antarctic regions. Although the mid-troposphere and lower stratosphere circulation over both areas is broadly similar (meridional heat and moisture transport toward the poles; subsidence of cold air from the stratosphere especially in winter), the surface environments create major differences in wind flows, cyclogenesis, distribution of extra-tropical lows, temperature, and precipitation. At both locations, the geographical distribution of land, water, and ice has a critical role in the seasonal and spatial climate pattern. Models suggest that both regions will warm in greenhouse scenarios. The main consequences for the Antarctic may be more precipitation and thinning of the surrounding ice shelf. For the Arctic, open water may replace the ice pack by the end of the twenty-first century, creating major changes in radiation budgets, circulation, precipitation, and temperature.

Table 5.4 *Geographical and climate comparisons between the Arctic and Antarctica*

Antarctica	Arctic
Large continent surrounded by water	Ice pack over water surrounded by land
High elevation continent (to 5000 m)	Low elevation (ocean/ice surface)
Ice and snow covered	Ice and snow covered
Surrounded by seasonally variable ice pack	Seasonal variability in ice pack size and thickness
Polar–equatorial temperature gradient = 70 K	Polar–equatorial temperature gradient = 50 K
High albedo from snow pack (up to 80%)	High albedo from snow cover (50–60%)
Minimum temperatures < −70 °C	Minimum temperatures −40 to −50 °C
Snow accumulation < 50 mm/y in interior, much higher along the coast	Precipitation 100–150 mm/y over basin, higher where local terrain supports
Strong, persistent surface inversions	Strong, persistent surface inversions
Strong, persistent katabatic winds from the interior to the coast	Katabatic winds highly variable spatially
Semi-annual oscillation (AAO) main controller of circulation	Semi-annual oscillation (NAO/AO) main controller of circulation
Circumpolar trough around coastal boundary	No well-defined circumpolar trough
Enhanced cyclogenesis in CPT	Cyclogenesis over polar ice or from mid-latitude lows
Little penetration of storms inland	Regular penetration of storms into polar basin
Very strong late winter/early spring PNJ, circumpolar vortex	Variable later winter/early spring PNJ, circumpolar vortex
No meridional incursions of warm air in winter/spring	Warm air meridional incursions at all levels in winter/spring
Very strong stratospheric ozone depletion, early spring	Weak stratospheric ozone depletion, early spring, occurs in "pockets"

5.15 Examples of polar websites

Antarctic Climate Evolution, www.ace.scar.org, describes proposals related to climate research, paleoclimatology and so forth associated with the multidisciplinary Scientific Committee on Antarctic Research (SCAR).

The Arctic and Antarctic Research Institute, www.aari.nw.ru/main_en.asp, focuses on the research activities of Russia's major polar scientific institute (AARI).

Alaskan Climate Research Center (ACRC), http://climate.gi.alaska.edu/ provides climate and climate change data for Alaska and the North.

The British Antarctic Survey, www.antarctica.ac.uk/BAS_Science/Index.html, studies climate change and variability, global warming, katabatic winds, and other aspects of Antarctic climate.

The Cooperative Institute for Arctic Research, www.cifar.uof.edu, supports major scientific research in the Western Arctic.

Dumont d'Urville and Dome C, www.gdargaud.net/Antarctica.index.html, French research activities in Antarctica, created by one of the researchers Guillaume Dargaud.

The National Oceanographic and Atmospheric Administration, www.cmdl.noaa.gov/obop/met, reports information from the United States background meteorological stations, in particular, the South Pole.

The World Meteorological Organisation, www.wmo.ch/web/www/Antarctica/antarctica.html, describes research in the polar regions associated with the World Climate Program.

5.16 References

Ackerley, S., Wadhams, P. and Comiso, J., 2003. Decadal decrease in Antarctic sea ice extent inferred from whaling records revisited on the basis of historical and modern sea ice records. *Polar Research*, **22**, 19–25.

Allison, I., Wendler, G. and Radok, U., 1993. Climatology of the East Antarctic ice sheet (100E–140E) derived from automatic weather stations. *Journal of Geophysical Research*, **98**, 8815–8823.

Ball, F., 1957. The katabatic winds of Adélie and King George V Land. *Tellus*, **9**, 201–208.

Bobylev, L., Kondratyev, K. and Johannessen, O., 2003. *Arctic Environmental Variability in the Context of Global Change*. Berlin: Springer.

Bokoye, A., Royer, A., O'Neill, N. and McArthur, L., 2002. A North American Arctic aerosol climatology using ground-based sun photometry. *Arctic*, **55**, 215–228.

Bridgman, H., 1997. Air pollution. In R. Thompson and A. Perry, eds., *Applied Climatology Principle and Practice*. London: Routledge, pp. 288–303.

Bromwich, D. and Parish, T., 1998. Meteorology of the Antarctic. In D. Karoly and D. Vincent, eds., *Meteorology of the Southern Hemisphere*. Meteorological Monographs Vol. 47, No. 49, Boston: American Meteorological Society, pp. 175–200.

Bromwich, D. and Stearns, C. (eds.) 1993. *Antarctic Meteorology and Climatology: Studies based on Automatic Weather Stations*. Washington, DC: American Geophysical Union.

Chapman, W. and Walsh, J., 1993. Recent variations of sea ice and air temperatures in high latitudes. *Bulletin of the American Meteorological Society*, **74**, 33–47.

Curtis. J., Wendler, G., Stone, R. and Dutton, E., 1998. Precipitation decrease in the Western Arctic, with special emphasis on Barrow and Barter Island, Alaska. *International Journal of Climatology*, **18**, 1687–1707.

Frauenfeld, O. and Davis, R., 2003. Northern Hemisphere circumpolar vortex trends and climate change implications. *Journal of Geophysical Research*, **108**, manuscript 4423.

Frey, K. and Smith, L., 2003. Recent temperature and precipitation increases in West Siberia and their association with the Arctic Oscillation. *Polar Research*, **22**, 87–300.

GEO, 2003. *Global Environmental Outlook 2003*. United Nations Environment Programme, Geneva, www.unep.org/GEO/geo3.

Giovinetto, M. and Bentley, C., 1985. Surface balance in ice drainage systems in Antarctica. *Antarctic Journal of the US*, **20**, (4) 6–13.

Hartmann, B. and Wendler, G., 2005. The significance of the 1976 Pacific climate shift in the climatology of Alaska. *Journal of Climate*, **18**, 4824–4839.

Holland, M., 2003. The North Atlantic Oscillation-Arctic oscillation in the CCSM2 and its influence on Arctic climate variability. *Journal of Climate*, **16**, 2767–2781.

Holloway, G. and Sou, T., 2002. Has arctic sea ice rapidly thinned? *Journal of Climate*, **15**, 1691–1701.

IPCC, 2001. *Climate Change 2001: The Scientific Basis. Contribution of Working Group I to the Third Assessment Report of the Intergovernmental Panel on Climate Change.* J. Houghton, Y. Ding, D. Griggs, *et al.*, eds., Cambridge: Cambridge University Press.

Jones, P., 1995. Hemispheric surface temperature variations: reanalyses and update to 1993. *International Journal of Climatology*, **7**, 1794–1802.

King, J. and Turner, J., 1997. *Antarctic Meteorology and Climatology*. Cambridge: Cambridge University Press.

König-Langlo, G., King, J. and Pettre, P., 1998, Climatology of three coastal Antarctic stations Dumont d'Urville, Newmayer, and Halley. *Journal of Geophysical Research*, **D103**, 10935–10946.

Köppen, W., 1936. Das geographische System der Klimate. In *Handbuch der Klimatologie*. Berlin.

Kuroda, Y. and Kodera, K., 2001. Variability of the polar night jet in the Northern and Southern Hemispheres. *Journal of Geophysical Research*, **D106**, 20703–20713.

2004. Role of the polar-night jet oscillation on the formation of the Arctic Oscillation in the Northern Hemisphere winter. *Journal of Geophysical Research*, **109**, manuscript D11112.

Makshtas, A., 1991. *The Heat Budget of the Arctic Ice in the Winter*. English edition, Cambridge: International Glaciological Society.

Mawson, D., 1915. *The Home of the Blizzard*. London: Heinemann.

Maxwell, B., 1992. Arctic climate: potential for change under global warming. In F. Chapin, R. Jefferies, J. Reynolds, G. Shaver and J. Svoboda, eds., *Arctic Ecosystems in a Changing Climate*. New York: Academic Press, pp. 11–35.

Meinardus, W., 1938. Klimakunde der Antarktis. In W. Koeppen and R. Geiger eds., *Handbuch der Klimatologie*, Band IV, 180pp.

Parish, T., 1982. Surface air flow over East Antarctica. *Monthly Weather Review*, **112**, 545–554.

Parkinson, C., 2002. Trends in the length of the Southern Ocean sea-ice season 1979–1999. *Annals of Glaciology*, **34**, 435–440.

Phillpot, H. and Zillman, J., 1970. The surface temperature inversion over the Antartic continent. *Journal of Geophysical Research*, **75**, 4161–4169.

Przybylak, R., 2003. *The Climate of the Arctic*. Dordrecht: Kluwer.

Radok, U., 1970. Boundary processes of drifting snow. *Studies of Drifting Snow*, Meteorological Dept., University of Melbourne. Vol. **13**, 1–20.

Rothrock, D., Yu, Y. and Maykut, G., 1999. Thinning of the Arctic sea ice cover. *Geophysical Research Letters*, **25**, 3469–3472.

Simmonds, I., 1998. The climate of the Antarctic region. In J. Hobbs, J. Lindesay and H. Bridgman, eds., *Climates of the Southern Continents*. New York: Wiley, pp. 137–160.

Simmonds, I. and Jacka, T., 1995. Relationship between the interannual variability of Antarctic Sea Ice and the Southern Oscillation. *Journal of Climate*, **8**, 637–647.

Turco, R., 2002. *Earth Under Siege From Air Pollution to Global Change*, 2nd edn. New York: Oxford University Press.

Turner, J., 2004. The El Niño-Southern Oscillation and Antarctica. *International Journal of Climatology*, **24**, 1–31.

Weller, G. and Wendler, G., 1990. Energy budgets over various types of terrain in polar regions. *Annals of Glaciology*, **14**, 311–314.

Wendler, G., Stearns, C., Weidner, G., Dargaud, G. and Parish, T., 1997. On the extraordinary katabatic winds of Adélie Land. *Journal of Geophysical Research*, **102**, 4463–4474.

Whitfield, P., Hall, A. and Cannon, A., 2004. Changes in the seasonal cycle in the circumpolar Arctic, 1976–1995: temperature and precipitation. *Arctic*, **57**, 80–93.

Zhang, X., Walsh, J., Zhang, J., Bhatt, U. and Ikeda, M., 2004. Climatology and interannual variability of Arctic cyclone activity: 1948–2002. *Journal of Climate*, **17**, 2300–2317.

Zwally, H., Comiso, J., Parkinson, C., Cavalieri, D. and Gloerson, P., 2002. Variability of Antarctic sea ice 1979–1998. *Journal of Geophysical Research*, 10.1029/2000JC00733.

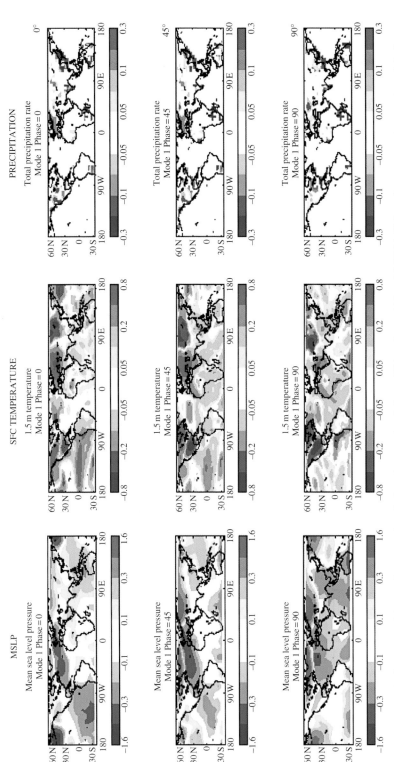

Plate 1 Canonical patterns of the spatiotemporal evolution of co-varying 2.2 year LFV peak quasi-biennial (QB) MSLP, surface temperature, and precipitation signals at 0, 45 and 90 degrees of phase over the domain 65° N to 35° S from an MTM-SVD analysis using data from 1900 to 1998. MSLP, surface temperature, and precipitation weights are shown as being redder for positive, and bluer for negative values.

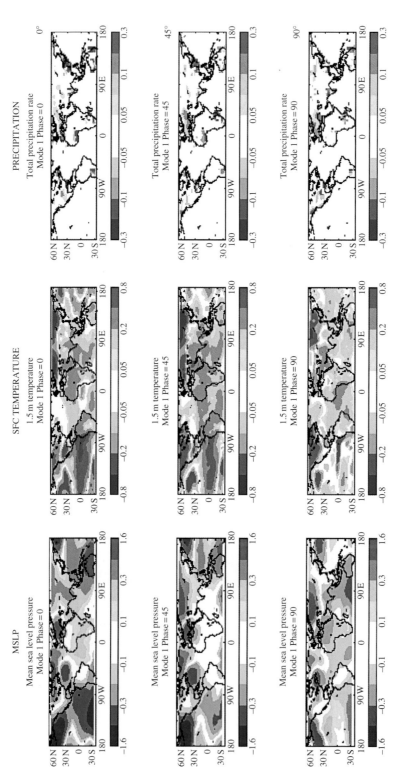

Plate 2 Canonical patterns of the spatiotemporal evolution of co-varying 3.6 year LF peak LF ENSO MSLP, surface temperature, and precipitation signals at 0, 45 and 90 degrees of phase over the domain 65° N to 35° S from an MTM-SVD analysis using data from 1900 to 1998. MSLP, surface temperature, and precipitation weights are shown as being redder for positive, and bluer for negative values.

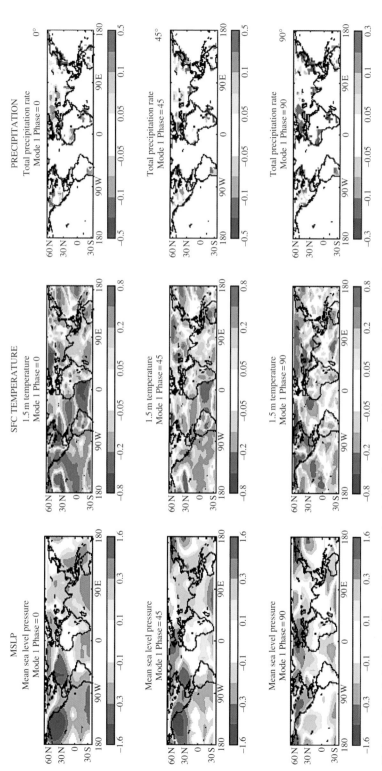

Plate 3 Canonical patterns of the spatiotemporal evolution of co-varying 11.6 year LFV peak quasi-decadal MSLP, surface temperature, and precipitation signals at 0, 45 and 90 degrees of phase over the domain 65° N to 35° S from an MTM-SVD analysis using data from 1900 to 1998. MSLP, surface temperature, and precipitation weights are shown as being redder for positive, and bluer for negative values.

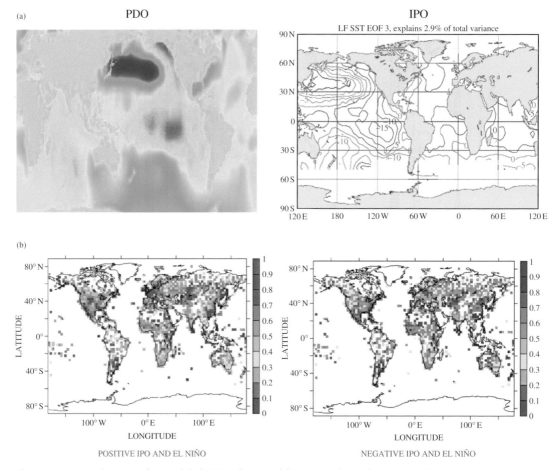

Plate 4 (a) Spatial pattern of near-global SSTs indicative of the PDO and IPO phenomena. The PDO diagram is from http://www.iphc.washington.edu/Staff/hare/html/decadal/decadal.html and the IPO panel is the second EOF pattern from the analysis of Folland *et al.* (1999). (b) Probability of exceeding median October–December rainfall over the globe during El Niño events with positive (left panel) and negative (right panel) IPO phase in the previous September. (Derived from the work of Meinke *et al.* 2001)

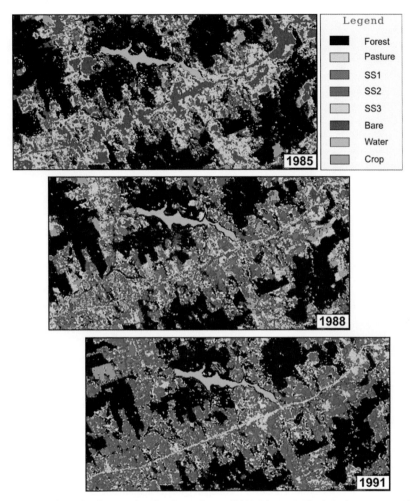

Plate 5 Land use/land cover classifications of a small (12.5 km × 5.5 km) study area near Altamira, Brazil, using computer analyzed August 1985, 1988, and 1991 Landsat TM.

Plate 6 This nighttime image of the Phoenix, Arizona, metropolitan area is from the Advanced Spaceborne Thermal Emission and Reflection Radiometer (ASTER) for October 3, 2003 at approximately 22:39:00 local time. A color ramp has been applied to highlight relatively hot and cool surfaces; the corresponding temperature scale is in °C. Urban materials, such as the asphalt street grid and Sky Harbor Airport exhibit relatively high surface temperatures, while heavily vegetated regions (mesic neighborhoods, riparian zones, agricultural fields) are relatively cool. Sparsely vegetated mountain ranges comprised of exposed massive bedrock and block talus also have relatively high surface temperatures compared to other natural surfaces. North is towards the top of the image. (Image source: William L. Stefanov, Johnson Image Analysis Laboratory, NASA Johnson Space Center, Houston, TX. Original unprocessed ASTER data (AST_08 surface temperature product) from NASA/ERSDAC)

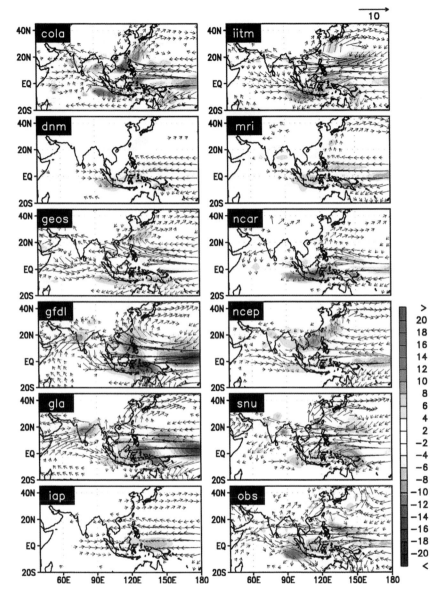

Plate 7 Rainfall (mm/day) and 850 mb wind (m/s) differences (JJA 1998 minus JJA 1997) for individual models participating in CLIVAR MMIP. The observation is shown in the right-hand bottom panel.

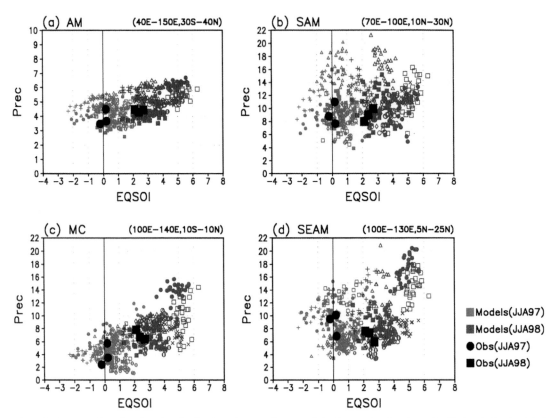

Plate 8 Scatter plots showing the distribution of anomalous precipitation vs. the Southern Oscillation Index for (a) the entire AM monsoon region, (b) the South Asian Monsoon region, (c) the Maritime Continent (MC), and (d) the South East Asian Monsoon region. The latitude–longitude boundary of each domain is indicated. Unit for precipitation is mm/day and for SOI is in mb.

Chapter 6
Post-glacial climatic change and variability

6.1 Introduction

An understanding of past climates, climate change, and variability within the Earth's climate system is essential to help understand current climate variability and to assist in the prediction of future climate change. Research over the past 50 years or so has demonstrated that changes in climate can be described for periods up to 420 000 years before present (BP). The very last few thousand years of record, the period since the end of the last major glaciation, the Holocene, is described by Petit *et al.* (1999) as "... the longest stable warm period recorded in Antarctica during the past 420 ky."

Jones and Mann (2004) emphasize that the last 12 000 years BP is a short enough period so that long-term effects on climate, such as changes in the Earth's orbit and major changes in global ice mass, have no impact. Natural forces over smaller temporal scales, such as variations in solar radiation output, dust and gases from irregular volcanic eruptions, ENSO variations (Section 2.8) and changes in ocean–atmospheric circulation, become major forcing factors. After about AD 1850, human influences become much more important, associated with excess greenhouse gas emissions, global warming, land use changes, and cooling from sulfate aerosol pollution.

Thompson *et al.* (1993) list several reasons why the study of climate during the Holocene, and especially over the past 1000 years, can provide an important contribution to understanding of climate variability. There is enough detailed information available from a range of sources to allow climate reconstructions. These can include rates of climate change, lags, leads, and annual and seasonal resolutions. The data show that periods of extremes existed, but also periods where climate could be considered "average" or "normal." Through modeling and other methods, the causes of Holocene climate change and variability can be explored, and the details of potential forcing functions can be established.

This chapter describes current knowledge about climate change and variability during the Holocene. After a short review of the methods used to determine past climate, the essay by Zielinski focuses on the Northern Hemisphere (NH), where the best data sets and strongest research effort have occurred. The time periods 7800–8200 BP, and AD 800–1850 (containing the Medieval Warm Period (MWP) and Little Ice Age (LIA)) are a focus for discussion. The chapter concludes with a description of climate change in the Southern Hemisphere (SH), and comment on comparison between the hemispheres. The evaluation of the post-1850 period is left to Chapter 9, where the impacts of global warming are assessed using global circulation models. Links to human behavior are explored in Chapter 8.

6.2 Determining past climate through the use of proxies

Detailed measurements of weather and climate are the best way to provide accurate interpretations of climate variability and change. However, measurements are rare earlier than AD 1850, and, those that do exist further back in time are not representative of data from today's methodologies. Past climates, therefore, must then be determined indirectly, through the use of proxy indicators that react or change as climate changes (Jones *et al.* 1998; IPCC 2001; Jones and Mann 2004). There are two main types of paleoclimate research: annual to decadal time series, which can be correlated against measured data to determine strength and accuracy; and assessments at poorer resolutions, over periods of 100 or more years, which provide a much coarser picture. More detailed time series provide the best information, with the coarser resolution proxy data providing background support.

Paleoclimatic reconstructions can occur for a number of atmospheric parameters. Temperature proxies are most often used, however, because the range of data is the largest, and temperature is possibly the most important indicator of climate change. Moisture is the next best choice, but data are more limited, and local and regional influences are strong confounding factors. Whatever the proxy or the parameter, researchers must recognize the limitations in the use of proxy data for past climate assessment. Some of the major proxy methods and their limitations are listed in Table 6.1. Trenberth and Otto-Bliesner (2003) list several complexities that must be considered. These include seasonality, oscillations (Chapter 2), environmental interferences, quality and scale of dating, potential biological factors, climate system reactions, and accurately establishing what the proxy actually represents.

Several authors emphasize the importance of a multi-proxy integrated approach which will allow the most accurate climate interpretations possible. Calibration and standardization are critical, allowing the quality of correlation and accuracy to be determined (Diaz and Bradley 1995). This can be difficult to accomplish with data that are highly spatially diverse, and that represent differing time scale accuracies. Given the quality of most proxy results, resultant multi-proxy temperatures are often plotted in annual or decadal averages, with a 30 to 50 year filter used to show major variations and trends.

Table 6.1 *Examples of major proxy methods to determine detailed past climate change and variability, their benefits and limitations*

Proxy method	Definition	Spatial extent	Time scale	Benefits	Limitations
Ice cores (O^{18}, gases, dust)	Cored ice from ice caps and glaciers	Greenland, Antarctica, some high-altitude and high-latitude glaciers	Several 100 000 years	Long records over multiple millennia; Wide range of information possible; Annual/seasonal analysis possible	Small area of Earth's surface; Cold climates only; Detailed dating can be imprecise; Accuracy reduced with depth; Affected by changes in moisture; Affected by rate of horizontal flow
Dendrochronology	Width and density of tree-rings	Poleward of 30° where cold season stops growth	Up to about 10000 years	Details about temperatures and moisture; Correlates well with measured data; Potential for O^{18} and C^{14} analysis; High-quality data record	Limited to high elevations or mid to high latitudes; Limited to trees with annual growth rings; Dependent on growing season months; Highly dependent on standardization method; Calibration only where cross-dating possible
Coral	Growth rings and chemistry from tropical coral massives	Tropics with shallow seas	Several hundreds of millennia possible	Tropical complement to tree-rings; Can provide precise dating; Continuous sampling possible; Provide teleconnection details (ENSO); Detailed isotope and elemental chemistry analysis	Tropics only; Limited spatial distribution; Long records are rare; Strong dependence on water temperature and salinity; Less accurate with depth; Seasonal cycle dominates

Table 6.1 (cont.)

Proxy method	Definition	Spatial extent	Time scale	Benefits	Limitations
Pollen	Pollen species from undisturbed lake and coastal cores	Mainly mid to high latitudes where trees and grass grow	Normally a few millennia	Widespread use Long records possible Can indicate both temperature and moisture	Dating can be uncertain Pollen identification can be difficult Depends on how vegetation responds to climate
Speleothems	Stalactites and stalagmites in cave environments	Cave environments: depends on water flow in sedimentary rock	Normally a few millennia	Wide range of proxies possible (chemistry, trace elements) Can indicate changes in water cycle; atmospheric circulation	Limited locations Dating can be uncertain Difficult to interpret climate variables
Varved sediments	Sedimentation of marine and lake organic remains	Areas with high sedimentation rates and strong seasonal changes	Normally a few millennia	Annual and seasonal resolution possible O^{18} analysis Faunal assemblies	Best in closed-basin glacial lakes Limited distribution
Historical records	Written records, diaries, phenology, crop harvests, etc.	Can be wide ranging, but usually mid to high latitudes	Up to about 1500 years	Wealth of information possible Wide range of potential sources Can provide highly detailed information	Patchy in space and time Can be anecdotal and lack accuracy Requires cautious interpretation Often emphasis on extremes May not represent seasonal and/or annual means

Greater details are available in Jones et al. (1998); IPCC (2001); Jones and Mann (2004).

6.3 ESSAY: Post-glacial climates in the Northern Hemisphere

Gregory A. Zielinski, *University of Maine*

6.3.1 Introduction

A commonly presented statement about the nature of NH climate following deglaciation is that climatic conditions have not changed greatly with time. This statement is only appropriate in a relative sense, because the many high-magnitude and rapid changes in climate during the last glacial period (i.e. Dansgaard–Oeschger Events; Figure 6.1) overwhelm the fluctuations that occurred since the Earth moved into the present interglacial period. Greatest evidence for the rapid climate change events during the last glacial comes from the very highly resolved climatic records available from the Greenland Ice Sheet ice-coring projects of the early 1990s (GISP2 and GRIP; American Geophysical Union 1998). However, detailed evaluation of the post-glacial record (i.e. the Holocene) in these same ice-core records (Figure 6.1), as well as in other types of proxy data, shows a great deal of variability in NH climate since the retreat of the large ice sheets. This variability becomes even more pronounced from a spatial perspective, as once the climatic impact and feedback mechanisms associated with large ice sheets are reduced and ultimately eliminated, post-glacial climates in the NH became very regionalized. Given such regionalization, a simple description of Holocene climatic conditions from a northern hemispheric perspective is not easily attained.

Nevertheless, there is evidence that factors which control Holocene climate in the NH operate on a roughly periodic time scale. As a result, there is an overarching theme to changes in climatic conditions over the last 11 500 years that provides, at least, a partial sense of consistency in Holocene climate. However, the magnitude of these changes is not necessarily the same for each event. Similarly, the way climatic conditions respond to changes in forcing factors through the Holocene varies, and thus the nature of the signal in proxy records for the same event may vary. This scenario further reflects the regionalization of post-glacial climate in the NH.

This essay summarizes post-glacial climatic conditions in the NH from a broad perspective with the beginning of post-glacial conditions chosen to be the end of the Younger Dryas or about 11 500 calendar years ago (e.g. Mayewski *et al*. 2004). The reason for taking the broad perspective approach is that there are many detailed and specific records of Holocene climatic change that would require much more explanation and space than available in this essay. Consequently, to give the best overall perspective on climatic conditions, a general summary of overall NH climatic conditions through the Holocene is presented, thereby highlighting the periodic nature of climatic

Figure 6.1 Greenland ice-core records of environmental change over the last 120 000 years, including the rapid, high magnitude changes from 11 500 to 110 000 years ago (i.e. Dansgaard–Oeschger Events). The top two records reflect global mean temperature change and δ^{18}O, a proxy for temperature change, from Greenland ice cores. The third time series depicts changes in the concentrations of sea salt species measured in the GISP2 ice-core glaciochemical record, as a proxy for intensity of atmospheric circulation. Bottom time series shows the NH$_4$ time series in the GISP2 ice core reflecting overall greater terrestrial biomass activity during periods of warmer climate. H designates the Holocene periods, comparatively quiet at this scale. (Modified from Mayewski and White 2002)

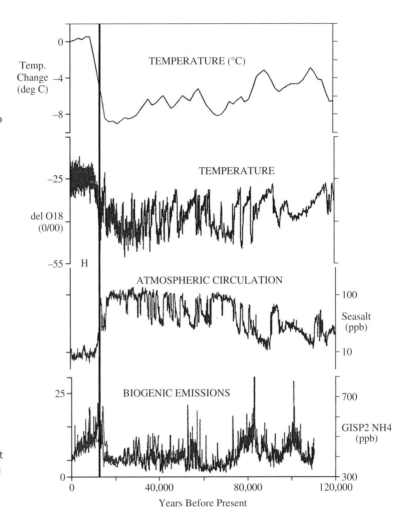

change. This will be accomplished initially by presenting several key records that homogenize and incorporate climatic conditions from throughout the NH. The best records that accomplish this are the ice-core records from Greenland. This will be followed by a presentation on Holocene climatic conditions from several other records, all of which are highly resolved. After presenting the "big picture" on overall Holocene conditions, a discussion will follow that looks at specific time periods during the Holocene. These individual periods are often noted in discussions about Holocene climate because they are known to be controversial or have some uncertainty in identifying the controls and spatial variability in climate. For instance, it has recently been observed in many paleoclimatic records that a distinct climatic event occurred about 8200–8400 years ago. This essay will summarize the thoughts on this event (i.e. the 8200-year event) and why it is such an important piece of the climatic puzzle in the NH. In addition, there has been much discussion recently on climatic conditions over the last millennium, such as the existence of the MWP and the consistency of the LIA. In fact, the use of these terms can be quite controversial. The last millennium is an important time period for evaluating the climatic system of the NH as there are more abundant proxy records to reconstruct climate, it contains the instrumental part of the record, and it contains two climatic periods (MWP and LIA) when the NH, at least in general, was under contrasting conditions. The final section briefly summarizes some of the dominant forcing factors on Holocene climate in the NH.

In addition to the use of the terms MWP and LIA, there are other climatic terms that still come into use at various times when discussing Holocene climatic variability. Figure 6.2 shows an early attempt to identify Holocene climatic stratigraphy as developed through the evaluation of discontinuous peat deposits. Some of these original terms are still used quite frequently in the discussion of post-glacial climate especially at the regional scale. However, in the decades since this original stratigraphy was developed, many highly resolved and continuous records have been constructed, thereby superceding the stratigraphy shown in Figure 6.2. The greatest modification has been in specific details and timing of particular events given the more abundant records that now exist with excellent dating control. Nevertheless, the overall climatic conditions identified in the stratigraphic column shown in Figure 6.2 give a broad idea of general changes in Holocene climate. Moreover, they provide an historical perspective of the evolution of the record of post-glacial climates in the NH.

6.3.2 General characteristics of Holocene climate

Ice-core records

Of the many types of proxy records available for the reconstruction of Holocene climate (e.g. Bradley 1999), ice cores are the only medium that produces details of past atmospheric conditions. Although ice cores have been collected from both polar regions, as well as from tropical and mid-latitude high alpine sites, the climatic records derived from the Greenland ice-coring projects of the 1990s set the standard by which subsequent climatic records must now be compared. The very high resolution available in these Greenland records (i.e. potentially to the sub-annual level during the Holocene) is one primary reason for the significant contributions they have made toward understanding our climate system. In addition, the length of record available at an annual resolution (i.e. last 110 000 years; Meese *et al.* 1997) is unprecedented at this time. Finally, the location of Greenland may be equally important to the nature of information available in various ice-core measurements used for reconstructing NH climate.

Figure 6.2 Terminology used in the Holocene climatic stratigraphy developed and modified from the 1950s to the 1970s. Atlantic and Boreal terms originally developed from a series of primarily peat records in Europe with several major (large jagged lines) and minor (small jagged lines) discontinuities. Radiocarbon dating modified the original time scale to show the range of climatic boundaries possible for the individual boundaries. Neoglaciation identifies the period of renewed alpine glacier activity following the retreat of glacial ice. (Modified from Bradley 1999)

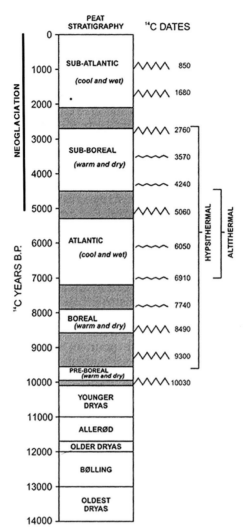

The Greenland Ice Sheet is a very large topographic barrier to the prevailing westerlies within the northern mid-latitudes. In addition, the position of the Icelandic Low, generally close to the southern border of the island, means that many air masses in the NH migrate toward Greenland. Some of these will ride over the top of the ice sheet while others will essentially collide with it and move around its perimeter. Regardless of the eventual path taken, these air masses bring gases, aerosols and insoluble particles to various portions of the ice sheet. These species, scavenged by precipitation (i.e. primarily snow) together with some dry fallout via settling by gravity, will be deposited on the snow surface. Their concentrations are preserved as the snow is converted to glacial ice with burial. Air found in the voids between individual snow grains is also trapped with glacial ice formation, thus providing a multi-parameter record of overall atmospheric characteristics through the Holocene. Thus, Greenland's high elevation and position in NH circulation patterns enable the GISP2 and GRIP records (American Geophysical Union 1998) to reflect hemispheric and global climatic conditions.

Individual parameters measured in ice cores provide a tremendous wealth of information about different aspects of changing environmental conditions including climatic conditions and climatically influenced environmental conditions. For example, physical characteristics of the ice (e.g. Alley *et al.* 1997a) including the presence of melt layers (e.g. Koerner and Fisher 1990) and yearly accumulation time series (e.g. Meese *et al.* 1997) provide critical information on changing summer temperature and precipitation conditions, respectively. Oxygen isotopes measured in the melted ice are well known as paleothermometers. In the GISP2 and GRIP ice cores the amount of temperature depression in the NH is on the order of 8 °C or more during the last glacial maximum (Figure 6.1). In addition to these parameters, various chemical species provide proxy records of changing environmental conditions (e.g. Mayewski *et al.* 1997).

The most versatile records from the GISP2 ice core that provide a multi-parameter record of NH climatic and environmental conditions are the various glaciochemical time series. As shown on Figure 6.3, the time series of these many species show that there is a cyclical nature to post-glacial climatic conditions in the NH. For instance, concentrations of Na^+ and Cl^- in the ice are related to the concentration of sea salt species in the atmosphere over the ice at the time of deposition. Atmospheric sea salt concentrations are directly related to parameters such as wind speed, with greater overall wind speeds leading to a greater amount of sea salt species over the Greenland Ice Sheet. Sea salt species also may be inversely related to sea ice extent as the less sea ice, the greater the amount of open water, the greater the amount of sea salt that may be transported to Greenland. However, in the case of the GISP2 record, the effect of greater sea ice extent during glacial conditions is overwhelmed by more vigorous wind speeds resulting in high Na^+ and Cl^- concentrations during colder climates (Figure 6.3). Greater concentrations of continental-derived dust species, such as Ca^{2+}, Mg^{2+}, K^+, also reflect increased wind speeds in the NH as well as greater aridity in these source areas. Furthermore, both marine and continental-derived species can be used to postulate changes in NH pressure systems, and thus overall circulation patterns. Thus, greater concentrations of these species also reflect the greater wind speed and aridity associated with cooler climatic conditions in the NH (Figure 6.3). There are many other ice-core parameters that provide detailed information on past environmental conditions (American Geophysical Union 1998), but those are not discussed in detail in this essay.

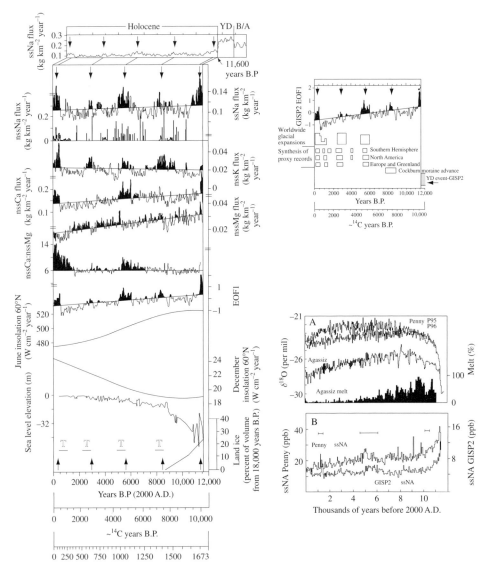

Figure 6.3   Time series of Holocene glaciochemical records and the first empirical orthogonal function (EOF1) derived from the GISP2 ice core (top portion of figure on left) showing the cyclicity of periods dominated by cool temperatures, overall aridity and vigorous circulation patterns during the Holocene (filled sections of curves above mean trend line). ss refers to a sea salt origin; nss refers to a non-sea salt origin. The bottom part of the figure on left shows the relationship of these parameters to insolation, sea level and land ice volume. The upper-right figure shows the relationship between these cold periods (as noted by the EOF1 time series) and periods of known alpine glacier advances and other proxy records of cooler climatic conditions. The lower-right shows the overall declining temperature record from the Penny Ice Cap δ^{18}O record and the Agassiz Ice Cap melt record (A). (B) shows the sea salt Na$^+$ time series for the GISP2 and Penny ice cores. Note the increase in the GISP2 sea salt Na$^+$ record during the last few centuries of the record (i.e. Little Ice Age) compared to the decrease in the Penny Ice Cap sea salt Na$^+$ record over the same time period. (From O'Brien *et al.* 1995; Fisher *et al.* 1998 and Koerner and Fisher 1990)

Although the individual chemical species reflect various environmental conditions as just presented, there are multiple sources for most of these species. For instance, there is also a Ca^{2+} component in sea salt and Na^+ may be derived from continental sources. Consequently, it is necessary to discern the source of the individual species, as well as the relative importance of each chemical species in the overall glaciochemical record. This task is accomplished through an empirical orthogonal function (EOF) analysis of the complete glaciochemical record (e.g. Mayewski *et al.* 1997). In fact, the time series of individual EOF components has added information to the overall environmental and climatic record available from the glaciochemical record. The time series of EOF1 in the GISP2 core provides additional evidence for the cyclical nature of Holocene climatic conditions in the NH (Figure 6.3), as this particular EOF explains the greatest amount of variability in all of the individual glaciochemical time series.

The glaciochemical record from the GISP2 ice core developed by O'Brien *et al.* (1995) provides a continuous high-resolution record of cyclical changes of cool climatic conditions during the Holocene (Figure 6.3). As shown by the time periods with high concentrations of individual chemical species and high EOF1 values, there are five periods during the Holocene when overall climatic conditions were characterized by cooler, drier conditions than other time periods, and with overall greater circulation pattern development and the resulting increase in wind speeds. The first period of such cooling following the Younger Dryas, the final cold pulse of the last glacial, occurs between 8800 and 7800 years ago. Three more periods of cool Holocene climatic conditions occur from 6100 to 5000, 2400 to 3100 and 600 years ago to almost the present. The primary conclusion drawn from these results is that NH climate, and probably global climate, has varied on approximately 2500-year cycles during the Holocene. This periodicity matches the 2500-year solar cycle as reflected in radiocarbon records from tree rings in the Holocene. In general, O'Brien *et al.* suggest that these time periods are characterized by an expanded polar vortex and possibly greater meridional flow in the NH. Subsequently, milder NH conditions, and thus a contracted polar vortex or weaker meridional flow, existed from 10 600 to 9300, 7900 to 6300, 2700 to 1500 and 960 to 610 years ago.

The use of EOF analysis on the GISP2 glaciochemical record (Mayewski *et al.* 1997) provides excellent evidence for the regionalization of climate during post-glacial times. EOF1, the most dominant component of the record, explains 50% of the variance in all glaciochemical species during the Holocene, but the amount of variance explained jumps to 70–80% during interstadial and stadial events within the last glacial period, respectively. Mayewski *et al.* suggest that EOF1 represents an expanded polar vortex during glacial times when the large ice sheets are in existence. They believe this expanded polar vortex highly influences the transport of all chemical species to Greenland, thus the very high level of variance explained. Without the prevalence of the expanded polar vortex during the Holocene, a greater variety of air masses and sources of chemical species were able to reach the Summit region of Greenland. A greater variety of prevalent circulation patterns in the NH undoubtedly would lead to a much greater degree of regionalization during post-glacial times.

Superimposed on the periodicity of Holocene climate is an overall general trend that is observed not only in the GISP2 glaciochemical record, but is seen in various records from the smaller ice caps of the Canadian Arctic. Interestingly, these records, and others we discuss later, provide even more evidence for the regionalization of NH climate once the influence of the large ice sheets is removed from the climate system. The GISP2 record essentially homogenizes NH conditions, and in doing so, a general straight line fit of the data suggests an overall warming of climate through the Holocene as noted by the overall decreasing trend in concentration values for most glaciochemical species and in EOF1 (Figure 6.3). Not all species display this trend, probably reflecting the multiple controls on the glaciochemical record in combination with the regionalization of Holocene climate in the NH.

On the other hand, the records from the smaller ice caps show an overall general cooling through the Holocene (Figure 6.3, lower right). The small ice caps in the Canadian Arctic contain more permanent ice than any place but the Greenland and Antarctica Ice Sheets, thus they also provide an excellent source of past environmental conditions. Evidence for the overall decrease in temperatures in the Canadian Arctic comes from the summer melt record for the Agassiz Ice Cap. The melt record reflects summer temperatures with an overall decrease in summer temperatures from the early Holocene to the present, as reflected in the decreasing number of melt layers in the core with time. Oxygen isotope records from the Penny and Devon Ice Caps, also in the eastern Canadian Arctic, show an extensive period of early Holocene warmth with the possibility of maximum warmth being attained more toward the middle of the Holocene. Mid-Holocene warmth, that is the Climatic Optimum, was the generally accepted norm in the early work on Holocene climatic conditions (Figure 6.2). Nevertheless, oxygen isotopes noticeably trend toward more negative values in the latest part of the Holocene in these same cores, that is the coolest period of the Holocene according to these records (Figure 6.3).

An additional piece of evidence for identifying the regionalization of post-glacial climate and the influence of regionalized climate on the record of post-glacial climatic change comes from comparing the sea salt Na^+ record between the GISP2 and Penny Ice Cap ice cores. The sea salt Na^+ time series in the Penny ice core shows a general decrease in concentrations going from high concentrations at the end of the Younger Dryas/beginning of post-glacial conditions to the lowest values at the end of the Holocene. The record for this same parameter in the GISP2 core shows the high concentrations at the end of the Younger Dryas with only a slight decrease in concentrations through the Holocene, except for the last few hundred years. In the case of the GISP2 record, sea salt Na^+ concentrations increase noticeably during recent centuries. Thus, the two ice cores show a difference in how they record the climatic change. Because ice-core records from the Summit region of Greenland reflect overall hemispheric conditions, the increased sea salt Na^+ over the last few hundred years probably reflects the cold conditions, more extensive sea ice, and more vigorous NH circulation during the LIA. On the other hand, the source area for sea salt reaching the Penny Ice Cap is probably more local, thus the decrease in sea salt Na^+ probably reflects greater sea ice in the Baffin Bay and Labrador Sea area during the LIA. Consequently, records from these smaller ice caps in the Arctic, such as the Penny Ice Cap (e.g. Fisher *et al.* 1998) better preserve regional climatic conditions than the Greenland records. In the case of the Penny Ice Cap record, there appears to be an overall general cooling through the Holocene, just as the Agassiz Ice Cap melt record suggests.

Other records

Evidence for these cyclical changes in post-glacial climate are found in other highly resolved records from the NH lending support to the conclusion that the Greenland ice-core records reflect hemispheric conditions. As shown in Figure 6.3, the GISP2 ice-core record matches the known glacial record; however, as noted in the figure, many of these glacial records are discontinuous as they were primarily based on morainal deposits. Consequently, establishing that the periodicity of these Holocene climatic shifts is hemispheric requires additional highly resolved and well-dated records. Several of these records are now presented as summarized by Mayewski *et al.* (2004).

In addition to the ice-core record, the most significant evidence for millennial-scale cycles of Holocene climatic conditions comes from the marine sediment record (Figure 6.4, records g–j; Bond *et al.* 1997). The evidence for such changes is found in radiocarbon-dated sediment cores from the North Atlantic that reflect ice-rafting episodes (Figure 6.4, record j). Ice-rafted debris (IRD) consists of various types of sediment found on the ocean floor that could have only reached their final depositional site, because of their large particle size, via drifting icebergs. As icebergs rotate, overturn, and eventually melt completely in their southward journey from glaciated areas surrounding the northernmost Atlantic Ocean, they release this debris. Specific indicators used to mark the presence of IRD are sections of core containing high concentrations of lithic grains (Figure 6.4, record g), high

Figure 6.4 Series of Holocene climatic records from the Northern Hemisphere with those at the top of the page from high latitudes and those at the bottom of the page from lower latitudes. Climatic and environmental significance of each record is noted in the figure. Light gray columns represent the timing of rapid climate change (RCC) events during the Holocene as identified by the authors, whereas darker gray shading on individual time series represents sections within individual time series that display anomalies. (From Mayewski *et al.* (2004), where details of the sources of the individual curves can be found)

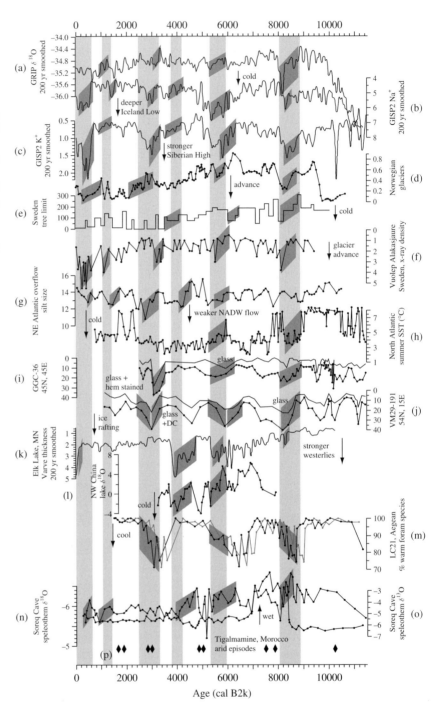

percentages of fresh volcanic glass from Iceland or Jan Mayen (Figure 6.4, record j), and high percentages of hematitic-stained grains (mostly quartz and feldspar) that originate from sedimentary deposits containing red beds, such as on Svalbard and in the eastern Canadian Arctic (Figure 6.4, record i). Although the presence of IRD in North Atlantic sediment cores has been used in the evaluation of rapid climatic change events during the last glacial, the work of Bond *et al.* (1997) shows that such events occurred throughout the Holocene, albeit of a lesser magnitude than equivalent-type events during the last glacial, as is the case for the ice-core records (Figure 6.1).

Bond *et al.* (1997) suggests that there were abrupt reorganizations in climate and ocean circulation in the North Atlantic to the point of bringing large numbers of icebergs into areas of the North Atlantic that do not see them today. Peaks in these periods of presumably cool conditions occur at 11 100, 10 300, 9400, 8100, 5900, 4200, 2800 and 1400 years ago. Any younger events would not be recorded because that part of the record was not collected in the sediment cores used in their study. The findings of Bond *et al.* suggest that there are periods of ocean surface cooling at an approximately millennial cyclicity (Figure 6.4, record h). Using the presence of planktonic foraminifera in the same layers that IRD was found suggests that the amount of cooling during these colder periods of the Holocene did not exceed 2 °C or 15 to 20% of the full Holocene to glacial temperature change. Nevertheless, the south-ward migration of ice-bearing surface waters would be on the order of 5° of latitude further south than present, thus penetrating into what is now the warm North Atlantic Current. The good agreement in the timing of peaks in cooler ocean surface temperatures with the time-series and periodicity of peaks in various glaciochemical species from the GISP2 ice core (Figure 6.4) suggest that Holocene changes in atmosphere and oceanic systems in the North Atlantic were coupled. Because this approximately 1500-year periodicity of climate change is found in other Holocene records, there is apparently a millennial-scale control within our climate system. Changes in NH climate during post-glacial times should, therefore, be highly influenced by this cycle of climatic change. However, the cause of this cyclicity is not known.

Given the ice-core and marine sediment records just discussed, there are general time periods in the Holocene that appear to reflect millennial-scale changes between overall cooler conditions with more vigorous circulation patterns and overall warmer conditions with less vigorous circulation pat-terns. During the cooler periods, atmospheric conditions were generally dustier, whereas the NH atmosphere was less dusty during the warmer periods. Timing of this cyclicity appears to be between about 1500 and 2500 years (i.e. millennial-scale changes).

Further evidence of the periodic fluctuation of climatic conditions during the Holocene comes from several other time series from terrestrial sources.

Glacier advances (Figure 6.4, record d) and more contracted tree-line limits (Figure 6.4, record e) in Scandinavia correspond to some of the same periods of cool conditions as observed in the ice-core and marine records. Very strong evidence for cooler conditions in North America during these same time periods comes from the varve record in Elk Lake, Minnesota, USA (Figure 6.4, record k). This varve record also indicates the presence of stronger westerlies during these same time periods, thus further evidence of enhanced circulation patterns during these cool events. Evidence in the lower latitudes of the NH for these periods of climatic change reflects not only similar climatic conditions, as far as overall temperature regimes during these events, but also similar moisture conditions. Reduced warm foram species are found in the Aegean Sea during the cold periods of the last 11 500 years (Figure 6.4, record m), whereas speleothem deposits from caves in Israel reflect the same periods of cooler and drier conditions in contrast to other time periods during the Holocene (Figure 6.4, records n, o). The ice-core record also suggests that these cooler periods were characterized by overall more arid conditions and by greater overall strength of circulation systems.

Despite the significant evidence supporting cyclical changes in Holocene climate, these records are not exact replicas of each other. Not every record indicates the same magnitude for each cooling event nor does each record show a cooling event at the same time one is found in many other records. For instance, the beginning of the cooling period that is generally from 3200 to 2500 years ago begins earlier in the North Atlantic marine sediments (Figure 6.4, record g) and in the Elk Lake varve record (Figure 6.4, record k) than in most other records shown on Figure 6.4. Further, the warm water forams in the Aegean Sea record (Figure 6.4, record m) indicate a warming of the waters through most of the Holocene time periods considered to be cold and dry. The Aegean record indicates cool conditions at the beginning of each of these periods, but unlike the other proxy records shown, the coldest time is the beginning of the event, not the middle or the end of the event. This spatial and temporal variability in the continuous record of Holocene climatic conditions further supports the regionalization of post-glacial conditions despite the evidence of a prevalent forcing factor that produces millennium-scale climatic change. However, it must also be noted that some of the differences in time-series trends are a direct function of different temporal and sampling resolutions.

6.3.3 Specific climatic events or periods

Within the scope of overall climatic conditions during the post-glacial period, there are several climatic events or periods that stand out as being important to our overall understanding of Holocene climate in the NH. The first event to be discussed is the 8200-year event (c. 7500 ^{14}C yr BP) The significance of this event is its high magnitude, as observed in many proxy records, and the apparent lack of a definitive cause of the event. Following this discussion will be a more specific evaluation of climatic conditions over the last millennium. In particular, the discussion will focus on the MWP and the LIA, two contrasting climatic periods over the last 1000 years. The significance of these two climatic periods centers on whether or not they should actually be called distinct climatic periods and if so, when did they occur and what caused them to occur. It is especially critical to understand our climate system over the last 1000 years if we are to predict what could happen in the future and the magnitude of the human influence on climate.

The 8200-year BP event

As presented above, there are several periods of cool climate conditions during the Holocene occurring roughly every 1500 years. However, the period that, in general, may be greatest from the perspective of the magnitude of the signal in proxy records, is the first of these events, that is, the 8200-year event. In fact, in the Greenland ice cores, the signal of this event in many of the individual parameters measured is about half that of the Younger Dryas. None of the other cool events are of that relative magnitude. Adding to its overall significance as a climatic event is the fact that evidence of the event is found in other proxy records in the NH, as well as globally. A summary of the key pieces of evidence that demarcate the event and its significance in the overall scheme of Holocene climatic change is taken primarily from the summary by Alley *et al.* (1997b).

Several key parameters in the Summit Greenland ice-core records provide evidence of climatic conditions in the NH during the 8200-year event. The overall signal in these parameters (Figure 6.5) generally covers the period from about 8400 to 8000 years ago, but the peak is typically at about 8250 years ago. The characteristics of the event are the same as other cold events, such as the Younger Dryas and stadials within the last glacial, as indicated by the following characteristics. Accumulation during the peak of the 8200-year event is about 20% lower than baseline conditions at 8400 and 8000 years ago indicating dry conditions on a local to regional scale around Greenland. Oxygen isotope ($\delta^{18}O$) values show a decrease of about 2 per mil over the same baseline, which translates into a cooling of $6 \pm 2\,°C$, based on the calibration of Cuffey *et al.* (1995). Concentrations in Na^+, Cl^- and Ca^{2+} are about 60% above this baseline indicating overall more vigorous circulation systems and the enhanced transport of sea salt species (Na^+, Cl^-) and dust from continental areas (Ca^{2+}). Greater Ca^{2+} concentrations also reflect more availability of dust for transport, and thus more aridity. Decreased methane production, because lesser amounts of the NH were in wetlands, and evidence for increased forest fire activity also support the contention of

Figure 6.5 Greenland ice-core records that show the pronounced cooling episode between 8000 and 8400 years ago (i.e. 8200-year event). All plots are shown with an approximately 50-year running mean. Timing of the Medieval Warm Period (MWP) and Little Ice Age (LIA) is also noted for these parameters. All records are from the GISP2 ice core except the methane record (dashed line), which is from the GRIP ice core. (Modified from Alley *et al*. 1997a)

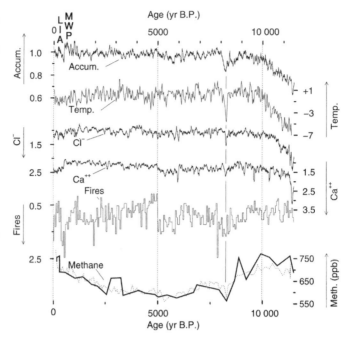

increased aridity at this time. These proxy data indicate that the 8200-year event, like other cold events in the Greenland ice-core records, was character-ized by cold, dry, and dusty conditions in the NH.

The significance of this event is further enhanced by evidence of distinct climatic change from other proxy records in the NH, thus further supporting the idea of a synchronous NH to global event. The dating and sampling resolution of these other records is not as high as that for the Greenland ice cores, as the timing of the peak of the event spans a millennium with a duration of 100 to 1000 years or more. Nevertheless, the temporal relation-ship is supportive of an overall hemispheric synchroneity for this event. For instance, cold conditions are suggested for northern Sweden and the eastern Canadian Arctic together with fresh, cool sea-surface conditions in the North Atlantic. Dry, windy conditions are suggested for the Laurentian Great Lakes region of North America, as well as in the monsoonal regions of Africa and Asia. These are the same general conditions indicated by the ice-core records from Greenland and by other proxy records shown in Figure 6.4.

Given the abundant evidence for this event within the cyclicity of Holocene climate, the more elusive problem associated with the 8200-year event is along the lines of why or what forced this event to occur? This is especially problematic because none of the other cool periods of the Holocene appear to be of an equivalent magnitude, at least in the Greenland ice-core records. One possible explanation is that the cause of this event is the same process as that which forces the ~1500-year cycles evidenced by the marine sediment record (Figure 6.4), as described above. An alternative cause may be the slowdown of the thermohaline circulation system from the influx of fresh water to the North Atlantic. However, such a "flooding" event is not know, thus there may have been an unknown fresh water event at this time or the thermohaline circulation system responded in a very large way to a very small fresh-water event. Ice-marginal lakes known to have existed at that time would provide a very small amount of fresh water to the North Atlantic system. Should this have been the case, the 8200-year event may be an excellent example of the non-linearity of the climate system. As pointed out by Alley *et al.* (1995), if the amount of fresh water put into the North Atlantic by those ice-marginal lakes was sufficient to slow down the thermohaline circulation, then the amount of meltwater that may be added to the North Atlantic from melting of NH ice with anthropogenic warming, is also of sufficient volume to slow down thermohaline circulation. Consequently, the nature of the 8200-year event has serious ramifications for future climatic change.

Medieval Warm Period

There are often suggestions of and reference to two prevalent climatic periods of very contrasting characteristics over the last millennium, the MWP and the LIA. Initially, the discussion will focus on the characteristics of the MWP and why it is argued that a MWP does not exist. Discussion about the LIA then follows.

The MWP was first labeled by Lamb (1995) as the period from about the ninth century to the fourteenth century AD when many places in not only the NH, but the world as a whole seemed to have felt renewed warmth (Figure 6.6). Parts of the period, such as during the eleventh and twelfth centuries, were nearly equivalent to that of the warmest millennium in post-glacial times (Hughes and Diaz 1994). Questions have arisen, however, that address the synchroneity of the warmer periods and the actual magnitude of the warmth as a whole. Evidence used to evaluate the existence of the MWP follows with the major source being continuous paleoclimatic records, such as those available in tree-rings and ice cores.

One of the most often cited lines of evidence for warm conditions, particularly in Europe and the northern Atlantic, is the establishment of Norse colonies in Greenland and over to Newfoundland (see Section 8.2). This evidence adds to an overall suggestion of warmer conditions, particularly in the tenth to thirteenth centuries, for most of northern Europe, southern Greenland and Iceland, and basically the North Atlantic, as a whole. More specifically, there is a persistent warm anomaly in northwestern Europe annual temperature reconstructions from 1190 to 1350. On the other hand,

Figure 6.6 Recent reconstructions of NH temperature for the last 1000 years. All time series have been smoothed to highlight variations on the 50-year time scale. References for each time series in box on the figure. The time period usually given for the Medieval Warm Period (MWP, ~ AD 900–1300) and the Little Ice Age (LIA, ~ AD 1450–1850) has been changed for this figure given the likely boundary at AD 1400 in the records shown. (Modified from Jones *et al.* 2001)

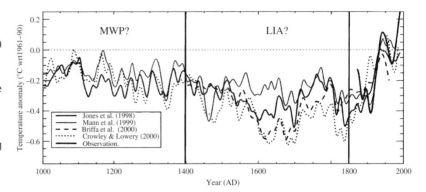

there is no evidence for any exceptional warmth in southern Europe until the mid fourteeth century. Interestingly, if one looks more at western Europe, there is evidence of warm springs and dry summers throughout most of the thirteenth century, evidence that may be used to contribute to the idea that there is a lack of temporal synchroneity in the MWP. However, even within northern parts of Europe there are contrasting climatic conditions. For instance, the positive summer temperature anomalies over the millennium mean in the Polar Urals from 1110 to 1350, as derived from tree-ring records, is opposite to the negative anomalies shown by summer temperatures in Fennoscandia during the same time period. On the other hand, there are positive anomalies in the Fennoscandian records from 971 to 1100 and 1350 to 1540.

Looking at tree-ring evidence in North America shows a similar discontinuity in regional climatic conditions for the MWP. For instance, high-altitude studies in the Canadian Rockies indicate climatic conditions favorable for glacial retreat in the tenth to thirteenth centuries, as well as the advance of forests into higher elevations of the Canadian Rockies. Summer temperatures in the Sierra Nevada show positive temperature anomalies from the millennium mean over a longer time frame than in the Canadian Rockies, that is, warm summers are prevalent from 1090 to 1450. The southwestern United States may have experienced climatic conditions that would favor greater crop growth (i.e. an increase in effective precipitation compared to other time periods) during the tenth to thirteenth centuries. This improved crop growth could be related to an enhanced Arizona monsoon between 700 and 1350. Summer enhanced monsoons could be related to an increase in solar receipt. However, in the southeastern United States, there does not appear to be any evidence of either prolonged wet spells or prolonged drought in the ninth through fourteenth centuries.

Overall, a consistent pattern of continuous and spatially consistent warming between about 900 and 1450 is lacking, but this does not preclude the use of the term MWP. Certainly, there are places, particularly in northern Europe, where the evidence is prevalent for warmer conditions in general over that time frame. Nevertheless, as is shown for the LIA below, Holocene climate is strongly characterized by the regionalization of climate and it should not be surprising that conditions will be inconsistent from one place to another at the same time. Moreover, a specific forcing component can produce different responses in climate given the non-linearity of the climate system.

Little Ice Age

The LIA was first used as a descriptive term referring to glacial activity over the last 4000 years (now often referred to as Neoglaciation; Figure 6.2), the formal term of LIA now often refers to the period from AD 1550 to 1850 (i.e. mid sixteenth to mid nineteenth century) (Jones and Mann 2004). Many records are available that support this contention, perhaps making the LIA the most spatially coherent of the climatic regimes in the NH during the latest parts of the Holocene. In fact, the most recent cool period of the cyclical events in the Holocene is the LIA (Figures 6.3–6.5). However, there still are temporal and spatial differences from area to area in the NH, reflecting the regionalization of Holocene climate, much like the evidence put forth to question the existence of the MWP. As was the case for the MWP, part of the variability in describing LIA climate is the different nature of various proxy records including the spatial and temporal resolution of the data. Records showing the variability in the LIA, as summarized by Jones *et al.* (1998) and Jones and Mann (2004), are now presented.

Greatest evidence for cooler climatic conditions in Europe covers the period from the early to middle part of the nineteenth century (around 1820) to the earliest part of the twentieth century (around the 1920s). Overall, the seventeenth century was also cold, but such conditions were not as prevalent as those in the sixteenth and eighteenth centuries. Coldest conditions in North America match those in Europe, as the nineteenth century was the coldest of the last millennium. There is also evidence of a cold seventeenth century, although places in the western United States may have been warm during that century. Climatic conditions in eastern Asia may deviate by the greatest amount from the general hemispheric climate, that is, assuming that general LIA climate follows the conditions observed in Europe and North America. For instance, there is little evidence for persistently cold conditions in the nineteenth century in eastern Asia, but there is evidence for cold climatic conditions in the seventeenth century. The late eighteenth and earliest nineteenth centuries seemed to be cold in eastern Asia, but these cold conditions do not appear to persist through the rest of that century unlike conditions observed in Europe and North America.

Given these general climatic conditions, it may be summarized that coldest climatic conditions in the NH occurred 1590s–1610s, 1690s–1710s, 1800s–1810s and 1880s–1900s. Between these periods of cooler conditions synchronous warm conditions during the LIA are found in the 1650s, 1730s, 1820s and more recently in the 1930s and 1940s. Nevertheless, the use of the LIA to describe climatic conditions over most of the last six centuries is generally valid.

6.3.4 Climate forcing

The variability in Holocene climate from both a temporal and spatial perspective is reflected in the factors that have caused these changes. As alluded to in the discussion about the periodicity of climatic fluctuations in postglacial times, there is an unknown factor that produces these approximately regular millennial-scale cycles. However, there are other forcing factors that have been shown to contribute to the variability in NH climate. Mayewski *et al.* (1997) used a series of band-pass filters to construct a series of band-pass components (i.e. a type of spectral analysis) on the complete 110 000-year glaciochemcial record from the GISP2 ice core. Although there are components of the record with periodicities beyond the 11 500-year Holocene record, Mayewski *et al.* were able to identify components of the climate system that force Holocene climate, such as the 6100-year periodicity that maintains itself through the Holocene. This component of the climate system appears to coincide with iceberg discharge and ocean surface water cooling events including those found during the Holocene that demarcate the LIA and other cooling events (Figures 6.3 and 6.4). It appears that this periodicity, as well as the less prominent 4500- and 3200-year periodicities in the record, could be a function of suborbital-scale changes in insolation, thus the likelihood of an orbital component that forces Holocene climate. In addition to an orbital forcing component, solar variability also plays a role in Holocene climate variability, as indicated by the presence of a 2300-year periodicity together with the 1450-year cycle. In fact, the 1450-year cycle may be related both to cycles in orbital parameters and to the solar cycle. However, feedback mechanisms, that is changes in atmospheric circulation associated with changes in sea ice extent and thermohaline circulation, may be produced with insolation and with solar variability (Mayewski *et al.* 1997).

Evaluation of factors that may force Holocene climate was also done for more recent time periods. Crowley (2000) modeled climatic conditions for the last millennium, including the MWP and the LIA, to determine the role that various climatic forcing mechanisms had in controlling changes over that time frame. Using an energy balance climate model verified through comparisons with controlled runs from a coupled model, Crowley found that natural multiple forcing factors such as changes in solar irradiance and volcanic eruptions (e.g. Zielinski 1995) account for 41 to 64% of the decadal-scale variability in climate over the last 1000 years prior to anthropogenic time (pre-1850). Recent warming above the suspected natural variability within the system suggests that greenhouse gas forcing is responsible for that recent warming (Jones and Mann 2004). Clearly, the regionalization of Holocene climate in the NH is a function of many forcing factors operating on different time scales. These factors become even more prevalent once the impact of the large ice sheets is removed with deglaciation.

6.3.5 Conclusions

Post-glacial climate (i.e. the Holocene or the last 11 500 years) in the NH has undergone a series of fluctuations between generally cool to cold conditions and warm or mild conditions. Although these fluctuations are not of the same magnitude as those that occurred during the last glacial (i.e. Dansgaard–Oeschger Events), they are significant in light of the overall warmer climate of the Holocene. These cyclical changes occur on the order of every 1500 to about 2500 years depending on the record used for reconstruction. Nevertheless, there does appear to be a millennium-scale component to Holocene climatic change. Periods when colder conditions were prevalent are 8400–8000, 5900–5100, 4200–3800, 3100–2400 and 600 BP.

However, superimposed on these cycles of Holocene climatic change is an overall general trend to climatic conditions that varies depending on the records observed. Greenland ice-core records suggest a slight warming trend through the Holocene, whereas records such as those from ice caps in the Canadian Arctic suggest an overall general cooling through the Holocene. These differences are probably a reflection of the regionalization of post-glacial climate in the NH once the influence of the large ice sheets is gone. Greenland records reflect general conditions in the NH while other proxy records may reflect more regional conditions. Similarly, other highly resolved records do not show the same magnitude of change in a particular cool event as the ice-core or marine records may show, or the timing of the environmental change may be different, again reflecting the temporal and spatial variability in environmental conditions with regionalization of NH climate during the Holocene.

Specific time periods during the Holocene further reflect the complex nature of the post-glacial climate. The 8200-year cooling event is found in many proxy records from the NH, although it is especially prevalent in the Greenland ice-core records. Its significance comes from the high magnitude of the signals that characterize it, a magnitude that is about half that of the

Younger Dryas. However, the cause for this event is not readily identifiable. In addition, the great deal of variability in climatic records that mark the MWP and the LIA within the last millennium is very indicative of the regionalization of Holocene climate in the NH. The many factors that force Holocene climate, including orbital, solar, volcanic, and greenhouse gas components, act on varying time scales further enhancing the regionalization of the post-glacial climate.

6.4 Southern Hemisphere climate reconstructions

The data to reconstruct past climates of the Holocene in the SH are much more limited than the NH (Mann and Jones 2003; Jones and Mann 2004). Since the SH is mostly ocean, proxies on land can only be drawn from Australia, Southern Africa, South America, New Zealand, and Antarctica. Shallow ocean-based proxies include coral and bottom sedimentation cores. Data before 1861 are especially sparse (IPCC 2001). Jones *et al.* (1998) list only seven major locations where detailed good-quality longer-term records currently exist. These include tree-ring records from Tasmania (Australia), and Lenca and Alerce (Argentina); coral records from the Great Barrier Reef (Australia), Galapagos Islands (Chile), and New Caledonia; and ice cores from Antarctica. To this list can be added tree-ring records from New Zealand, and ice-core information from Quelccanya, Peru (Mann and Jones 2003), and speleothem information from Southern Africa (Tyson *et al.* 2000).

As a result, conclusions about climate change and variability in the SH cannot be as well defined or as representative as those for the NH. However, the data are useful to provide some balance in the discussion for the globe, and individual site results are comparable to their NH counterparts.

6.4.1 The 8200-year BP event in the Southern Hemisphere

The 8200-year BP event defined for the NH by Zielinski (Section 6.3.3, Figure 6.5) also appears in the SH proxy records but not as strongly. Speleothem results for New Zealand (Williams *et al.* 1999) suggest that the period 11 000 to 8500 BP was a warm, wet period, but subsequently temperatures declined. There were small advances in New Zealand mountain glaciers between 8200 and 7800 BP (IPCC 2001). In Southern Africa, however, evidence for abrupt cooling during this period is lacking (Lindesay 1998).

Stager and Mayewski (1997) provide data from the Taylor Dome ice core in Antarctica which suggested a "climatic reorganization" that lasted about 200 years. Increased sodium (Na^+) concentrations in the ice layers dated to this period indicate an abrupt change to a more meridional circulation pattern, with stronger atmospheric mixing. Polar circulation was weak (Mayewski *et al.* 2004). Changes to ocean circulation and terrestrial conditions lead to an expansion of the sea ice surrounding Antarctica, and a shift in the circumpolar vortex toward the equator. The extended ice volume enhanced the ice–ocean temperature gradient, further strengthening the meridional circulation. Impacts in other areas of the SH included increased aridity and seasonal rainfall patterns in the tropics, weaker summer monsoons, and some spatial changes to mid-latitude precipitation patterns, such as increased precipitation in Chile. The details and reasons for the recovery after 7800 BP are not known.

6.4.2 Southern Hemisphere temperature reconstructions after AD 900

Figure 6.7 presents a composite of SH temperature variations from AD 900 onward, based on the average of seven widely distributed proxy records from various seasons. The SH results are compared to those for the NH. A 50-year Gaussian filter emphasizes any overall trend and variations. There are several interesting aspects to the temperature patterns. Aside from the strong warming in both hemispheres beginning in the twentieth century, most likely caused by human impacts (Jones *et al.* 1998), the variability associated with natural forcing consists of small fluctuations (within $\pm 1\,^{\circ}\mathrm{C}$) around the 1901–50 mean used as a baseline comparison. Overall the NH temperatures remain somewhat lower, and vary more, compared to the baseline, while the SH temperatures are slightly higher than the standard, suggesting a greater ocean influence. Higher NH temperatures around AD 1100 suggest evidence

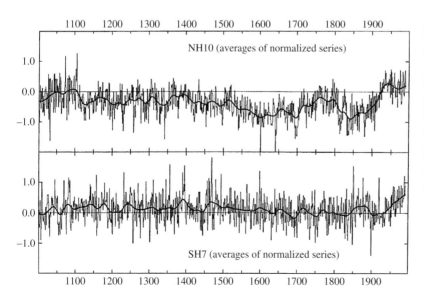

Figure 6.7
Reconstructions of average temperatures from various proxy data sets for the NH (10 data sets) and the SH (7 data sets). The heavy line is a 50-year Gaussian filter showing broad trends. (1998, Edward Arnold Publishers Ltd., Jones *et al.*, *Holocene*, **8**: 455–471, www.hodderarnoldjournals.com)

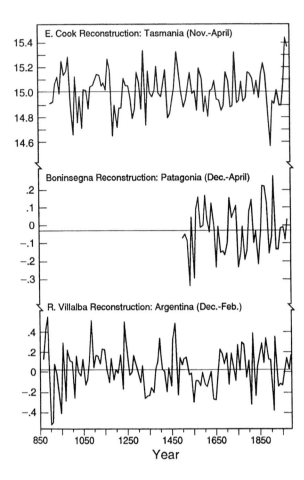

Figure 6.8 Tree-ring reconstructions of summer temperatures in Tasmania (°C) and two locations in Argentina (anomalies from the average, °C). (Diaz and Bradley 1995: with permission from National Academies Press)

of the MWP, which is not reflected in the SH data. The LIA between 1550 and 1850 is strongly indicated in the NH data, but only weakly in the SH series (see detailed discussion in Section 6.4.3). The overall difference between the pre-1900 maximum and minimum components of the curves is less than 1.5 °C.

One major problem with the SH composite result is that, individually, the data sets that comprise average patterns such as in Figure 6.7 do not show much similarity (Diaz and Bradley 1995). This situation is illustrated in Figure 6.8, which contrasts tree-ring reconstructions in Tasmania and Argentina. The detailed variations show little agreement, even in the two Argentinean records, and there are no apparent trends. Similar disagreements occur between records for New Zealand (Williams *et al.* 1999). This problem emphasizes that care needs to be taken in interpreting multi-proxy records, and that regional effects and large low-frequency variance can interfere with obtaining representative results on a hemispheric scale.

6.4.3 Did the Southern Hemisphere experience the MWP and the LIA?

Jones and Mann (2004) argue that the MWP and LIA were defined based mainly on European and North Atlantic climate periods, but could not be called global phenomena. However, several other authors disagree. For example, Soon and Baliunas (2003) suggest that there is global evidence for glacial retreat during the MWP and significant glacial advance during the LIA. However, the start and end dates are often not well established, and the timing varies in different parts of the world. What evidence exists that either period occurred in the SH, given the inconclusive indicators in Figure 6.7? There seems to be greater evidence for a global LIA than for a MWP.

Figure 6.9a shows temperature records from tree-rings in New Zealand from Cook et al. (2002). Figure 6.9b compares oxygen-18 records from ice cores in Greenland, Peru, China, and Antarctica (Thompson et al. 1993), and Figure 6.9c presents speleothem records from Southern Africa (Lindesay 1998). Table 6.2 lists the coldest and warmest pre-1850 summers, decades, and centuries in each hemisphere, based on the analysis of Jones et al. (1988). These records indicate the complexities of defining MWP and LIA periods outside of Europe.

Figure 6.9a indicates a period of strong cooling (13.5 °C) around AD 1000, which does not appear in the NH records. Mayewski et al. (2004) state that small advances in New Zealand glaciers occurred, eastern Southern Africa was dry and cool, and mid-latitude Chile was dry. The tropical regions were also considerably drier than today. After 1000 temperature increased to 15.5 °C by about 1160, in Tasmania. The warmer period lasted until about 1300, with some decadal fluctuations, which is a similar time period to the MWP in Europe. The MWP-type period in Tasmania began about 300 years later, and seems to be shorter in duration. Evidence for a colder period occurs between about 1520 and 1680, again a similar but shorter time scale to the LIA in Europe. Cook et al. (1995) state that a similar result emerges from tree-rings from Tasmania, and is supported by carbon-14 measurements.

The ice core O^{18} results in Figure 6.9b show a broadly similar variation in timing, and that both a MWP-type warming and LIA-type cooling occurred in Peru and Antarctica. Both the South Pole and Peru data sets clearly show a much greater number of warm than cold years between 1100 and 1550. The timing of warm year periods differs, and Antarctica had more and stronger cold years in between the warm years. In comparison to the two NH records, warm years in the SH were more abundant during this period. Between 1550 and 1850, all records except Siple Station in Antarctica show more cold than warm years, but the SH records are more consistent. At this time, the Quelccaya records also showed the greatest precipitation accumulation (Markgraf 1998). Mayewski et al. (2004) indicate that both polar regions were wet and windy, East Antarctica was cold, the tropics remained wet, glaciers advanced in New Zealand, Chile was moist,

Figure 6.9 Three representative proxy data sets for locations in the SH: (a) tree-rings in New Zealand (Cook *et al.* 2002: permission from American Geophysical Union); (b) Antarctic O^{18} data from ice cores compared to those in Greenland, China, and Peru (reprinted from Thompson *et al.* 1993, with permission from Elsevier); and (c) speleothem records from southern Cape Province, Southern Africa (Lindesay 1998, Figure 6.14, Copyright John Wiley and Sons Ltd., with permission), illustrating regional variations in temperatures over the past 2000 years. Higher O^{18} values indicate colder temperatures.

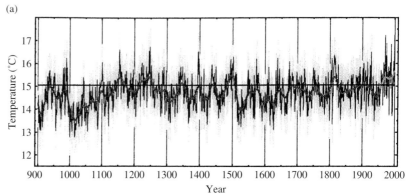

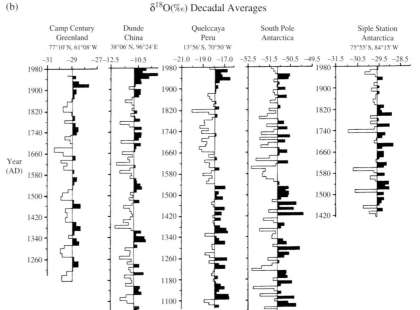

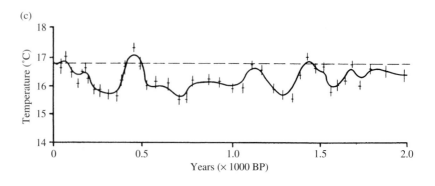

but South Africa cool and dry. Siple, Antarctica, is a coastal location, and may have been influenced by Antarctic coastal circulation, which is dominated by the ocean (see Section 5.5). The graphs in Figure 6.9b differ considerably in detail, reflecting regional influences.

Reviews of a range of literature by Markgraf (1998) establish a complex picture of climate variability and change during this period in South America. Evidence from northern Patagonia in Argentina supports the alternation of warm and cool periods from the Peruvian ice core (Figure 6.9b). Galapagos Island sea surface temperatures were colder than normal between the seventeenth and nineteenth centuries. Tree-ring evidence established that northern Patagonia temperatures were warmer for 1080–1250, and 1720–1750, but cooler for 1270–1660. The coldest years were 1340, and 1520–1650. Precipitation generally was lower during warmer years, but this varied by spatial location and season. Winter precipitation on central Chile (approximately 33° S) was highest for 1220–1280 and AD 1450–1550, and lowest for 1270–1450 and 1580–1680. Summer precipitation at 44° S in Argentina increased for 1600–1620 and 1670–1690, associated with cooler conditions, but was lower in warmer conditions for 1720–1740 and 1770–1790. The precipitation distribution was usually inverse between the two locations.

In Southern Africa (Lindesay 1998; Tyson *et al.* 2000), speleothem records show that warmth was centered around 1500–1520, and cooler and drier conditions around 1700. Figure 6.9c also shows a period of cooler climate between 600 and 800, and between 1000 and 1500. After 1000 the country became drier, although summer rainfall in Natal increased after 1580. The Southern African interior was up to 3 °C warmer than today during the MWP, and about 1 °C cooler during the LIA. There is considerable spatial variability in both temperature and precipitation variations.

The hemispheres are compared by pre-1850 extremes in Table 6.2. The original data (Jones *et al.* 1998) contained the period after 1850, when human influences become important, and therefore the temperature anomalies reflect differences with a 1961–90 baseline. The warmest NH summers are grouped over a 30-year span around 1100, reflecting the MWP. In the SH they occur between 1356 and 1469, which in the NH was a transition period toward the LIA. The coldest summers in the NH occurred in the seventeeth century, but also in the middle and toward the end of the MWP. The period around 1300 must have been cold in the SH.

Decade and century results are perhaps more reflective of the generality of the MWP and LIA descriptors. The warmest NH decade was in the middle of the MWP, and the coldest decades in the seventeenth and nineteenth centuries. The MWP is clearly defined as the centuries between 1000 and 1300. In the SH, the warm centuries began and ended a little later, between 1100 and 1400. The cold centuries extended between 1500 and 1800, but the eleventh century was also cold in the SH, in contrast to the NH.

The results in Figure 6.9 and Table 6.2 suggest there is some evidence for both LIA and MWP occurring in both hemispheres, but this is obscured in Figure 6.7

Table 6.2   *Pre-1850 periods (p) of extreme temperatures in the Northern and Southern Hemispheres based on proxy data, for summers, decades, and centuries. The value (v) is the difference in °C with the 1961–90 baseline period. (After Jones et al. 1988)*

| Period of time | Northern Hemisphere | | | | Southern Hemisphere | | | |
| | Warmest | | Coldest | | Warmest | | Coldest | |
	p	v	p	v	p	v	p	v
Summers	1106	0.54	1601	−1.21	1469	0.86	1257	−0.85
	1074	0.49	1641	−1.16	1468	0.78	1750	−0.84
	1103	0.27	1032	−0.96	1356	0.72	1302	−0.83
			1342	−0.94	1396	0.70		
			1695	−0.93				
Decades	1101–10	−0.02	1691–1700	−0.68	1461–70	0.19	1691–1700	−0.45
			1831–40	−0.68	1391–1400	0.16	1751–60	−0.41
			1601–10	−0.64	1171–80	0.04	1051–60	−0.39
			1641–50	−0.61			1011–20	−0.35
			1811–20	−0.61			1441–50	−0.35
Centuries	11th	−0.16	17th	−0.52	14th	−0.12	17th	−0.25
	14th	−0.23	19th	−0.48	12th	−0.13	18th	−0.22
	12th	−0.25	16th	−0.40	15th	−0.15	19th	−0.20
	13th	−0.30	15th	−0.31	13th	−0.16	11th	−0.19
			18th	−0.30			16th	−0.16

for the SH by averaging. The data are still too sketchy to make any firm conclusions, and regional influences may dominate some results. The timing of the two periods, both in extent and in start and finish, seems to vary spatially, both between and within hemispheres. The values in Table 6.2 suggest that the climate variations are stronger overall in the NH, which perhaps is not surprising given the much stronger ocean mitigating influences in the SH.

6.4.4 Reasons for Holocene climate variations in the Southern Hemisphere

Cook *et al.* (1995) state that there are no proven explanations for the Holocene climate variations in the SH. A combination of several possibilities is likely, and the discussion by Zielinski in Section 6.3.4 is applicable. During colder periods, the expansion and contraction of the circumpolar vortex, leading to more meridional circulation and larger wave numbers, can cause important climate changes. Expansion of the westerlies and shifts in the subtropical anticyclone can create drier climates in central Chile (Petit *et al.* 1999) but wetter climates in the mid-latitude zones (Markgraf 1998), cooler temperatures over southern Australia (Cook *et al.* 1995), and changes in spatial temperature and precipitation patterns in Southern Africa (Stager and Mayewski 1997). Lindesay (1998) states that the northern displacement of the westerlies usually brings warmer, wetter conditions to Southern Africa. Overall, sea ice extent expands northward, and cyclonic activity increases. Goodwin *et al.* (2004) demonstrate that increased Na^+ concentrations measured in ice cores show that transport of air toward Antarctica was enhanced. There is a strong anti-correlation with the AAO (see Section 5.4) and a high correlation with Rossby wave number 3 circulation in the SW Pacific and Indian Ocean region (see Chapter 4).

During warmer periods, the circumpolar vortex is closer to the pole, zonal circulation dominates, and variations in the climate pattern are reduced. Goodwin *et al.* (2004) suggest that their Na^+ data indicate a reduction in circulation variability since 1600, more positive AAO values, and stronger circumpolar zonal vortices in both the troposphere and stratosphere.

The details of what causes these changes are unknown. Jones and Mann (2004) and Mayewski *et al.* (2004) describe the importance of changes in solar radiation (associated with sunspots), and the extent and frequency of volcanic eruptions, as major forcing factors. Oscillations in internal ocean circulation and deep water dynamics (Cook *et al.* 1995) may also be important. Changes in regional circulation dynamics, such as ENSO (Section 2.8), would also play a role. IPCC (2001) states that the ENSO and its teleconnections were very different in the mid-Holocene compared to today. Between AD 700 and 1000, El Niño events were at the highest frequency in the Holocene period, gradually dropping off over the next few centuries (Markgraf 1998). Answering the questions "why?" and "how?" past climate change and variability occurred are important areas for future climate research.

6.5 Chapter summary

Assessment of past climate change and variability, and the forcing functions that cause the changes, is a complex challenge. For the periods before AD 1850, proxy analysis must be used. Fortunately, a range of very good proxies, such as ice cores, coral cores, tree-rings, and speleothems, is available. Unfortunately, there are difficulties with scale, length of record, calibration, and consistency. Jones *et al.* (1998) emphasize that often detailed proxy data have little in common with each other. Regional impacts may interfere with global representativeness. Mayewski *et al.* (2004) list several reasons for these differences, including the complexities of the climate record, abrupt local or regional changes, and spatial irregularities associated with multiple controls.

There are, however, enough data sets to begin to establish a broad global-scale picture for the Holocene, the most recent 11 500 years. The NH data records are more complete than those for the SH. A series of fluctuations between warm and cool periods, which on a hemispheric average do not exceed 2 °C, occur in all of the data sets. The reasons for these fluctuations before AD 1850 have been defined as variations in solar radiation (linked to sunspots) and volcanic activity leading to changes in dust concentration in the stratosphere and lower troposphere. Other forcing factors may be changes in ocean circulation and atmosphere–ocean interactions and variations between the ice packs and the ocean. The expansion and contraction of the circumpolar vortex in each hemisphere has an important role. After AD 1850, human influences such as greenhouse warming must be included to properly explain the variations.

The 8200-year cooling event appears to be global, but is more evident in the NH. The global existence of the MWP and LIA is debatable, but many data sets show some indication of appropriate warming or cooling, although starting and ending dates, and the temporal scale, vary. Further assessment of current data and the development of new data sets will lead to increased knowledge and understanding about past climate variability and more accurate predictions for the future.

6.6 Examples of paleoclimate websites

The World Data Center for Paleoclimatology (WDC-A) is based at the US National Oceanographic and Atmospheric Administration in Boulder, Colorado, USA, and has lots of excellent programs and data sets: www.ngdc.noaa.gov/paleo/paleo.html.

The Byrd Polar Research Center at Ohio State University (USA) focuses on paleoclimatic research from ice cores in polar regions and Peru: www-bprc.mps.ohio-state.edu/Icecore/frontpage.html.

The Quaternary Palaeoenvironments Group at the University of Cambridge (UK) researches a range of paleoenvironmental aspects over the last 2.6 million years: www-qpg.geog.cam.ac.uk/.

The University of Washington (USA) Quaternary Research Center studies Earth systems interactions over the past 2 million years, including high-resolution climatic change in Antarctic ice cores: depts.washington.edu/qrc/index.cgi.

The Climate System Research Center at the University of Massachusetts (USA) focuses on the climate system, climate change, climate variability, and global change issues: eclogite.geo.umass.edu/climate/climate.html.

The Climate Change Research Center at the University of New Hampshire (USA) studies global climate change and forcing mechanisms, linking data sets from different locations around the world: www.ccrc.sr.unh.edu.

6.7 References

Alley, R. B. *et al.*, 1995. Changes in continental and sea-salt atmospheric loadings in central Greenland during the most recent deglaciation: Model-based estimates. *Journal of Glaciology*, **41**, 503–514.

Alley, R. B. *et al.*, 1997a. Visual-stratigraphic dating of the GISP2 ice core: Basics, reproducibility and application. *Journal of Geophysical Research*, **102**, 26,367–26,382.

Alley, R. B., Mayewski, P. A., Sowers, T., *et al.*, 1997b. Holocene climate instability: A large event 8000–8400 years ago. *Geology*, **25**, 402–406.

American Geophysical Union, 1998. Greenland Summit Ice Cores. Reprinted from *Journal of Geophysical Research*, **102**, 26,315–26,886.

Bond, G. *et al.*, 1997. A pervasive millennial-scale cycle in North Atlantic Holocene and glacial climates. *Science*, **278**, 1257–1266.

Bradley, R. S., 1999. *Paleoclimatology*, 2nd edn. San Diego: Harcourt Academic Press, 644pp.

Cook, E., Buckley, B. and D'Arrigo, R., 1995. Interdecadal temperature oscillations in the Southern Hemisphere: Evidence from Tasmanian tree rings since 300 BC. In *Natural Climate Variability in Decade-to-Century Time Scales*. Washington DC: National Research Council, pp. 523–532.

Cook, E., Palmer, J. and D'Arrigo, R., 2002. Evidence for a 'Medieval Warm Period' in a 1,100 year old reconstruction of past Austral summer temperatures in New Zealand. *Geophysical Research Letters*, **29**, 1667–1671.

Crowley, T. J., 2000. Causes of climate change over the past 1000 years. *Science*, **289**, 270–277.

Cuffey, K. M., Clow, G. D., Alley, R. B. *et al.*, 1995. Large Arctic temperature change at the Wisconsin-Holocene glacial transition. *Science*, **270**, 455–458.

Diaz, H. and Bradley, R., 1995. Documenting natural climate variations: How different is the climate of the twentieth century from that of previous centuries. In *Natural Climate Variability in Decade-to-Century Time Scales*. Washington DC: National Research Council, pp. 17–31.

Fisher, D. A. *et al.*, 1998. Penny Ice Cap cores, Baffin Island, Canada, and the Wisconsin Foxe Dome connection: Two states of Hudson Bay ice cover. *Science*, **279**, 692–695.

Goodwin, I., van Ommen, T., Curran, M. and Mayewski, P., 2004. Mid latitude winter climate variability in the South Indian and southwest Pacific regions since 1300. *Climate Dynamics*, **22**, 783–794.

Hughes, M. K. and Diaz, H. F., 1994. Was there a "Medieval Warm Period", and if so, where and when? *Climatic Change*, **26**, 109–142.

IPCC, 2001. *Climate Change 2001: The Scientific Basis. Contribution of Working Group I to the Third Assessment Report of the Intergovernmental Panel on Climate Change*. J. Houghton, Y. Ding, D. Griggs, *et al.*, eds., Cambridge: Cambridge University Press.

Jones, P. D., Briffa, K. R., Barnett, T. P. and Tett, S. F. B., 1998. High-resolution palaeoclimatic records for the last millennium: Integration, interpretation and comparison with General Circulation Model control run temperature. *Holocene*, **8**, 455–471.

Jones, P. D., Ogilvie, A. E. J., Davies, T. D. and Briffa, K. R., 2001. Unlocking the doors to the past: Recent developments in climate and climate impact research. In P. D. Jones, A. E. J. Ogilvie, T. D. Davies and K. R. Briffa, eds., *History and Climate: Memories of the Future*. New York: Kluwer Academic/Plenum Publishers, pp. 1–8.

Jones, P. D. and Mann, M., 2004. Climate over the past millennia. *Reviews of Geophysics*, 42:RG2002, doi:10.1029/2003RG00143.

Koerner, R. M. and Fisher, D. A., 1990. A record of Holocene summer climate from a Canadian high-Arctic ice core. *Nature*, **343**, 630–631.

Lamb, H. H., 1995. *Climate, History and the Modern World*, 2nd edn. London: Routledge.

Lindesay, J., 1998. Past climates of Southern Africa. In J. Hobbs, J. Lindesay and H. Bridgman, eds., *Climates of the Southern Continents*. Chichester, UK: Wiley, pp. 161–206.

Mann, M. and Jones, P., 2003. Global surface temperatures over the past two millennia. *Geophysical Research Letters*, **30**, 1820–1824.

Markgraf, V., 1998. Past climate of South America. In J. Hobbs, J. Lindesay and H. Bridgman, eds., *Climates of the Southern Continents*, Chichester, UK: Wiley, pp. 151–264.

Mayewski, P. A. and White, F., 2002. *The Ice Chronicles*. Hanover, NH: University Press of New England.

Mayewski, P. A. *et al.*, 1997. Major features and forcing of high latitude Northern Hemisphere atmospheric circulation over the last 110,000 years. *Journal of Geophysical Research*, **102**, 26,345–26,366.

Mayewski, P. A. *et al.*, 2004. Holocene climate variability. *Quaternary Research*, **62**, 243–255.

Meese, D. A. *et al.*, 1997. The Greenland Ice Sheet Project 2 depth-age scale: Methods and results. *Journal of Geophysical Research*, **102**, 26,411–26,424.

O'Brien, S. R., Mayewski, P. A., Meeker, L. D., *et al.*, 1995. Complexity of Holocene climate as reconstructed from a Greenland ice core. *Science*, **270**, 1962–1964.

Petit, J. *et al.*, 1999. Climate and atmospheric history of the past 420,000 years from the Vostock ice core, Antarctica. *Nature*, **399**, 429–436.

Soon, W. and Baliunas, S., 2003. Proxy climatic and environmental changes of the past 1000 years. *Climate Research*, **23**, 89–110.

Stager, J. and Meyewski, P., 1997. Abrupt early to mid-Holocene climatic transition registered at the equator and the poles. *Science*, **276**, 1834–1835.

Thompson, L. *et al.*, 1993. "Recent Warming": ice core evidence from tropical ice cores with emphasis on central Asia. *Global and Planetary Change*, **7**, 145–156.

Trenberth, K. and Otto-Bliesner, B., 2003. Toward integrated reconstruction of past climates. *Science*, **300**(5619), 589–591.

Tyson, P., Karlen, W., Holmgren, K. and Heiss, G., 2000. The Little Ice Age and medieval warming in Southern Africa. *South African Journal of Science*, **96**, 121–126.

Williams, P., Marshall, A., Ford, D. and Jenkinson, A., 1999. Palaeoclimatic interpretation of stable isotope data from Holocene speleothems of the Waitomo District, North Island, New Zealand. *Holocene*, 9, 648–657.

Zielinski, G. A., 1995. Stratospheric loading and optical depth estimates of explosive volcanism over the last 2100 years derived from the GISP2 Greenland ice core. *Journal of Geophysical Research*, **100**, 20,937–20,956.

Chapter 7
Urban impacts on climate

7.1 Introduction

Anthropological research tells us that humans originally congregated in central locations at important transport junctions, such as rivers, to trade goods, communicate, and to obtain information. Such locations eventually grew to villages, then to towns, and then to cities as human populations expanded, and the technology increased for buildings, roads, and so forth. Along with the expansion of population and the growth to cities came modifications to the environment, including changes to the local and regional climate. Depending on the state of development, cities are often noisy, smelly, polluted, densely packed, and overpopulated, with major difficulties associated with waste, water, the atmosphere, and human health. However, as Figure 7.1 shows, cities have many benefits to populations, including providing shelter, employment, services, technological development, education, cultural and social interactions, and income. Therefore cities have become essential to the lifestyle and well-being of human beings.

Spatially, cities cover less than 1.5% of the Earth's land surface. However, the latest summary of the state of cities around the world from the United Nations Environment Programme (GEO 2003) states that currently about 47% of the world's population live in cities. This is almost double the percentage in 1975 (27%). Projections during the first half of the twenty-first century suggest an average urban population growth of 2% per year, and that by 2050, 65% of the global population will live in cities.

Much attention has been paid to the development of megacities, those with populations of 10 million or more. Figure 7.2 shows there are 18 megacities, with the majority in Asia. Here, urban growth rates are among the fastest in the world. Tokyo/Yokohama is the largest city with over 26 million in population, followed by Mexico City, Sao Paulo (Brazil), and Bombay (India). Virtually all the megacities are located in developing countries, with the exception of Tokyo, New York, and Los Angeles. According to GEO (2003), however, the majority of urban dwellers live in medium to small sized cities and towns, locations that do not attract much global attention.

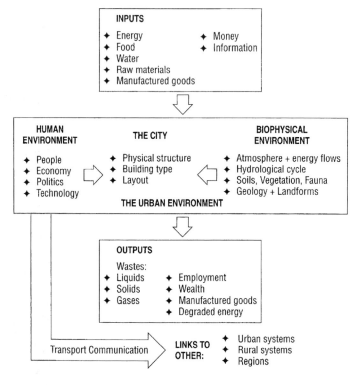

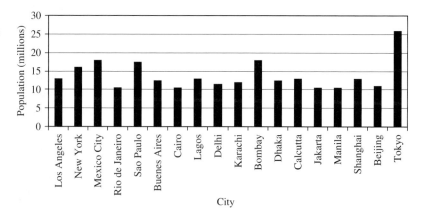

With their increasing populations, wide diversities, and important problems, it
is little wonder that urban environments are of such concern to global, regional,
and local decision makers. Cities have a very important role in regional climate,
affecting temperature, wind, and precipitation. Urban impacts on global climate
are less clear, although they contribute to increased cloud cover frequency, are a

Table 7.1   *Flowchart showing the range of spatial scales in an urban environment that contribute to urban climate. Each unit interacts to contribute to a range of processes that create alterations to urban climate, some of which are listed in the right-hand column (based on Arnfield 2003)*

Urban unit (smallest to largest size)	Interacting processes (occur between each unit level)
Roof and walls	
↓	Shortwave solar energy interception
Buildings, paved areas, green space	Longwave radiation release
	Artificial heat release
↓	Increased urban temperatures
Urban canyons	Increased roughness
↓	Reduced wind speed
City blocks	Altered wind direction
↓	Urban circulation development
Neighborhoods	Convective cloud enhancement
↓	Increased downwind precipitation
Land use zones	Excess water runoff
↓	Reduced evaporation/transpiration
Urban area	

major source of anthropogenic greenhouse gas emissions, and create or contribute to aerosol plumes that can transport many thousands of kilometers down wind (Oke 1997). Critical, however, to urban impacts on climate is the importance of urban unit scale and spatial variability (see Table 7.1).

7.2 Highlights in the history of urban climate research

Research and analysis of cities and their impacts on the atmosphere are not new (see www.urbanclimate.net for a useful historical bibliography). Air quality, rather than climate itself, was defined as a problem first. John Evelyn in 1661, in his pamphlet *Fumifigum*, describes the sulfur and smoke from the city of London. He identified air pollution as a major health problem, a fact not formally recognized by the medical profession until 300 years later! The first known urban climate study was published in 1833 by Luke Howard, an alchemist who also provided the first description of cloud types. Using methodologies and equipment the quality of which would not be acceptable today, Howard established differences in climate between London and the surrounding countryside. In 1925, the first circulation pattern associated with an urban center was identified in Munich.

Table 7.2 *A summary of urban properties compared to rural properties, and urban climate effects (after Oke 1997)*[a]

Property	Urban–rural comparison
Roughness length	*Rural*: 0.01–0.5; *Suburban*: 0.6–1.0; *Urban*: 1.5–2.5
Albedo	*Rural*: 0.12–0.20; *Suburban*: 0.15; *Urban*: 0.14
Emissivity	*Rural*: 0.92–0.98; *Urban*: 0.94–0.96
Thermal admittance ($J\,m^{-2}\,s^{\frac{1}{2}}\,K$)	*Rural*: 600–2000; *Suburban*: 800–1700; *Urban*: 1200–2100
Anthropogenic heat ($W\,m^{-2}$)	*Rural*: Nil; *Suburban*: 15–50; *Urban*: 50–100 (winter up to 250)
Condensation nuclei (cm^{-3})	
Aitken	*Rural*: 10^2–10^3; *Urban*: 10^4–10^6
Cloud	*Rural*: 2500; *Urban*: 10^3–10^4

Variable	Urban change compared to rural
Turbulence	10–50% greater
Wind speed	5–30% less in strong wind; increased in weak winds with UHI
Wind direction	1–10° variation, more in canyons
UV radiation	25–90% less
Solar radiation	1–25% less
Infrared radiation	5–40% greater
Evaporation	~50% less
Convective heat flux	~50% greater
Heat storage	~200% greater
Air temperature	1–3 °C greater annual average; up to 12 °C hourly average greater
Humidity	Less in summer daytime; higher in summer night, winter day
Cloud	More, especially downwind of city
Fog	More or less depending on aerosol numbers; local environment
Total precipitation	Greater, especially downwind of city
Thunderstorms	More
Snow	Less, turns to rain
Tornadoes	Fewer

[a] Assumes mid-latitude city with about 1 million population; summer season except where indicated.

In 1954, T. J. Chandler produced the first major description of the London heat island, including spatial distribution maps. During the early 1960s, the highly controversial La Porte (Indiana, USA) precipitation anomaly was identified, and blamed on urban effects downwind of Chicago. In 1964, D. B. Turner produced the first major air pollution dispersion model for urban areas, based on Gaussian principles. In 1968, the World Meteorological Organisation (WMO) hosted the first international symposium on urban climates (and building climatology). By 1969, there was enough information available for J. Peterson to publish a US government survey of known literature on urban climates.

The first major urban climate experiment, the METROpolitan Meteorological EXperiment (METROMEX), occurred in St Louis (USA) in the early 1970s. Interest and the literature in urban climates were expanding rapidly. In 1972, P. Rao introduced the use of remote sensing from satellites for urban heat island analysis. In that year, the American Meteorological Society held its first conference on the urban environment. In 1974 and 1975, the first detailed work on urban canyons began. Further reviews on urban climate during the 1970s were produced by T. Oke for the WMO. In 1981, Helmut Landsberg's landmark book, *The Urban Climate*, was published by Academic Press, followed in 1985 by *The Urban Atmosphere* by B. Atkinson.

Since then there have been major improvements to urban climate research and understanding. This chapter explores our current understanding of the impacts of urban areas on climate, most of which are broadly summarized in Table 7.2. Following literature trends, it provides a major focus on the urban heat island (UHI). A description of the energy balance, urban heat island development, and aspects of methodology occurs in an essay by Sue Grimmond. Sections on wind, urban canyons, moisture and precipitation, air pollution, and remote sensing follow. Examples from specific cities are used in support of the discussion, especially Melbourne, Australia, Mexico City, and Göteborg, Sweden (see Background Box 7.1). The chapter ends with some comments about ways to mitigate urban impacts on climate.

Background Box 7.1

Cities used as major urban examples

Melbourne, Australia

Melbourne has a population of about 3.6 million people, and is located in southern Victoria, Australia, at the head of Port Phillip Bay ($37°49'$ S, $144°38'$ E). Melbourne is under the influence of the mid-latitude westerlies in winter and subtropical circulation in the summer. The city is surrounded on three sides by higher elevations, but these are not consistent in altitude and there are gaps. To the south, the Bay exerts some influence on the regional climate, but not enough to overcome the influence of the city itself.

Mexico City, Mexico

Mexico City is located in a mountain environment at $10°25'$ N, $99°10'$ W. It is considered to be a tropical city, but is located at an altitude of 2250 m. In a valley, surrounded by mountains, and under the synoptic control of the semi-permanent subtropical high in the eastern North Pacific, Mexico City provides the classic environment for major air pollution episodes and urban climate impacts. With a population of about 20 million people, and major emissions from traffic and industry, Mexico City is rated by UNEP as one of the three most polluted cities in the world

(GEO 2003). There are two major seasons. The cool season (November to April) is dry, and synoptic circulation is generally weak. Anticyclonic weather dominates and surface inversions are frequent, creating a stable boundary layer atmosphere. The warm season (May to October) is wet. The trade winds bring instability and convection, with rain occurring most frequently in the afternoon and evening.

Göteborg, Sweden

Göteborg is a small northern European city located on the west coast of Sweden (57°62′ N 11°45′ E). Its location close to the Arctic Circle means it has about 5 hours of daylight during the winter and around 15 hours of daylight in the summer. Surrounded on three sides by a mix of large and small valleys and higher elevations in between, Göteborg has a population of about 750 000. The central part of the city is divided into the eastern section, containing old, densely packed buildings of between four and six stories; and the western section, which is newer, with more variety in building heights and also landscape features, such as parks. Climatically, Göteborg is firmly located in the mid-latitude westerly circulation belt. Alternating high pressure systems and fronts represent the synoptic weather pattern.

7.3 ESSAY: Variability of urban climates

Sue Grimmond, *King's College London*

7.3.1 Introduction

Urban areas are locations where there is a dramatic human influence on the natural environment. Urban construction materials have different thermal conductivities and capacities; the geometry of buildings and their spatial arrangement trap radiation and pollutants and create a very rough surface that influences air flow and dispersion; the heat released by human activities, from vehicles, industry, etc., supplements natural sources of energy; and engineering structures (pipes) remove water from the surface and modify natural topography and drainage networks, thereby altering runoff and humidity regimes (Oke 1987). The net effect is profound changes to the radiative, thermal, moisture, and aerodynamic characteristics from the pre-existing landscape, which alter natural budgets of heat, mass, and momentum, resulting in the development of distinct urban climates. The most widely recognized urban climate effects include urban heat islands (the common observation that cities are warmer than their surroundings); urban induced wind circulation; and precipitation enhancement downwind of urban areas (see detailed reviews in Landsberg 1981; Oke 1997; Lowry 1998).

Cities and their inhabitants also have a broader impact, influencing directly and indirectly climates beyond their boundaries. The "footprint" of urban areas extends downwind as a consequence of the transport of pollutants, and downriver as a consequence of storm water and sewerage flows; regionally as a consequence of the construction of dams to provide water, and intensive agriculture to provide food to sustain city inhabitants; and globally through enhancement of greenhouse gases and aerosols, significant fractions of which are emitted in urban environments.

In this essay, attention is directed primarily to the climatic processes and effects within and near cities. It is important to recognize that there is no one urban climate or effect. Urban areas are not always hotter, drier, or windier than the pre-existing or surrounding landscape, a consequence of the variability of cities in terms of their surface cover and roughness, energy use and emissions, and regional setting (cities in arid settings have different apparent effects on the pre-existing or regional climate to those constructed in tropical rainforests or coastal wetlands). This essay will consider some of the critical issues that must be considered when documenting urban climates and identify key factors that influence the magnitude of effects.

7.3.2 The variability of cities and the importance of scale

Globally, cities are highly variable, not only in terms of size and population, but in terms of the key properties of the surface that influence energy, mass and momentum exchanges and thus climate. Both within and between cities, roughness (the size, shape, and separation of buildings and vegetation) and surface cover (the radiative, thermal, and moisture properties of all facets of the urban fabric and their spatial arrangement) vary. Commercial, industrial, suburban, downtown, areas of different cities are not the same (see by way of example Figure 7.3). Contrast, for example, central areas of European and North American cities, or downtown residential neighborhoods with suburbs on the fringes of cities, the products of urban sprawl.

Key to interpreting and/or modeling urban effects on atmospheric processes is a conceptual understanding of the spatial scales at which urban geometry and surface materials vary, and thus the appropriate scales at which to study climatic effects. Based on Oke (1984), three spatial scales, micro, local, and meso are commonly recognized based on the natural range of scales in urban morphology (Figure 7.4). At the micro scale (10^1–10^2 m), where issues of dispersion around buildings or within a canyon are relevant, important spatial differences in processes occur in response to variability in building dimensions or canyon height-to-width ratios and orientations. At the local scale (10^2–10^4 m), processes represent the integrated response of an array of buildings, vegetation, and paved surfaces. At this scale, spatial

Figure 7.3 Photographs of contrasting urban areas, classified by density. (Grimmond and Oke 2002)

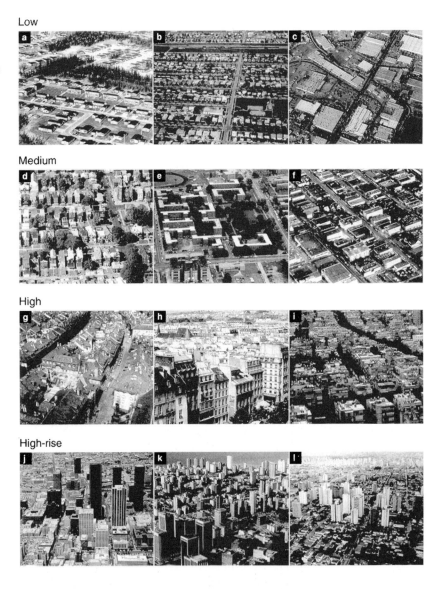

Low

Medium

High

High-rise

variability across a city reflects different neighborhoods, with various combinations of built and vegetated cover and morphometry. At the meso scale (10^4–10^5 m), the city is considered in its entirety, and is differentiated from its regional surroundings, areas of forest, agriculture, etc.

Vertically the urban atmosphere can be divided into the urban canopy layer (UCL) (which lies below the mean roof level), the urban boundary layer (UBL) from the mean roof level (upwards), and the surface layer (approximately the lowest 10% of the boundary layer) (Figure 7.4). The surface layer

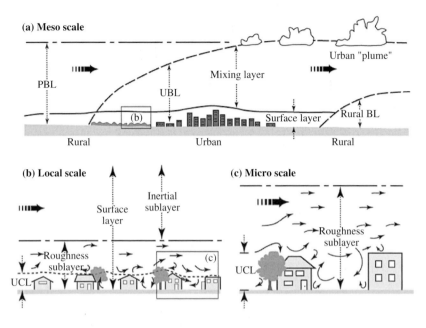

Figure 7.4 Schematic of the urban boundary layer (UBL) including its vertical layers and scales. PBL stands for planetary boundary layer, and UCL for urban canopy layer. (Revised by Oke and Rotach after a figure in Oke 1997) (Source: Piringer *et al.* 2002: with permission from Elsevier)

can be divided into the roughness sublayer (RSL) and the inertial sublayer (ISL). In cities, the presence of large roughness elements (buildings, trees, etc.) means that the RSL, which includes the UCL, is deep. Within this layer the turbulent fluxes of momentum, energy, moisture, and pollutants are height dependent, thus any measurements are representative of micro-scale processes. To document local-scale (neighborhood) processes, instruments have to be mounted above the RSL and within the ISL.

The importance of appropriately considering scale is well illustrated by studies of urban air temperature (Figure 7.5). Temperature anomalies associated with urban areas, from villages to megacities, have long been studied; Luke Howard's (1833) investigation of London is often cited as one of the first studies of urban climate. These urban temperature effects are of interest not only to urban climatologists but also to global change researchers trying to document long-term regional and global temperature trends and thus wishing to remove any urban "contamination" from long-term records, and those interested in the impact of urban areas on energy consumption, in particular related to air conditioning demand, and the implications for enhanced global carbon dioxide concentrations given the use of fossil fuels for power generation.

When meso-scale (whole city) urban heat islands (UHI) are studied in detail (Figure 7.5a and b), it is apparent that rarely does temperature increase continuously from the outskirts of the city into the central business district. Rather, local-scale thermal influences are evident. Parks,

Figure 7.5 Variability of air temperature at different scales. (a, b) Hypothetical representation of the spatial features of the canopy layer urban heat island in a mid-latitude city with "ideal" (calm, clear) conditions. Spatial pattern (a) along cross section AB and (b) in relation to plan outline of the city (Oke 1982). (c, d) Measurements in an urban canyon (H/W 1.06). (c) Temporal variation of surface temperature on the canyon floor together with the air temperature at 0.5 m. (d) Isotherm distributions across a canyon for two 10-minute periods (Nakamura and Oke 1988: permission from Royal Meteorological Society).

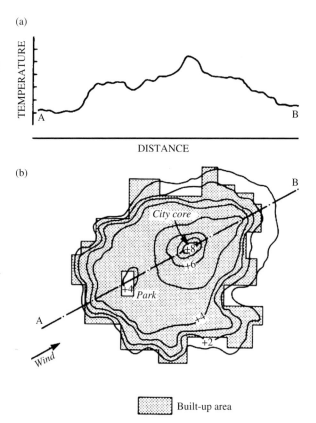

(a)

TEMPERATURE

A   B

DISTANCE

(b)

City core

Park

+6

+4

+4

+2

A

B

Wind

Built-up area

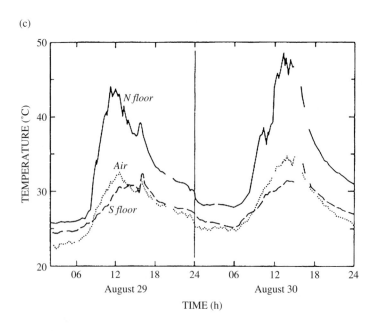

(c)

TEMPERATURE (°C)

50

40

30

20

N floor

Air

S floor

06   12   18   24   06   12   18   24

August 29   August 30

TIME (h)

(d)   **Figure 7.5** (cont.)

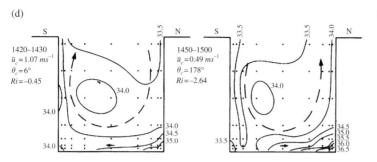

for example, may generate "cool islands," caused by longwave radiative cooling (because of more open geometry) and enhanced latent heat flux. Industrial or dense residential areas often generate enhanced heat islands. At the micro scale, the scale at which most people directly experience urban climate effects, differences in air temperature within an urban canyon (Figure 7.5c and d) or across a park may be greater than the variability across the entire urban area.

Different UHI phenomena can be distinguished vertically (Figure 7.6). Air temperature sensors at weather-screen level (UCL) record something different from those at higher elevation above the surface, in the UBL, and from that derived from remotely sensed surface temperatures. Increasingly, UHIs are being studied using remotely sensed surface temperature data. These satellite-derived ground temperatures do not always reveal the same spatial and temporal patterns as air temperature surveys: UHI intensities tend to be greatest by day and smallest at night (almost the reverse of the UCL temperature patterns); and spatial temperature patterns tend to be linked closely to land use (parks, parking lots, etc.) by day but not at night (Roth *et al.* 1989). Some of these differences can be linked to the difference between a radiative surface temperature and a near-surface (UCL) air temperature, where the latter depends on surface energy exchanges as well as heating and cooling of an air volume (see Section 7.8 for more detail).

Even when surface temperatures are all that is of interest, depending on their view angle, satellites may preferentially measure horizontal surfaces (roofs or roads) or wall facets with a preferred aspect. Thus the geometry of urban areas will complicate interpretation of remotely sensed surface temperature data due to the presence of vertical surfaces and shadows. Work by Voogt and Oke (1997), using ground and airborne infrared radiometers, has taken into account the total active surface area of an urban environment (all the roofs, walls, and road surfaces), to document the complete surface temperature. Comparisons between complete surface temperatures and those derived from satellite images with different view angles are significant.

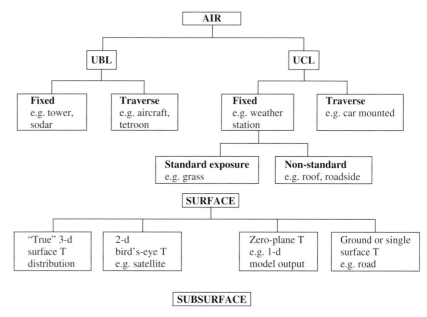

Thus, satellite-based observations of surface temperatures must be interpreted with care, if conducted in the absence of consideration of geometric conditions of the surface and the viewing conditions.

7.3.3 Identifying urban effects

Documenting urban effects is often difficult. Only rarely are extensive pre-urban measurements of the climate of a region available against which urban observations can be compared. More commonly, urban data are collected at a point and compared with those at a nearby rural "reference site" and urban effects are defined in terms of urban–rural differences.

Returning to the urban effect on air temperature (T), in this framework, the UHI can be defined as:

$$\Delta T_{u-r} = T_u - T_r$$

Where, the subscripts u and r refer to the urban area and the non-urban reference site, respectively. Close attention must be paid to reference sites used in such comparisons. The reference station, although outside the urban area, may lie downwind and be subjected to urban influences. Moreover, many rural areas are significantly influenced by human activities, for example, agricultural development, irrigation, etc. Such sites do not necessarily provide a surrogate for conditions in the city pre-urban development.

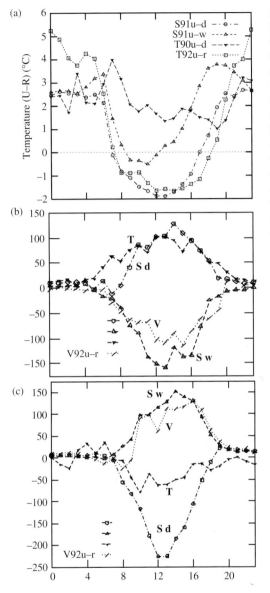

Figure 7.7 Difference between urban and non-urban (a) temperature, (b) latent heat flux, and (c) sensible heat flux. Comparison for Sacramento, California, USA (S91, S) where the urban (u) is a residential neighborhood with irrigated vegetation and a dry (d) unirrigated long grass area and a wet (w) irrigated sod farm; Tucson, Arizona (T90, T) where the urban (u) area is a residential neighborhood with some irrigated vegetation and the natural desert (d); and Vancouver, British Columbia (V92, V) which is a residential neighborhood (u) with vegetation that was not being irrigated during this period because of drought but would typically be an agricultural field of soy beans (r).

To illustrate this point, results from a study by Oke and Grimmond (2002) are presented (Figure 7.7). While conventional wisdom states that urban areas are warmer than their rural surroundings, the urban–rural differences may in fact reverse, depending on the land cover/use of the region that surrounds the urban area (Figure 7.7). For example, in Sacramento, California, a naturally arid area, extensive agricultural irrigation results in greater contrasts between rural areas than between rural and urban areas. Depending on the rural

Figure 7.8 Land use change for Phoenix region, Arizona, USA, showing the shift from desert (natural) and agriculture (human modified) to urban from 1912 through 1995. Notice how the land use changes around the climate stations through time. (After Brazel *et al.* 2000)

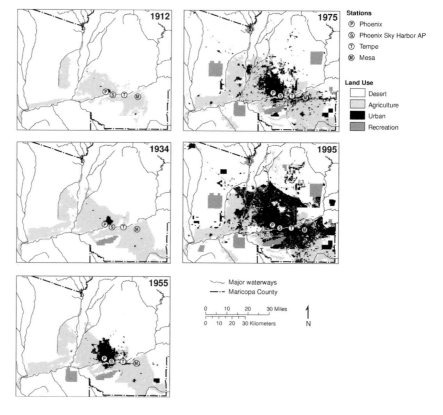

reference, the urban area may appear hotter (irrigated reference) or cooler (desert reference) than the surrounding rural sites. Moreover, in studies that look at the development of the UHI with time, it is important to recognize that the surface and thermal properties of both cities and their rural surroundings change through time. Brazel *et al.* (2000), in a study of Phoenix, Arizona, document landscape changes (Figure 7.8) both within and around the city (irrigation and intensification of agriculture) that impact urban–rural differences in temperature.

7.3.4 The surface energy balance

Ultimately urban climate effects are due to differences between the city and its pre-existing landscape surface characteristics. Thus understanding, prediction, and mitigation of urban climate effects are intricately tied to knowledge of surface–atmosphere exchanges in urban environments. A powerful framework for studying urban climate effects, and their relations to the materials and morphology of the urban surface, is the surface energy balance (SEB), defined for an urban areas as:

$$Q^* + Q_F = Q_H + Q_E + \Delta Q_S + \Delta Q_A \ [\text{units}: \text{W m}^{-2}]$$

where Q^* is the net all-wave radiation (the net balance of the incoming and outgoing radiative fluxes), Q_F is the anthropogenic heat flux (the energy released by human activities), Q_H is the turbulent sensible heat flux (the energy that heats the air), Q_E is the latent heat flux (the energy taken up/released with the phase change of water, i.e. with evaporation and condensation), ΔQ_S is the net storage heat flux (the energy that heats and is stored in the urban fabric and volume), and ΔQ_A is the net horizontal heat advection (the lateral movement of energy into or out of an area).

Given advances in instrumentation, measurements of surface energy balance fluxes increasingly are being made in urban areas to complement measurements of more standard meteorological variables, for example, temperature, wind speed, humidity, to gain greater insight into the processes that underlie the generation of urban climates. In these studies, careful attention has to be directed to the siting and operation of the instruments to ensure reliable and representative data are collected. Recent work has demonstrated that turbulent sensible, latent, and storage heat fluxes all are important terms in the SEB of most cities. Each of the heat fluxes varies both spatially and temporally. Under low-wind conditions, storage heat flux is most important at downtown and light industrial sites (at least 50% of daytime Q^*), and the sensible heat flux is most important at residential sites (40 to 60% of daytime Q^*) (Grimmond and Oke 2002). At all sites there is distinct hysteresis in the diurnal course of the storage heat flux; much more of the net radiation is used to heat the urban fabric in the morning. In addition, the sensible heat flux remains positive after the net all-wave radiation turns negative at night. This has important implications for stability (mixing) of the urban atmosphere and therefore for urban air quality. At residential sites, latent heat flux, if sustained by garden irrigation and/or frequent rainfall, is also significant (20 to 40% of daytime Q^*). Cities are not just impervious building materials (concrete and asphalt sidewalks, roads and parking lots, rooftops, etc.); in many urban areas, trees and other vegetated surfaces cover a significant area (commonly up to 40% of the plan area of a city in North America) and surface detention ponds are common. Surface cover, notably the fraction of the surface vegetated and irrigated, has been shown to exert an important control on Q_E and the relative heat partitioning between sensible (Q_H) and latent (Q_E) heat.

Here select data derived from Oke and Grimmond (2002) are presented to illustrate the importance of both urban and rural surface properties in SEB comparisons. Whilst most cities have an increased sensible heat flux

---

Background Box 7.2

The energy balance in Mexico City (Oke _et al._ 1999)

The structure of a city has a major impact on the energy balance. Measurements in the dry season (early December 1993) over a rooftop in a densely built-up location, provide results vastly different from those in rural areas. During the day, absorption of incoming shortwave radiation and anthropogenic heat sources (industry, traffic, etc.) are critical to the spatial variation of the energy budget. Daytime net radiation (Q^*) peaks at about 400 W m^{-2}, a reduced value due to urban pollution and also lower diffuse radiation in a thinner winter atmosphere. Outgoing longwave radiation under clear skies is enhanced. The buildings act as a major sink, storing 58% of average daytime Q^* (on the order of 155 W m^{-2}). The afternoon convective (sensible) heat flux is reduced to 38% of Q^*, where in other cities it would dominate the daytime energy budget.

At night, Q^* is strongly negative, on the order of -120 W m^{-2} after sunset, and -90 W m^{-2} near sunrise. However, the release of daytime heat from storage in the urban surfaces overcomes this deficit, releasing up to 1.3 times the Q^* deficit, and is often the only source of energy at night. Spatial variations depend on urban geometry, building density, and the level of stored energy release. A weak convective heat flux is thus maintained overnight.

The dominance of the storage and sensible fluxes and the small amount of vegetation cover mean that evaporation is very small, on the order of 4% on Q^*. At night, evaporation is virtually zero. This creates a Bowen ratio of 0.58, considerably larger than the 0.19 to 0.40 range determined for other mid-latitude cities.

The dominance of storage in the energy budget enhances the development of the UHI, especially at night. Nighttime UHIs average 4–5 °C and can occur in both seasons, although maximum UHIs >7 °C only occur during the calm clear conditions of the dry season. The highest UHIs occur at daybreak, associated with maximum longwave radiation loss and a cleaner atmosphere, whereas in many other cities maximums occur before midnight.

---

compared to the countryside, it can be less if the city is irrigated, or if urban heat storage is particularly large in a city center (e.g. Mexico City, see Background Box 7.2; Oke _et al._ 1999). Evaporation is usually reduced by the "waterproofing" of the urban surface but urban–rural comparisons can be complicated by urban irrigation, stronger surface–mixing layer coupling in the city, intra-urban advection, and the spatial variability of rural moisture (see the example of Sacramento conditions described above) (Figure 7.7b). Any such reversals of expected patterns have the potential to produce counter-intuitive urban–rural effects on parameters such as stability and mixing depth, or climate elements such as temperature and humidity. For example, it is possible for $Q_{H\,u-r}$ to be negative (see Figure 7.7c), resulting in less boundary layer heating in cities surrounded by dry landscapes. Whether this results in a shallower UBL depends on the relative roles of heat flux and mass convergence (due to dynamic processes, roughness, and barrier effects) in producing uplift. The Sacramento results show that choosing a rural site, which is a reasonable analogue of pre-urban conditions, gives radically different outcomes to that using a site characteristic of the modern agriculturally managed landscape (Figure 7.7).

7.3.5 Investigating urban climates

Conducting measurements in urban areas requires all the quality assurance and quality controls of any measurement campaign. Often, routine observations become more difficult in urban area and other issues need to be addressed. Moreover, many cities are located in complex settings: along a coast or shoreline, the confluence of river valleys, on a hill, in a sheltered basin. Each of these settings carries with it an additional set of local climate influences that complicate the study of urban areas.

A critical question in any urban climate study, intricately tied to the scale of interest, is where does it make sense to locate instruments to obtain a representative measurement? Not only does the height of the instruments relative to the roughness elements (Figure 7.4) have to be considered, but also the spacing of the elements around the sensor location and the orientation of streets and buildings (which will affect radiation geometry and diurnal and seasonal heating patterns and wind speeds) need to be taken into account. If the typical micro-scale unit is the urban canyon (Figure 7.4), the 1–2 m height of instruments in a standard weather screen is less than the height of buildings and much urban vegetation. In cities, there is often the added complication that instruments have to be mounted higher above the surface to avoid vandalism or to be out of the way of traffic or pedestrians. The field of view of urban-installed sensors may be less open than what would be recommended for a standard climatological/meteorological station. All of these things will introduce bias that the user of the data must be aware of when interpreting data.

An alternative approach to studying urban climates involves the use of models. These range from physical hardware models to study micro- and local-scale effects, for example wind tunnel studies of air flow or studies of the thermal effects of parks (Spronken-Smith and Oke 1999), to numerical models to simulate energy, mass, and momentum exchanges. Many larger scale numerical models represent urban areas by just changing the thermal materials of the "soil," representing the urban area as concrete. Comparisons of such model output with urban-based observations show this approach to be too simplistic and many of the features of urban climate are not captured. Increasingly, more complex models are being developed, which incorporate the morphology and complexity of surface materials in urban environments and their results are showing significant promise. For example, the Town Energy Balance (TEB) model of Masson (2000) treats the surface as an urban canyon without orientation. With such models, incorporated into a meso-scale model it is possible to have areas with different height to width characteristics across a city. It is possible to begin to incorporate the spatial variability of urban morphology within and between cities. At smaller scales, three-dimensional dispersion models now take into account the flow processes with the RSL (Figure 7.4), and three-dimensional energy balance models capture the variability of sky view factor within urban canyons. With more powerful computers, models at all scales will capture more completely the fundamental exchange processes generating urban climates with greater spatial resolution, accounting more fully for the geometry and diversity of construction materials of cities, providing better simulations of urban climates as a result.

7.3.6 Concluding comments

In many instances urban climate effects are similar to, and maybe even greater than, those changes predicted from global climate models (GCM). However, there is no one urban effect; cities are diverse as are their settings. Great care must be taken in observing urban climates, with particular attention to the scale of interest and the scale of the observations, and in generalizing results to other locations. Greater insight into urban climates is being gained by studies that measure and model the fundamental surface–atmosphere exchanges of energy, matter, and momentum in these settings.

Given that cities worldwide are reaching an unprecedented size, in terms of their number, area, and population, the magnitude and extent of urban effects are increasing, with profound implications for energy consumption and human health and well-being. Most growth in urban populations, approximately 90%, is occurring in countries of the developing world, where growth rates, notably in Africa and Asia, are approximately 4% per year. In Europe and North America, the issues associated with urbanization are different. On these continents, the greatest urban growth took place a century ago; by 1995 more than 70% of the population was living in urban areas (World Resources 1996) (this proportion exceeds 90% in certain countries, for example, Australia). Much of the population shift now underway involves movement away from concentrated urban areas to extensive, sprawling metropolitan regions, or to small- and intermediate-size cities. This is especially manifest in the United States. This means the imprint of urban areas on climate is increasing.

In all urban areas resource use and emissions are increasing with global consequences. Urban environmental problems, related to air quality, thermal stress, issues of water demand and quality, all of which are linked directly or indirectly to urban climate, are emerging as major environmental concerns at the start of the twenty-first century. The study and understanding of fundamental causes of urban climates is critically important to predict, plan for, and mitigate negative inadvertent effects.

7.4 Wind, cloud cover, and pressure

7.4.1 Wind

The urban environment has significant interactions with airflow. The city modifies the vertical and horizontal wind structure in several ways, creating alterations in the forces and pressures that make up the wind. The variations in the height of the buildings increase the surface roughness, creating extra turbulence, and the potential for stronger convection. Wind speed is generally lower in the city (see Table 7.2) due to friction, but under light winds, urban streets and canyons can act as wind tunnels, increasing local airflow strength. Wind direction is deflected considerably in the maze of buildings and streets, often making definition very difficult. The buildings themselves will respond in varying ways to the stress created by the wind. These alterations make using a 10 m mast to measure wind inappropriate.

The variation of wind speed with height in the lower troposphere is often described through wind profiles or roughness calculations. For example, the exponential wind profile depends on the speed of the wind with height ($u(z)$), the airflow at a defined base level (z_0), the zero-plane displacement of the airflow (z_d), the height and wind speed at a defined reference level (z_{ref}, u_{ref}), and an exponent (α) which indicates the roughness of the surface (Plate 1995):

$$u(z)/u_{ref} = (z_0 - z_d/z_{ref})^\alpha$$

Unfortunately, use of wind profiles alone can only provide a coarse indication of vertical wind structure over a city, because the roughness varies considerably and aerodynamic responses are irregular depending on the type, size, and shapes of surface features (urban morphology). Grimmond and Oke (1999), in a review of methodologies used to assess the relationship between roughness geometry and aerodynamics over a city, summarize how difficult it is to obtain representative results. They concluded, for example, that acceptable estimates of z_d and z_0 are almost non-existent.

When the wind is strong, its interaction with the urban structures is mechanical, and internal friction and roughness are the main reasons for the wind variations (Plate 1995). Buildings of various heights will tend to slow and mix the wind (Figure 7.4), creating uncomfortable and unexpected variations in wind speeds for people at ground level. Under these conditions, the roughness sub-layer created by the city can extend to 2.5 to 3 times the height of the buildings, especially where a group of buildings of varying heights exists. Coriolis force is reduced in the city due to the friction, and the convective interactions between the urban airflow and synoptic airflow can affect weather fronts, clouds, and precipitation (Oke 1997). The variations in the surface morphology mean that there is no consistently reliable method to incorporate urban surface roughness into a wind model (Arnfield 2000).

When the synoptic wind is weak, the extra thermal effects (artificial heat sources; release of heat by surfaces, etc.) in an urban environment become much more important to the wind than roughness and friction. Thus UHI formation from thermal effects best occurs at night under anticyclonic circulation, cloud cover less than 2 oktas, wind speed less than $3 \, \mathrm{m \, s^{-1}}$, and a large negative Q^*. Regional circulations, such as drainage flows from surrounding higher eleva-tions, and micro-advection processes, become influential (Eliasson 1994). The stability of the atmosphere and the strength of any inversion present define the depth of the mixing layer, and it thus becomes an important controller. The wind in the planetary boundary layer may be separated from that in the UBL by an inversion (Plate 1995). Further separations between airflows at UCL and UBL level can allow local circulations to dominate (such as between the buildings in Figure 7.4). In smaller coastal cities, such as Göteborg, such separations can be critical to UHI development (Eliasson 1994).

Depending on the physical size of the city, its population, and its density and height variation of buildings, one or more UHIs may develop when the wind speed is small. UHIs also can affect the wind, altering local pressure fields, and creating changes in wind direction (by up to 20°) due to increased convective mixing from the heat releases (Oke 1995). While the general structure of thermal turbulence characteristics may be similar to the rural environment, the urban center modifies the stability characteristics of the lower atmosphere and can significantly increase the convective turbulence.

7.4.2 Meteorological conditions and urban temperatures

The relationship between meteorological conditions and the UHI is generally well known. The city of Melbourne is used as an illustration. Research results reported by Morris and Simmonds (2000) and Morris *et al.* (2001) provide a detailed evaluation of the relationship between UHI development and strength and meteorological conditions (wind speed, synoptic pressure anomaly, cloud cover) between 1973 and 1991 in Melbourne. Comparing a central business district site with three rural airport sites, the UHI detected over the 20-year period averaged 1.13 °C and ranged from −3.16 to +6 °C. Negative temperature periods suggested the existence of an urban cool island (UCI – warmer rural than urban temperatures). On a seasonal basis, the average UHI was 1.29 in summer, 1.13 in spring, 1.02 in autumn, and 0.98 in winter. The differences between spring/summer and autumn/winter were statistically significant. Diurnally there was little change in the mean UHI between 10:00 p.m and 6:00 a.m. This study allowed an evaluation of the total range of weather conditions on UHI, rather than just a focus on extreme conditions.

The dominant frequency of UHI values occurred between 0 and 2 °C (on about 75% of the days). The average pressure anomaly for the 0–1 °C UHI was statistically significant at -2.12 hPa, representing a low pressure center between 28 and 48° S (SE of Tasmania). For the 1–2 °C UHI, the average pressure anomaly was a slightly positive (non-significant) 1.73 hPa, associated with the high pressure center over and just east of Tasmania. Wind speed averaged 3–4 m s^{-1} and cloud cover 4.5 to 5.6 oktas. The influence of cloud cover and wind speed on UHI development and strength was mixed during these periods, with cloud cover tending to have a dominant influence under calmer conditions, and wind speed the dominant influence under clear or slightly cloudy skies.

UCI conditions were dominated by high pressure over eastern Australia, centered over the Tasman Sea (32 to 48° S). The pressure anomalies averaged 5.06 hPa for UCI < 1.0 °C (significant), and 3.9 hPa for $-1 \leq$ UCI < 0. A cooler urban environment was associated with the prevention of warmer (prefrontal) NW airflow, which reached the rural airport sites, but was prevented from entering the city by the urban area. This situation was associated with higher than average minimum temperatures. Often, with a change in the synoptic situation, these periods were immediately replaced by UHI conditions, associated with the release of heat storage from the urban area. UCI conditions were more frequent between March and July, the winter half of the year.

At the other extreme, the 17% of the data set exhibiting a strong UHI (>2 °C) also showed highly significant mean pressure anomalies (3.6 and 4.9 hPa respectively). The mean location of the pressure center was over Bass Strait, just east of Melbourne but somewhat north of Tasmania. These conditions are typical of the expected UHI development from the literature under high pressure and a stable atmosphere, and the meteorological data supported this conclusion. Airflow was often warm, and weak (~ 1.5 m s^{-1}) from the N or NE, with cloud cover less than 3 oktas, creating optimum UHI conditions. Under these situations, the UHI averaged 2.7 °C. Wind speed was the more important controller of UHI development. These periods occurred with greatest frequency between October and March, the summer half of the year.

A UHI did exist under a cloud cover of 8 oktas and a wind speed of >5m s^{-1}, averaging about 0.5 °C. The main controller was cloud cover, with diminishing cloud leading to increases in UHI. Overall, although both the wind speed and cloud controllers were not strong statistically, cloud cover tended to be more important. The contributions from each tended to be independent of the other.

7.4.3 The urban heat island circulation (UHIC)

When synoptic conditions are calm or very weak, a local urban circulation, the UHIC, can form, controlled by the difference in temperature between the urban center and the rural surroundings (Haeger-Eugenssen and Holmer 1999). The higher temperatures in the city induce a pressure differential, and air from the surrounding rural areas is attracted toward the urban center (Oke 1995). This air becomes modified in speed and direction by the urban morphology. Once it reaches the UHI center, the increased heating and turbulence causes it to rise. At the boundary layer inversion, it then spreads outward, back to the country-side, where it cools and sinks. Variations in the number and strength of convection cells can add to the mixing and turbulence, creating an intermittent and variable UHIC.

The UHIC forms most strongly in the daytime because solar heating favors greater vertical convection (Oke 1995). In Mexico City, there is a relationship to high ozone levels, and the stable synoptic-scale meteorology that creates these. Overall, there is no significant correlation with change in boundary layer thermal structure with height, or the difference in temperature between the bottom and top of the inversion. Shallow surface inversions, however, are much more likely to create stronger UHIs than deeper inversions, but shallow inversions may interfere with UHIC development (Jauregui 1997). Clouds and rain showers will reduce the urban–rural temperature contrast, reducing the UHIC.

When UHIs were stronger than $2.5\,°C$, an urban circulation system was established over Göteborg on winter nights (Eliasson and Holmer 1990). Development also depended on a $Q*$ of $-90\,W\,m^{-2}$ or less, and wind speeds $<3\,m\,s^{-1}$. A strong inversion (at least $4\,°C$), located at an average altitude between 40 and 60 m, separated the regional airflow from that close to the surface. The UHIC occurred in stages. Early in the evening, there was significant differential cooling between the city center and the rural areas, with the latter cooling very rapidly compared to the former. The UHIC started about two hours after the UHI was established (4 to 6 hours after sunset), bringing a sudden increase in urban cooling as the circulation transferred sensible heat from the rural environment. Cooling rates in both the urban and rural environments then equalized (at about $0.5\,K\,h^{-1}$) and the UHIC became self-sustaining until morning. Any change in one component of the circulation/heat differential system (for example wind speed) required compensatory changes in the other components to maintain UHIC equilibrium.

In the eastern part of Göteborg, the surface airflow is toward the center of the city, but in the western part of the city, the urban circulation flows through toward the ocean. Its spatial extent is determined by the valleys to the north, and other, more coastal, airflows to the south. The urban circulation does not extend more than 13 km outside the city center. It can be difficult to distinguish from topographic airflows. UHIC development at night can bring cooler air into the city, reducing the urban–rural temperature gradient, but creating an equilibrium between the city and its rural surroundings that allows the UHIC to continue for several hours (Haeger-Eugensson and Holmer 1999).

Once developed the UHIC can interact with, and interfere or support, other local airflows in very complex ways (Oke 1995; Haeger-Eugensson and Holmer 1999, and others). Cold air drainage from higher elevations around a city, channeled though valleys, may reduce UHI and UHIC development. If airflows from valleys are weak, the UHI may prevent these airflows from entering the city, creating airflow stagnation and high pollution episodes. If the sides of the valleys extend above the inversion, the valleys may create UHIC channeling, influencing cooling rates and wind direction. In coastal areas, the UHIC can act against sea breeze development, retarding the strength and depth of its inland flow. However, when the convective zone of the UHIC and the sea breeze merge together, the sea breeze can be strengthened.

Light winds can also carry the urban influence to the atmosphere downwind, over the countryside (Figure 7.4a). The resulting urban "plume" may extend over the rural boundary layer, creating a series of complex meso-scale inversions with height. The extent of urban plume transport and development depends on the regional terrain and the atmospheric conditions.

7.5 Urban canyons

An urban canyon is defined as a street or flat area bounded on two sides by buildings, with an, at least partially, open top to the sky. The building sides are normally defined as vertical with minimal variation. The top of the urban canyon usually defines the boundary between the UCL and the lowest part of the UBL. The exchange between the UCL and UBL of water vapor, heat, and pollutants, etc., occurs mainly through turbulence associated with wind interactions and canyon structure (Arnfield and Mills 1994a). Combinations of canyons are a micro-scale structure, and can create local-scale climates which influence the UHI, energy balance fluxes, and wind, depending on their orientation, depth, and materials (Arnfield and Mills 1994a, b; Bonan 2002).

Solar energy entering the canyon though the open top may be absorbed and reflected several times, depending on the angle of entry, the geometry of the buildings, and the building surface type. Urban canyons tend to trap solar energy during the day, and release this energy as longwave radiation and sensible heat over the diurnal cycle. There is a direct relationship between the amount of sky viewed at the top of the canyon and the amount of energy trapped (Oke 1981). The sky view factor (Ψs) is defined as the geometric relationship between the building height (H) and the street width (W), which determines the angle of view (a) from the bottom to the top of the canyon. If an infinitely long street is assumed, $\tan(a) = 2H/W$ and $\Psi s = \cos(a)$. The smaller the central city Ψs, the stronger will be the maximum air UHI. Figure 7.9 shows little difference in

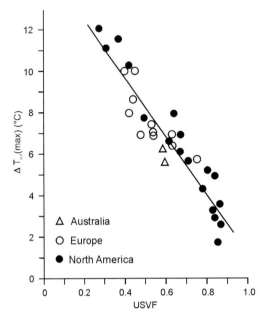

maximum air UHI and Ψs between European, North American, and Australian cities, but there are no reported results for cities in developing countries, such as China.

In Göteborg, with an average UHI of 2 to $4.5\ °C$, which is strongest in the summer, the role of urban canyons in establishing the strength and spatial distribution of the air UHI was investigated in depth, using Ψs and temperature measurements on various canyon surfaces (Eliasson 1994, 1996). In the old city, Ψs ranged from 0.25 to 0.45, but in the western section, were higher, 0.5 to 1.0 (in open parkland). Overall the canyon Ψs had warmer UHI of about $1\ °C$ in all seasons, but the relation is very complex.

There was a significant correlation ($p < 0.05$) between Ψs and surface temperature, but the correlation between Ψs and air temperature was poor. Detailed temperature measurements in the canyons established that the surface temperature variations (partially due to differential rates of cooling) do not translate very well into the air temperature above 0.1 m in height (Figure 7.5). Whether within a canyon, or in a traverse across the city central area, the urban air temperature distribution in central Göteborg remained relative uniform ($<3\ °C$ variation).

Urban canyons also create major changes in both the wind speed and wind direction of the prevailing airflow (Johnson and Hunter 1999). If the prevailing wind is parallel to the direction of the canyon, then down-canyon airflow occurs. The major influence will be friction, variations in canyon orientation, and gaps in the canyon sides (side streets and alleyways). If winds are light and variable, it is difficult to establish a relation between the wind flow within the canyon, its geometry, and the prevailing wind profile. These conditions create stagnation and trapping of air pollutants.

It is the cross-canyon airflows that make complex changes to the prevailing wind structure. Depending on the angle of the cross-wind flow and the wind speed, vortices within the canyon develop, which are partially controlled but also independent of the prevailing winds. The prevailing wind will strike the opposite (windward) wall first, deflect down toward the street, and then move up the opposite (leeward) wall. Thus the flow at the floor of the canyon becomes directly opposite that of the prevailing wind. Wind speed on the leeward wall is about 40% of that on the windward wall. Johnson and Hunter (1999) and Arnfield and Mills (1994a) suggest that vortices can occur when the height to width ratio (H/W) of the urban canyon exceeds 0.4, in wind speeds as low as $1-2\,\mathrm{m\,s^{-1}}$. Below a H/W of 0.4, any interaction with the wind between adjacent buildings tends to be restricted to isolated roughness or wake interference influences. When the prevailing wind direction is at an angle (for example 45°) to the canyon direction, spiraling vortices can develop and extend down the length of the canyon, in a combination of cross-canyon and parallel-canyon airflow. When canyons are very deep ($H/W > 1.0$), two vortices can develop: an upper vortex that is controlled by the ambient airflow; and near-surface circulation that is opposite in direction (Arnfield and Mills 1994a).

The wind direction and speed, the turbulence associated with the vortices, Ψs, the types of building and surface materials, and the exchange of air through the canyon top all influence the components of the energy balance. There are major differences in effect between individual cities and between individual canyons. The incoming solar energy, its absorption and reflection through the canyon, and interactions with atmospheric and artificially released water vapor influence the exchanges of heat and water vapor between the air and the canyon surfaces (Arnfield and Mills 1994a). The wind assists this process by generating turbulence and mixing. Air temperatures in the canyon may therefore show little variation, as compared to sunlit versus shaded surface temperatures.

Arnfield and Mills (1994b) review several studies that evaluate the energy balance of urban canyons, and the effects of wind. In general, daytime Q^* mainly goes into sensible heat flux, which can then be transferred by turbulence and convection out through the canyon top. There is some storage of Q^* in the surface material, but, depending on surface material, very little is used for latent heat of evaporation. At night, Q^* is mainly negative, and is balanced by heat release from the canyon surfaces. The advective impact on Q^* can be either positive or negative, depending on the time of day and the wind speed. The variations in radiative heat in different parts of the canyon, reflected in the SVF and the angle of the sun, mean that there is variation in the heat balance in different parts of the canyon at different times of day.

7.6 Moisture and precipitation

Urban environments create major alterations to the water balance. The solid impervious surfaces in the city encourage water runoff and prevent infiltration into the subsurface. Infiltration is an important feature of the rural environment. According to Bonan (2002), on average in cities, 83% of precipitation runs off the urban surfaces (mainly into sewers and gutters, depending on the development level of the city), and only 17% reaches the subsurface. The amount of evaporation was not considered. In heavy rain situations, peak discharges are increased, runoff speed is enhanced, and the volume of runoff is much higher than in a rural environment. Flooding can be more prevalent if the sewage and storm water runoff system cannot handle the excess water. The solid urban surfaces also ensure that evaporation in the city remains considerably lower than in the rural environment, and the lack of trees and parks creates reductions in transpiration as well (see Figure 7.7).

As a result of the urban surface plus the existence of the UHI, there are differences between urban and rural atmospheric moisture levels (absolute humidity). Humidity variations in the city are caused by varying evaporation rates, varying levels of condensation, advection of air from outside the city, extra turbulence and anthropogenic moisture release from different activities such as combustion or watering the lawn (Oke 1997). Holmer and Eliasson (1999) describe an urban moisture excess (UME) in city atmospheres at night, and a moisture deficit during the daytime, for temperate climate cities. The UME also varies by season; for example winter UME is usually smaller than summer UME, but colder overall climate and artificial moisture sources may cause exceptions. Excess heat in the urban atmosphere means that urban air can hold more moisture as water vapor at night, and condensation processes such a dewfall may be less likely in areas with high Ψs and low vegetation amounts.

The greatest difference between urban and rural atmospheric moisture levels occurs during the afternoon and early evening. In St Louis, USA, for example, a rural site averaged $14.5\,\mathrm{g\,kg^{-1}}$ and the urban site $14\,\mathrm{g\,kg^{-1}}$, but from 0700 to midnight, the urban–rural difference exceeded $1\,\mathrm{g\,kg^{-1}}$ (Karl *et al.* 1988 in Bonan 2002). In Mexico City, the urban center tends to be more humid at night (by about $6{-}8\,\mathrm{g\,kg^{-1}}$); the rural area more humid during the day (by about $1{-}3\,\mathrm{g\,kg^{-1}}$) (Jauregui and Tejeda 1997). Stronger winds reduce urban–rural humidity differences. Several other cities exhibited urban–rural moisture differences of between 1 and $3\,\mathrm{g\,kg^{-1}}$ (Holmer and Eliasson 1999), but with considerable individual and diurnal variation.

Holmer and Eliasson (1999) suggest that there can be a linear relationship between UME and UHI development, with the former lagging behind the latter by 1 to 5 hours. Ideally a UME should enhance UHI because a warmer, more humid atmosphere should add $20{-}40\,\mathrm{W\,m^{-2}}$ to sky-emitted longwave radiation.

Since urban atmospheres are also polluted, and particles can have a similar effect, it is difficult to separate particulate and moisture influences.

Whether UME or moisture deficit, urban areas act to increase turbulence, convection, cloud formation, and rainfall. Lowry (1998) and McKendry (2003) suggest that cities induce stronger convection zones, especially associated with the UHI, which support enhanced cloud formation and thunderstorm development. Variations in urban topography play a secondary role. But clouds are also an important controller of the strength of the UHI. An increased cloud cover generally reduces the strength of the UHI. Urban areas also create an excess of large and giant nuclei, many of which are hygroscopic, enhancing cloud condensation processes. On occasion, too many nuclei can be produced, creating too many cloud droplets and preventing efficient cloud droplet growth to raindrops.

Precipitation and severe weather can be enhanced, especially associated with frontal systems. Enhancement usually occurs at some distance downwind of the city, associated with the urban plume (Figure 7.4a), because time is needed for the raindrop formation process to develop. One of the earliest, and most controversial, urban rainfall enhancement descriptions was the LaPorte, Indiana, anomaly. Here major increases in rainfall were blamed on excess particulates and convection, blown downwind from the city of Chicago (Lowry 1998 and others). Summertime rainfall downwind of St Louis also showed increases blamed on urban effects (Brazel and Quatrocchi 2005). Not only did 15% more rainfall occur on the east of the southeast side of St Louis, but also storms with more than 25 mm of rain were 50% more frequent. The historical precipitation records show significant increases which parallel the growth of Mexico City (Jauregui and Romales 1996). For example, the frequency of rain periods greater than 20 mm h^{-1} has increased, especially between 1200 and 1800, in the warmer part of the day. While there is no evidence of an urban influence upwind of the city, increased rainfall downwind of the prevailing winds is clearly evident. The spatial distribution of rain shows a significant relation with the core of the UHI, where additional convective heating is available. However this relation is complex and non-linear, and there is considerable spatial variation within the urban area from storm to storm. Broadly similar results have been found for several other US cities, depending on the storm type. Urban enhancement of precipitation has the greatest effect on the strength and frequency of smaller rainstorms.

Bonan's (2002) summary describes weekly variations in urban rainfall associated with the urban cycle of pollutants. During the working week, assuming no major change in synoptic situation, pollutants accumulate in the atmosphere, reaching a maximum on Thursday, Friday, and Saturday. This can lead to increased rainfall toward the end of the working week and on the weekends, while Monday and Tuesday show little indication of rainfall enhancement over the city or downwind.

Lowry (1998) argues that, since precipitation is discontinuous, there is considerable uncertainty about urban effects. Complications include the role of local

Figure 7.10 TRMM satellite summary of downwind precipitation enhancement for five cities in the southern United States. The arrow indicates the prevailing wind, and the cloud the enhanced area of precipitation. (After Shepherd *et al.* 2002) Included for comparison is St Louis from earlier ground-based analysis. (From Brazel and Quatrocchi 2005)

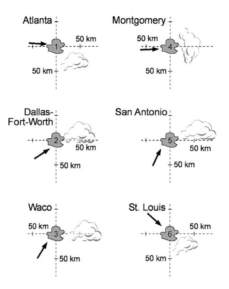

terrain, adjacent water bodies and shorelines, problems associated with point measurements being unrepresentative, inconsistent spatial distribution analysis, and lack of knowledge about urban interactions with different synoptic weather types. However, more recent evaluations of precipitation distributions downwind of urban areas using satellites suggest urban influences are real. For example, Shepherd *et al.* (2002) used the precipitation radar on the Tropical Rainfall Measuring Mission (TRMM) satellite to assess the spatial distribution of rainfall over three years for five cities in the southern United States. They found an average increase in warm-season monthly rainfall rates of 28% at locations 30 to 60 km downwind of the urban center, compared to upwind areas (Figure 7.10). Maximum rainfall rates downwind increased by 48 to 116%.

7.7 Effects of air pollution

Throughout this chapter, there has been regular referral to air quality in cities and its relation to the components of urban climate. GEO (2003) provides a summary of major air pollution problems in cities around the world. While air pollution problems have abated somewhat in developed cities, such as in the United States and Europe, they are increasingly serious in rapidly developing cities, such as in India, China, and Southeast Asia. The main sources are industry and the ever-increasing traffic fleet.

There is a wide range of air pollutants in the urban environment. Gases are mainly created through the burning of fossil fuels and evaporation from liquids, and are emitted from the surface to the atmosphere as primary pollutants. Examples include sulfur dioxide, nitrogen oxides, carbon monoxide, and volatile organic hydrocarbons, which are mainly of urban and regional concern, and

Table 7.3 *Major interactions between air pollutants (mainly fine particulates and ozone) and urban climate components*

Interaction	Comment or impact
Scatter shortwave energy	Reduction of available SWR for surface absorption; increased diffuse SWR, reduced UHI?
Absorb shortwave energy	Reduction of available SWR for surface absorption; reduced UHI?
Absorb longwave radiation	Increased UHI
Increased nuclei for cloud formation and precipitation	Increased urban storms and rainfall; increased fog
Inclusion in UHIC	Reduced visibility; increased pollution episodes
Inclusion in urban plume downwind	Transfer of pollution impacts to rural areas
Photochemical reactions	Increased atmospheric temperature; reduction of available SWR

greenhouse gases such as carbon dioxide and methane. Gases can also be created through secondary reactions in the atmosphere, between primary gases and atmospheric components such as ultraviolet radiation and moisture. Tropospheric ozone, the major component of photochemical smog is the most prevalent example.

A second category of pollutant is particulate matter. Particles in the atmosphere exist across a wide range of sizes. Sources range from windblown dust from bare soil surfaces, to sea spray, to volcanic emissions, to biomass and fossil fuel burning. Of most concern to the respiratory health and welfare of the urban population is particulate matter below 10 μm in diameter (PM10) and that below 2.5 μm in diameter (PM2.5). The latter represents the size range called fine particulates.

It is PM10, PM2.5, and photochemical smog that cause the greatest problems in urban atmospheres today, and also have the strongest influence on urban climate. Table 7.3 lists the major interactions of air pollutants with components that create the urban climate. In Mexico City, the impact of air pollution (mainly ozone and fine particulate matter) on incoming all-wavelength radiation is significant, with an average reduction of 21.6% in the city compared to a rural location (Jauregui and Luyando 1999). There is little difference in reduction between the dry and wet seasons. Although rainfall will clean the atmosphere in the wet season, the increased humidity and hygroscopic action create haze which enhances short-wave scattering. Attenuation tends to be highest during weekdays (35%) and least on weekends (25%), reflecting a major change in the traffic regime.

Shortwave radiation depletion in Mexico City is related positively to relative humidity and air pollution concentrations, but inversely to wind speed and temperature (Jauregui and Luyando 1999). Periods of maximum reduction occur when winds are less than $2\,\mathrm{m\,s^{-1}}$ and relative humidity is high. These conditions mainly occur around midday. High temperatures reduce the relative humidity, and higher wind speeds mix and disperse the pollutants. The overall reduction in shortwave radiation has created lower mean annual maximum temperatures in rural areas on the downwind side of the urban center, associated with the weak transport of the pollution layer. These conditions are broadly representative of many other cities around the globe, but in larger cities such as Mexico City, especially in the most built-up areas, the UHI influence is dominant over any cooling created by the air pollution layer.

Air pollutants may either enhance or diminish urban climate development and impact (see Table 7.3). During the day, when fine particulates and ozone scatter and absorb incoming shortwave radiation, its availability at ground level is reduced. However, through absorption, particulates and ozone also increase the radiative heating of the boundary layer (Arnfield 2000). This increases stability and reduces sensible heat flux, but potentially also increases downward longwave radiation toward the surface. At night, warm urban surfaces release excess longwave radiation to the atmosphere, which can be absorbed by the atmospheric pollutants, and then re-released back toward the surface, enhancing the UHI.

7.8 Remote sensing and the UHI

The use of remote sensing to assess the spatial and temporal variations in UHI and other urban characteristics has some major benefits in the study of urban climates. Remotely sensed information has been used to define land use variations, to evaluate the strength of the UHI, and to help understand the process of urban surface–atmospheric exchanges. The current state of remote sensing in urban climate studies is extensively reviewed by Voogt and Oke (2003). There is considerable potential for detailed UHI evaluation (Figure 7.6), but also some major complications to overcome.

Infrared thermal imagery from various satellite sensors (ATLAS, LANDSAT, AVHRR and so forth) is used for surface heat-release assessment, in wavelengths within the 8–12 μm band (the "atmospheric window" in the infrared, where longwave radiation under clear skies escapes to space with minimal gaseous absorption). The visual result, presented as a series of squares (or pixels), can provide an attractive picture of the variation in surface temperatures in an urban environment. An example is presented in Figure 7.11 (Plate 6) for Phoenix, Arizona. The thermal pattern in Figure 7.11 suggests that there are urban–rural temperature differences on the order of 5 °C, depending on the surface.

Thus major advantages are that spatial distributions can be assessed, and areas of "hot" or "cold" thermal emissions detected (Roth et al. 1989). A series of satellite images taken over time can allow an assessment of how diurnal changes

Figure 7.11 This nighttime image of the Phoenix, Arizona, metropolitan area is from the Advanced Spaceborne Thermal Emission and Reflection Radiometer (ASTER) for October 3, 2003 at approximately 22:39:00 local time. A color ramp has been applied to highlight relatively hot and cool surfaces; the corresponding temperature scale is in °C. Urban materials, such as the asphalt street grid and Sky Harbor Airport exhibit relatively high surface temperatures, while heavily vegetated regions (mesic neighborhoods, riparian zones, agricultural fields) are relatively cool. Sparsely vegetated mountain ranges comprised of exposed massive bedrock and block talus also have relatively high surface temperatures compared to other natural surfaces. North is towards the top of the image. (Image source: William L. Stefanov, Johnson Image Analysis Laboratory, NASA Johnson Space Center, Houston, TX. Original unprocessed ASTER data (AST_08 surface temperature product) from NASA/ERSDAC) For color version see Plate 6.

of temperature vary in relation to the surface energy balance. However, the use of remote sensing has also created important complications to the determination and interpretation of UHI.

The remote sensor is not measuring the urban–rural contrast in air temperature (AUHI), but the "upwelling thermal radiance" (Voogt and Oke 2003) from the surface structures at ground level (SUHI). Thus the satellite "sees" a different temperature regime than a ground-based sensor, and integration between the two can be difficult. Also, SUHI is greatest during the day, while under ideal conditions (clear skies, high pressure, light winds) AUHI is greatest at night. During the day, land-cover distribution is very important to SUHI. At night the importance of land cover diminishes in the development of AUHI (Roth *et al.* 1989), and surface themal release becomes more important. For example, for Huntsville (Alabama, USA) irradiance from surface land uses during the day ranged from $70\,\mathrm{W\,m^{-2}}$ for the commercial district to $55\,\mathrm{W\,m^{-2}}$ for water and rural areas, based on 5-meter resolution from ATLAS. At night, except for the water surface, irradiance from all surfaces ranged from 43 to $47\,\mathrm{W\,m^{-2}}$ (Lo *et al.* 1997).

A major part of the reason for this difference is the complexity of heat transfer between a solid surface and a fluid atmosphere. Radiative and convective heat transfers occur initially from a surface layer into a volume containing the atmosphere. Under strongly convective conditions, the volume will be very large, and the heat flux will mix and dilute rapidly, with little resultant impact on atmospheric temperature. Under stable atmospheric conditions at night, with clear skies and very light winds, the volume can be much smaller, but heat transfer will still be diluted. If there is a near-surface inversion, the UCL layer can be decoupled from the UBL above (Figure 7.5), and mixing to the upper layers may be prevented.

The thermal sensor does not see the entire urban surface. Rather, depending on surface structure and morphology, the angle of the Sun, and the sensor viewing geometry, the thermal emissions detected can be spatially limited. Depending on the three-dimensional structure of the urban landscape, the sensor may only see rooftops, treetops, flat parking lots, etc. While the measurements from the satellite sensor should be more representative of horizontal surfaces than the vertical sides of buildings (Nichol 1996), the thermal results are likely to be unrepresentative of the full urban surface.

Table 7.4 lists many of the important considerations that must be addressed for accurate use of remote sensing data. Corrections for the atmospheric interferences and surface structure are essential or considerable interpretation error may result. Matching the scale of the thermal observations with that of the surface features provides much more representative results. For example, AVHRR satellite imagery has a coarse resolution of about 1.1 km. Studies trying to explain SUHI and AUHI distributions using AVHRR have not been very successful. However, Nichol (1996) used LANDSAT's thematic mapper, with a

Table 7.4 *Important considerations in the accurate use of satellite thermal imagery for urban climate analysis (after Roth* et al. *1989)*

Resolution of satellite sensor (and image)
Noise equivalent associated with wavelengths used
Angle of satellite image collection (angle of scan, geometric distortion)
Weather (cloud, wind, etc.)
Impact of intervening atmosphere (scattering, absorption, refraction, re-radiation)
Image enhancement needs and procedures
Emittance or emissivity of surface
Geometry of surface (three-dimensional roughness)
Surface land use
Surface directional effects (anisotropy)
Size and orientation of urban canyons
Surface layering (how much of the actual surface is obscured)

resolution of 120 m, to find a good relationship between SUHI and AUHI in Singapore. While surface temperatures were higher and had greater variability compared to air temperatures, the diurnal air temperature pattern did follow the surface temperature changes. Figure 7.11 (Plate 6) demonstrates the resolution detail a high-quality sensor can provide.

Voogt and Oke (1997) attempted to overcome the problems created by the use of SUHI to define AUHI, by defining a "complete surface temperature" between the surface and the air. Using a combination of ground and airborne thermal sensors in Vancouver, Canada, they estimated the impacts of emissions from all surfaces and all directions, removing the inherent satellite viewing biases described earlier. The result was an area-weighted overall temperature (Tc) that included inputs from rooftops, sides of buildings, under trees, etc. Tc must be calculated, and is not directly observable, and depending on the urban environment, can be very complicated to determine. Voogt and Oke suggested that hemispheric measurements of outgoing longwave radiation might be a reasonable surrogate for Tc. While Tc did provide a significant improvement compared to SUHI in the description of AUHI, especially around noon, there were still important differences between the two values.

Both Roth *et al.* (1989) and Voogt and Oke (2003) suggest that, as a result of the complications mentioned above, many different types of UHI can be defined (Figure 7.6). While Voogt and Oke in 1997 stated that the linkages between air and surface temperatures are far too complex to define properly, six years later, their viewpoint was a little more positive (Voogt and Oke 2003). While qualitative descriptions and simple correlations based on remote sensing are still very much the norm, there have been some improvements. Advances in satellite

sensor technology have created major improvements in resolution and detail of
the infrared images, but not yet to the level of small-scale variations (roof
geometry for example). There has been better use of multi-directional emissions
to improve the accuracy of thermal results, but these are hampered by poor
emissivity information from various surfaces. Progress on finding a relationship
between SUHI and AUHI suggests that if meso-scale advection is reduced,
especially at night, the two values correlate better. Much more research needs
to be completed in this area. Voogt and Oke (2003) expect better resolution
thermal scanners in the future, higher quality and less expensive surface representations, and the development of a portable high-resolution thermal scanner
that can be used at the surface to supplement the satellite data.

7.9 Mitigation of the UHI

Mitigation measures designed to reduce UHI and its impacts have been focused
on the use of trees and green space, and reducing absorption of solar radiation by
increasing surface albedos (Bonan 2002; McKendry 2003). The research completed in Göteborg, Sweden, clearly established the benefits of urban parks in
reducing the overall UHI (Upmanis *et al.* 1998). In general, the larger the park,
the stronger the temperature difference (up to 6 °C) with the urban center. The
thermal boundary between the park and the city was occasionally sharp (within
200 m) but more often the transition occurred over several hundred meters
(0.3 to 0.4 °C per 100 m). Park influences on the temperature of the built-up
area can extend to over 1 km, depending on size, but even small parks can have
an influence (to a few tens of meters from the park boundary). Within the larger
parks in particular, cooling rates at night can vary spatially, probably associated
with variations in Ψs and evaporation rates.

Trees and green spaces in cities help shade buildings, reducing surface
temperatures and the need for air conditioning. They provide a surface for
interception and the soil allows infiltration of rainfall. Trees can reduce pollution
by physically cleaning the atmosphere, and they act as buffers to wind. Through
transpiration they increase urban evaporative cooling. Trees and parks also
provide a very pleasant social and cultural environment for the people living
and working in the city.

There are some disadvantages, however, which must be accepted. Trees and
parks take up room, potentially limiting development in parts of the city. They
require surface preparation and regular maintenance, which adds costs to local
government and individual budgets. Some trees release natural hydrocarbons,
which can add to the mix of anthropogenic pollutants that create photochemical
smog. If UHI temperatures, and wind speeds, are reduced, then UBL inversion
levels over the city could be reduced, potentially creating more frequent air pollution episodes. For cities in the northern parts of the Northern Hemisphere, the UHI
may have some important benefits during winter, and reduction may not be desired.

In parts of Los Angeles (USA), cool community strategies have been introduced to provide a more comfortable climate for residents (McKendry 2003). Rooftops and flat areas have been covered with lighter colors, enhancing the albedo and reducing absorption. Trees have been planted to enhance shading. Model results for Los Angeles indicate that if these simple measures were adopted for the whole city, the UHI could be reduced by 3 °C, ozone exceedences reduced by 12%, and the use of air conditioning reduced by half. Cities located around the Mediterranean Sea in Europe have been using white buildings to assist cooling for centuries. Similar simple approaches in other cities around the world would have considerable overall benefit to the climate and the urban community.

7.10 Chapter summary

The urban environment has a major influence on local and regional climate, which, given the growth of cities now and into the future, is creating an increasing influence on the global scale. For example, cities are major contributors to the Asian Brown Cloud, a conglomerate of pollutants that covers much of South and Southeast Asia (GEO 2003). In the springtime, this material can be transported via the middle troposphere across the Pacific to the west coast of North America.

Arnfield (2003) provides a useful summary of the relationship between UHI in cities and climate, which is reproduced in Table 7.5. However, unique aspects of altitude, geographical location, and urban morphology apply to each city, and influence the spatial and temporal variability of urban climate features, especially the UHI. Included in Table 7.5 as well are general comments about the other impacts that cities can have on the local and regional climate, which summarizes the discussion above.

7.11 Examples of urban websites

CEROI, Cities Environment Reports on the Internet, summarizes the state of the environment in different cities around the world, www.ceroi.net.

CSRUR, Consortium for the Study of Rapidly Urbanizing Regions, investigates urban climate impacts at the meso, local, and micro scales, www.urbanheat.org.

GHCC, Global Hydrology and Climate Center, US National Aeronautics and Space Administration (NASA) provides a useful range of information on urban climate and hydrological problems, www.ghcc.msfc.nasa.gov/ghcc_home.html.

IAUC, International Association for Urban Climate, provides information on urban climate research and activities, including major international conferences, www.urban-climate.org.

LTER, Long Term Ecological Research Network, a major international project investigating ecological processes over long temporal and broad spatial scales, including urban impacts in the environment, www.lternet.edu.

Table 7.5   *General summary of the impacts of cities on the local and regional climate, based on Arnfield (2003) and the discussion in this chapter*

Air urban heat island strength
Reduced by increasing wind speed
Reduced by increasing cloud cover
Greatest during anticyclonic conditions
Best developed in summer
Greatest at night
Increases with increasing urban population, size, building density
UHI does not always exist
UCI can occur during the day and in parks
Rates of heating and cooling in the city are generally lower than the
   countryside
Cumulative impacts of urban canyons are important
UHI enhances convection and cloud formation

Urban effects on other parameters
Wind speed is reduced through friction and turbulence
Light winds may be increased though canyon channeling
Coriolis force is reduced, causing changes in wind direction
Urban canyons change the wind direction through channeling
Urban canyon vortices are important to energy exchanges and
   cross-canyon flow
Sensible heat and storage heat dominate the energy balance in canyons
Evaporation is minimal, runoff is maximized on urban surfaces
Actual humidity is lower, especially in afternoon/evening
Added particulate matter enhances cloud formation
Extra convection and turbulence enhance downwind precipitation

STADTKLIMA, a German-based urban climate website, provides a very useful bibliography on urban climate, www.urbanclimate.net.

UNEP, United Nations Environment Programme, summarizes the state of the world's cities and provides web links to urban projects, www.unep.org/themes/urban.

7.12 References

Arnfield, A., 2000. Micro- and mesoclimatology. *Progress in Physical Geography*, **24**, 261–271.
   2003. Two decades of urban climate research: a review of turbulence exchanges of energy
      and water and the urban heat island. *International Journal of Climatology*, **23**, 1–16.
Arnfield, A. and Mills, G., 1994a. An analysis of circulation characteristics and energy budget
      of a dry, asymmetric, east-west urban canyon: I. Circulation characteristics. *International
      Journal of Climatology*, **14**, 119–134.

1994b. An analysis of circulation characteristics and energy budget of a dry, asymmetric, east-west urban canyon: II. Energy budget. *International Journal of Climatology*, **14**, 239–261.

Bonan, G., 2002. Urban ecosystems. Chapter 14 in *Ecological Climatology Concepts and Applications*. Cambridge: Cambridge University Press, pp. 547–586.

Brazel, A. and Quatrocchi, D., 2005. Urban climatology. In J. E. Oliver, ed., *Encyclopedia of World Climatology*. Dordrecht: Springer.

Brazel, A., Selover, N., Vose, R. and Heisler, G., 2000. The tale of two climates – Baltimore and Phoenix urban LTER sites. *Climate Research*, **15**, 123–135.

Bridgman, H., Warner, R. and Dodson, J., 1995. *Urban Biophysical Environments*. Melbourne: Oxford University Press.

Eliasson, I., 1994. Urban-suburban-rural air temperature differences related to street geometry. *Physical Geography*, **15**, 1–22.

1996. Urban nocturnal temperatures, street geometry and land use. *Atmospheric Environment*, **30**, 379–382.

Eliasson, I. and Holmer, B., 1990. Urban heat island circulation in Göteborg, Sweden. *Theoretical and Applied Climatology*, **42**, 197–196.

GEO, 2003. *Global Environmental Outlook 2003*. United Nations Environment Programme, Geneva, www.unep.org/GEO/geo3.

Grimmond, C. S. B. and Oke, T. R., 1999. Aerodynamic properties of urban areas derived from analysis of surface form. *Journal of Applied Meteorology*, **38**, 1262–1292.

2002. Turbulent heat fluxes in urban areas: Observations and a local-scale urban meteorological parameterization scheme (LUMPS). *Journal of Applied Meteorology*, **41**, 792–810.

Haeger-Eugensson, M. and Holmer, B., 1999. Advection caused by the urban heat island circulation as a regulating factor on the nocturnal heat island. *International Journal of Climatology*, **19**, 975–988.

Holmer, B. and Eliasson, I., 1999. Urban-rural vapor-pressure differences and their role in the development of urban heat islands. *International Journal of Climatology*, **19**, 989–1010.

Howard, L., 1833. *Climate of London*, 1st edn.

Jauregui, E., 1997. Heat island development in Mexico City. *Atmospheric Environment*, **31**, 3821–3832.

Jauregui, E. and Luyando, E., 1999. Global radiation attenuation by air pollution and its effects on the thermal climate in Mexico City. *International Journal of Climatology*, **19**, 683–694.

Jauregui, E. and Romales, E., 1996. Urban effects on convective precipitation in Mexico City. *Atmospheric Environment*, **30**, 3383–3389.

Jauregui, E. and Tejeda, A., 1997. Urban-rural humidity contrasts in Mexico City. *International Journal of Climatology*, **17**, 187–196.

Johnson, G. and Hunter, L., 1999. Some insights into typical urban canyon air flows. *Atmospheric Environment*, **33**, 3991–3999.

Landsberg, H. E., 1981. *The Urban Climate*. Int. Geophys. Series 28. New York: Academic Press.

Lo, C., Quattrochi, D. and Luvall, J., 1997. Application of high-resolution thermal infrared remote sensing and GIS to assess the urban heat island. *International Journal of Climatology*, **18**, 287–304.

Lowry, W. P., 1998. Urban effects on precipitation amount. *Progress in Physical Geography*, **22**, 477–520.

Masson, V., 2000. A physically based scheme for the urban energy budget in atmospheric models. *Boundary-Layer Meteorology*, **94**, 357–397.

McKendry, I., 2003. Applied climatology. *Progress in Physical Geography*, **27**, 597–606.

Morris, C. and Simmonds, I., 2000. Associations between varying magnitudes of the urban heat island and the synoptic climatology on Melbourne, Australia. *International Journal of Climatology*, **20**, 1931–1954.

Morris, C., Simmonds, I. and Plummer, N., 2001. Quantification of the influences of wind and cloud on the nocturnal heat island of a large city. *Journal of Applied Meteorology*, **40**, 169–182.

Nakamura, Y. and Oke, T., 1988. Wind, temperature and stability conditions in an east-west oriented urban canyon. *Atmospheric Environment*, **22**, 2691–2700.

Nichol, J., 1996. High resolution surface temperature patterns related to urban morphology in a tropical city: a satellite-based study. *Journal of Applied Meteorology*, **35**, 135–146.

Oke, T. R., 1981. Canyon geometry and the nocturnal heat island: comparison of scale model and field observations. *Journal of Climatology*, **1**, 237–254.

1982. The energetic basis of the urban heat island. *Quarterly Journal of the Royal Meteorological Society*, **108**, 1–24.

1984. Methods in urban climatology. In *Applied Climatology*, Zürcher Geographische Schriften, **14**, 19–29.

1987. *Boundary Layer Climates*. London: Methuen.

1995. The heat island of the urban boundary layer: characteristics, causes and effects. In J. E. Cermak *et al.*, eds., *Wind Climate in Cities*, Dordrecht: Kluwer, pp. 81–107.

1997. Urban climates and global environmental change. In R. D. Thompson and A. Perry, eds., *Applied Climatology: Principles and Practice*. London: Routledge, pp. 273–287.

Oke, T. R. and Grimmond, C. S. B., 2002. Urban-rural energy balance differences. European COST 715 [Co-operation in the Field of Scientific and Technical Research] Surface Energy Budget in Urban Areas. *Proc. COST-715 Expert Meeting, Antwerp, Belgium, 12 April 2000*. European Commission, Report EUR 19447, 52–65.

Oke, T. R., Spronken-Smith, R., Jauregui, E. and Grimmond, C. S. B., 1999. Recent energy balance observations in Mexico City. *Atmospheric Environment*, **33**, 3919–3930.

Piringer, M., Grimmond, C. S. B., Joffre, S. M., *et al.*, 2002. Investigating the surface energy balance in urban areas – recent advances and future needs. *Water, Air and Soil Pollution: Focus*, **2**, 1–16.

Plate, E., 1995. Urban climates and urban climate modelling: an introduction. In J. E. Cermak *et al.*, eds., *Wind Climate in Cities*. Dordrecht: Kluwer, pp. 23–29.

Roth, M., Oke, T. R. and Emery, W. J., 1989. Satellite derived urban heat islands from three coastal cities and the utilization of such data in urban climatology. *International Journal of Remote Sensing*, **10**, 1699–1720.

Shephard, J., Pierce, H. and Negri, A., 2002. Rainfall modification by major urban areas: observations from spaceborne rain radar on the TRIMM satellite. *Journal of Applied Meteorology*, **41**, 689–701.

Spronken-Smith, R. A. and Oke, T. R., 1999. Scale modelling of nocturnal cooling in urban parks. *Boundary-Layer Meteorology*, **93**, 287–312.

Upmanis, H., Eliasson, I. and Linquist, S., 1998. The influence of green areas on nocturnal temperatures in a high latitude city (Goteburg, Sweden). *International Journal of Climatology*, **18**, 681–700.

Voogt, J. A. and Oke, T. R., 1997. Complete urban surface temperatures. *Journal of Applied Meteorology*, **36**, 1117–1132.

    2003. Thermal remote sensing of urban environments. *Remote Sensing of the Environment*, **86**, 370–384.

World Resources, 1996. *The Urban Environment*. A combined publication of the World Resources Institute, United Nations Environment Programme, United Nations Development Programme, and The World Bank, Oxford: Oxford University Press.

Chapter 8
Human response to climate change

8.1 Introduction

How human beings and societal organizations respond to climate change is an area of considerable debate. McGovern (1991) points out that unambiguous cases of climate impact are rare, and therefore simple correlations between climate change and human activities are not valid. Climate change and variability are included in a range of impacts that can influence human activities, the way humans live, and how lifestyle decisions are made. Other aspects include the level of technological advancement, societal structure, level of economic development, the influence of warfare, the importance of religion, and so forth. Detection of the role of climate variations and stresses in cultural reactions can be very difficult if some of these other factors dominate the lifestyle and decision-making process. It is, however, essential to consider climate along with these other impacts influencing society, if a complete understanding of cultural reaction is to emerge.

It is impossible in one chapter of a book to explore all the potential relationships between climate and human activity. We therefore focus on the period since about AD 500 (therefore all dates referred to are AD) when human activities were starting to dominate natural processes in parts of the world. At the beginning of this period, agriculture began to define cultural landscapes, creating a more sedentary lifestyle, and enhancing an increase in population. Humans also began to have a bigger impact on the local environment, through activities such as forest clearing, draining and filling wetlands, irrigation, and the creation of dams. Our examples come from Europe and the North Atlantic Ocean area.

Figure 8.1 provides a reconstruction of NH temperatures since 1000, based on evaluation from a range of proxy and measurement data (see Chapter 6 and Figure 6.6 for details). The data show considerable variation, especially since the fourteenth century. This chapter provides three examples of society response and adaptation to climate variations and change during the time period shown in Figure 8.1. Section 8.2 describes the Viking settlements in Greenland from 800 to 1450 as an example of the migration period. Section 8.3 discusses adaptation and crisis in Europe during two periods, the fourteenth century, at the beginning

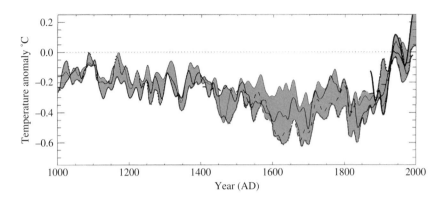

Figure 8.1 "Envelope" of reconstructions of NH temperature variations between AD 1000 and 2000, from five interpretations of proxy and measured data on 50-year time scales. The data are calculated as anomalies from the 1961–90 global average. The envelope obscures the details from each of the interpretations, but emphasizes the temperature variations over time. The greatest range in the estimates occurs in the middle of the Little Ice Age, from about 1550 to 1800. (Adapted from Jones *et al.* 2001; for further details see Figure 6.6)

of a significant climate change period; and the seventeenth century, when colder, stormier conditions were dominant. Section 8.4, the essay by Changnon, explores the relationship between climate variations and economics in the United States in modern times.

Diaz *et al.* (2001) emphasize that one of the fundamental impacts of climate change is on food production. Over the short term, over perhaps one to five years, climate variations can disrupt food supplies and create famine, causing considerable social unrest, until the society adjusts or the climate improves again. Over the long term, perhaps 10 years or more, climate change can threaten the viability of civilizations through loss of food and water resources, creating the need for radical change, or even leading to the breakdown of the society (Burroughs 1997). These climate impacts can be exacerbated by human activity, and are uniform neither spatially nor temporally across the globe. Much is still unclear about how and to what degree both large and small societies are vulnerable to climate changes and variations.

8.2 The Viking settlements in Greenland, AD 800–1450

8.2.1 Introduction

Between 800 and 1450 was a period of major change and movement in populations around the globe, encompassing the rise and fall of some great civilizations. The Mayan Classic period on the Yucatan Peninsula in Mexico occurred between 300 and 900, followed by the Toltecs in the tenth to twelfth centuries. In Cambodia, the Khmer Empire was founded in the ninth century and peaked during 1100 with the building of Angkor Wat. The Koryo Dynasty ruled in Korea between 935 and 1250. Germanic and Slavic tribes moved across Europe between 500 and 900, finally settling in Britain, Italy, and what is now Eastern Europe. In the thirteenth century, the Mongols conquered all of China and made major incursions across Asia and into Eastern Europe. Two great ocean migrations occurred during this period. The Polynesians sailed from Samoa and

Tonga into the central Pacific, populating major island groups from Hawaii to Easter Island to New Zealand. The Vikings explored the North Atlantic, settling Iceland, Greenland, and briefly, the east coast of what is now Canada (in around 1000).

Details of the climate and its variations in the North Atlantic during this period come from a variety of proxy sources (see Section 6.2). A good example is sediment cores from Igaliku Fjord, Southern Greenland, which provide diatom assemblies (Jensen *et al*. 2004). The MWP, a period of warmer, drier climate, occurred between 800 and 1250 in the North Atlantic area, with scattered cooling events between 960 and 1140. Between 1250 and 1550 occurred a period of transitional cooling, and the development of stormier, wetter conditions. This culminated in the LIA, between 1580 and 1850, a period of harsher and stormier conditions. Support also comes from Ogilvie *et al*. (2000), who review oxygen-18 evidence from ice cores from the Greenland ice cap. These show a period between 900 and 1200 with temperatures 1–2 °C warmer than subsequently.

Much has been published on the Viking movements westward across the North Atlantic (Figure 8.2), and the reasons for the rise and then demise of outlier colonies, such as in Greenland. Information from a wide range of fields, including archaeology, geology, paleoenvironmental, paleoclimatic, and historical records allows valid deductions about these colonies. Given the detail known about the change in climate after 1250, it is easy to blame harsher climatic conditions for the eventual loss of the Greenland settlements. But this explanation is much too simple and ignores the complexities of societal structure and interaction.

Figure 8.2 Viking migrations across the North Atlantic and the settlement sites in Greenland.

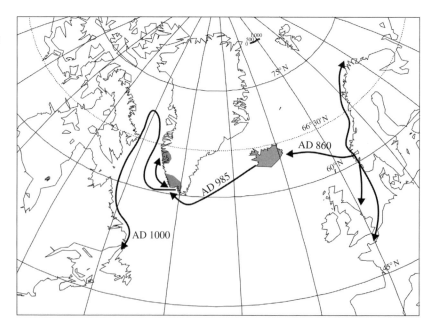

8.2.2 Migration

Reasons suggested for the beginnings of the Viking migrations are varied, and probably all have some relevance. McGovern (1991) and Vail (1998) suggest internal political conflict, population pressure, improved technological developments, scarcity of land, food and resources, and an improving climate, which encouraged decisions to venture further from home. The Vikings were hardy seafarers, who built excellent ships for long-distance ocean voyaging. Their navigation skills were the best available during this period of history. They seemed to have a spirit of adventure and discovery which allowed exploration of the unknown distant horizon. Figure 8.2 shows the geographical extent of movement between 800 and 1100 in the North Atlantic region.

By 874 the Vikings had settled Iceland, and in 985, Erik "The Red" Thorvaldsson formed a settlement on the southern tip of Greenland, termed the "eastern" settlement. Soon thereafter, a second, "western" settlement was formed, further north along the west coast (see Figure 8.2). By 1100, the population of both settlements had grown to between 4000 and 6000, and the colonies had an active trade with the Norwegian homeland. After about 1450, however, all contact with the Greenland colonies ceased. The major question is, why?; and then how important was climate change? In order to answer this question, an understanding of the Norwegian way of life and structure of society is needed, and how this society interacted with the environment.

8.2.3 Settlement and lifestyle in Greenland

The warmer, drier conditions of the MWP were ideal for sea transport. Analysis from the marine sediment cores indicates that this period was dominated by the warmer waters of the Atlantic Ocean, creating open water, limited sea ice, and few icebergs (Ogilvie *et al.* 2000; Jensen *et al.* 2004). The intensity of circulation was relatively low and dominated by zonal flow, creating a more stable climate regime. Polar–equatorial temperature gradients were weak, and the circumpolar vortex remained close to the Arctic (Lamb 1977). While there were periods of intermittent cyclonic activity and cooler conditions, overall there was little to hinder long-range migration across the North Atlantic.

The Vikings settled in the more protected areas on the inland edges of the fjords on the southwestern side of Greenland. Here, they found a landscape very much like home. These areas encouraged their agriculture-based lifestyle. The colonists imported and maintained cattle, sheep, goats, and pigs. The environment was marginal, even in good years, but initially there was enough vegetative support for survival and growth from year to year. McGovern and Pendikaris (2000) emphasize that the Greenland environment is not linked to the North Atlantic Current, as is that of Norway. As a result, even during the MWP period, summers were shorter, winters colder, and the vegetation much more fragile.

Unfortunately, the European hard-hooved animals began to have an immediate negative impact on the local environment (Vail 1998). Pastures were damaged or destroyed through erosion, for which there was no recovery. Even the use of seasonal grazing at higher elevations in summer brought no relief from this problem, and in fact it spread. Erosion interfered with the quality of and amount of hay, which was required for winter fodder for the animals. Inside protection in large sod barns, and enough fodder for up to nine months a year, was needed if the animals were to survive the harsh winters. However, the number of cattle and their quality were considered to be an indication of wealth in the Norwegian society, and therefore their presence was essential.

McGovern (1991) reviews the lifestyle of the colonists from archaeological records. The poorer colonists were unable to maintain a full agricultural lifestyle. However, the sea was readily accessible through the fjords, and had fish and seasonal migrations of seals. In spring, community seal hunts were held. Distances to the migrations were often a problem, and the hunts were dangerous, resulting in regular loss of life. Seals provided a range of essential items, from fur skins, to meat for food, to bone for fish hooks and sewing needles. Seals were not important to the upper strata of Greenland society until late in the settlement period.

Archaeological evidence described by McGovern (1991) and Vail (1998) shows that the Greenland colonies were organized according to wealth and status, based on the Norwegian model. There were severe economic inequalities. At the very bottom were the poor peasants, normally under feudal tenancy to the landowners. Their lifestyle was meager, work was hard, and they were required to contribute more than 20% of their limited products per year to the landowners and the church. The landowners controlled much of the wealth in the colonies, and had their own hierarchy. They were subservient to the Christian church. The church owned vast tracts of the best land and had the best cattle herds. The structure of the society was conservative, rigid, inflexible, and authoritarian. As an example, caribou in Greenland took on the status of deer in Europe. Caribou could only be hunted by the upper classes, and were not available to supplement food and equipment supplies of the peasants. While this helped preserve the caribou from extinction, a source of food was cut off for the working classes.

As the colonies became settled, and the landowners more wealthy, the church became more dominant in society. By the end of 1100, Norway was unified under one king, and under the Christian religion. The church, in combination with the landowners, controlled the trade and commercial economy. A big source of revenue was dried salted cod, called stockfish, which was used as a surrogate currency. According to McGovern and Pendikaris (2000) stockfish was an indicator of wealth, and could be traded for many things, including barley to brew beer. Another source of trade income was walrus ivory tusks and hides, and polar bear hides, obtained during hunting expeditions to the Arctic during summer. In 1167, the Greenland colony traded a live polar bear cub for a church bishop. Even under good climate conditions, these hunts were costly, both in men and in ships. A considerable amount of colonial income was traded for church construction material and trappings. At the height of colonial development, churches were the largest structures, with bells, stained glass windows, the finest wood, and other adornments similar to those in Europe.

8.2.4 Little Ice Age transition, 1250–1550

From about 1250 onwards in the North Atlantic region, a gradual cooling became more evident, with greater fluctuations in seasonal weather and climate from year to year and decade to decade (Figure 8.1). A significant increase in the number of cold seasons occurred (Figure 8.3), interspersed with occasional mild seasons, creating much higher variability in climate conditions then the previous four centuries. Vail (1998) lists results from the Greenland Ice Project, which show that cold years dominated in the periods 1308–18, 1324–9, 1343–62, and 1380–4, for example. Jensen *et al.* (2004) suggest that more stormy conditions with increased winter wind speeds occurred in a stronger general circulation. The Viking settlements at the heads of the fjords found little protection against the increasingly harsh climate conditions.

Increases in sea ice and iceberg frequency occurred as well. Lamb (1977) reports that sea ice appeared regularly off the coast of Iceland after 1002. The warmer Atlantic waters were increasingly replaced by colder Arctic water from the East Greenland current (Jensen *et al.* 2004) along the coast and in the fjords of Greenland.

The change in climate resulted in shorter summer growing seasons and considerably longer, harsher winters. Heavier rainfall and colder temperatures interfered with hay harvests, creating more winter seasons where animal starvation occurred. The eventual drop in average temperature by 2–3 °C compared to the MWP translated into a 50–60% loss in pasture productivity. Three to four centuries of

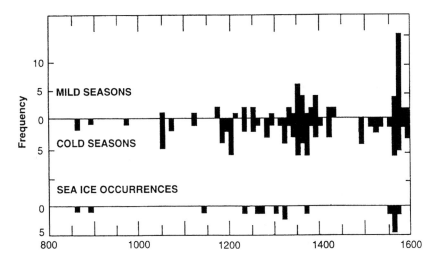

Figure 8.3 Occurrence of milder or colder seasons, and sea ice reports, from Iceland between AD 800 and 1600. (From Ogilvie *et al.* 2000)

erosion with limited recovery, plus demands from an increasing population, meant not enough food for everyone. The availability of seals as an alternative source of food and material, more important as the transitional period wore on, became highly erratic. Sealing became more dangerous, and there was increasing competition from the native Inuits, who were moving south as the climate became worse.

The stormier conditions and increased icebergs interfered with transatlantic trade. As McGovern (1991) states, succinctly, "... icebergs are bad for ships." The ships lost could not be replaced because there was no wood available. Nor could the increasing numbers of men lost, both at sea and on the Arctic summer hunts. Vail (1998) states that by this time the markets in Norway had begun to look south for goods, and the shipments from Greenland were not as profitable. Church doctrine may also have interfered with trade from Greenland.

How did the Greenland society structure respond to the threats to agriculture, the loss of trade income, and from the harsher climatic conditions? The rigid structure tried to maintain the status quo, and was unable to adjust to the change in environmental conditions. The tenants were the first to lose their tenuous lifestyle. Always marginal at best, the holdings did not survive the winters, and both animals and the population died. Without the income from the tenants, the landholders suffered, and without the income from both, the church lost its support, despite by this time controlling two-thirds of the best land in the colonies. The increasing loss of farmland through wind and water erosion, the changing economics of trade, and the increase in loss of men and boats all contributed to the demise of the colonies. The reduced interaction with Norway and Iceland also enhanced deterioration. The society could not adjust to the worsening conditions. The western settlement was lost around 1343, and by the middle of the fifteenth century, there was no more contact between the Greenland settlements and the outside world.

8.3 Climate change and adaptation in Europe during the Little Ice Age

This section focuses on climate variability and societal adaptation in Europe during the fourteenth century transition from MWP toward LIA, and during the seventeenth century, when the LIA was at its strongest. Information from a combination of proxy, economic and written data creates a picture of societal change and the role of climate (Pfister *et al.* 1996, and others).

8.3.1 The fourteeth century

Climate change and variability

Figure 8.4a provides details of fourteenth century isotope ratios from a GISP2 ice core in Greenland, a reasonable representation of the climate variability in Europe (Ogilvie *et al.* 2000). Figure 8.4b depicts the frequency of cold or

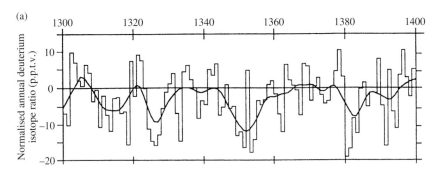

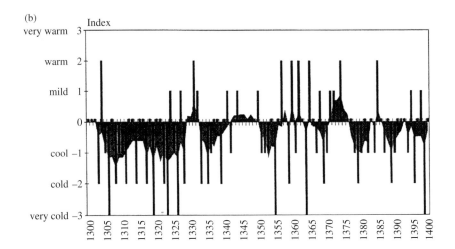

Figure 8.4 Fourteenth century: (a) Deuterium isotope ratio data from a GISP2 ice core in Greenland (Ogilvie *et al.* 2000) and (b) winter severity index for Europe (Pfister *et al.* 1996).

mild winters for the century (Pfister *et al.* 1996). Table 8.1 complements Figure 8.4 by highlighting years of climate variation and stress, and some of the major human activities in Europe. The fourteenth century was a period of increasing unrest across Europe, associated with a higher frequency of poor harvests and warfare.

Overall, the frequency of colder winters (on the order of 1–1.5 °C colder than today on average) and cool wet summers increased (Lamb 1977). Pfister *et al.* (1996) divided the century into four segments. Between 1303 and 1328, there was a series of cold, severe winters. The winter of 1322–3 was recorded as the coldest of any in the last 300 years. Many of the summers during this period were wetter than normal. From the end of the 1320s to around the mid 1350s, temperatures were "average" for the century. Western Europe saw wetter summers after about 1340, but eastern Europe was comparatively dry. While cold seasons dominated, there were no excessively severe winters. Between 1354 and 1375 winter temperatures were highly variable. Winters were very severe in 1354–5 and 1363–4, and lakes and rivers froze in northern Italy and on

Table 8.1   *Climate variations and major events in Europe during the fourteenth century*

Pre-1300–50	English wars with Scotland
1303–28	Erratic climate, mix of cold and warm years, increased flooding and severe storms
1304	Major sea flooding in coastal areas, north and west
1304–7	Very severe winters, most of Europe
1310	Floods in eastern Europe
1310–17	Higher level of volcanism around the globe
1314–17	Cool wet summers, poor harvests, famine
1315	Floods in eastern Europe
1316	Price of wheat in London eight times higher than 1313; harvests reduced by 60%
1316	Rain and mud stop Louis X (France) attacking Flanders
1319–22	Very poor harvests and very cold weather
1322–3	Worst cold intensity in winter in 300 years
Late 1320s to early 1350s	"Average" weather, cold winters dominate but no severe winters
1337	Beginning of Hundred Years War between England and France
1340–50	High level of volcanism around the globe
1348–50	Black Plague years, population of Europe reduced by 25–33%
1354–5	Very severe winter
1355–75	Highly variable winters
1359–60	Very mild winter
1362	Major sea flooding in coastal areas, north and west
1363–4	Very severe winter; rivers freeze in north Italy and the French Mediterranean coast
1375–6	Poor harvests in England
1376–1400	"Average" temperatures for the Little Ice Age but yearly fluctuations
1398–9	Very cold winter

Based on information from Lamb (1977), Pfister *et al.* (1996),
Burroughs (1997), and Brown (2001).

the French Mediterranean. One the other hand, the winter of 1359–60 was very mild and dry. Summer wetness extended across most of Europe during the later 1350s and early 1360s. Between 1376 and 1400, there was a return to "average LIA" winter temperatures, with milder fluctuations from year to year. Wetter summers dominated western Europe at the close of the century.

Lamb (1977), Pfister *et al.* (1996), and Brown (2001) describe several probable reasons for the increased harshness of climate (see further details in Section 6.3.4). Between 1280 and 1350, a period of sunspot reduction (the Wolf Minimum) occurred. While this was not as significant as the Maunder

Minimum at the height of the LIA (see Section 8.3.2), there is a correlation between harsher climates and low sunspot activity. The NAO (Section 2.2) most likely reversed compared to the MWP, enhancing high pressure and outbreaks of cold air from the north. The circumpolar vortex began to shift south as the equatorial–polar temperature gradient increased. This lead to increased periods of meridional circulation and blocking. Changes in circulation of deep bottom water in the Atlantic may have had significant impacts on sea surface temperature levels and distribution, forcing a change in the circulation pattern over Europe. There was also a considerable increase in volcanism across the globe, especially from 1310 to 1317 and 1340 to 1350, compared to earlier centuries. The resulting increase in particulate matter in the upper troposphere and lower stratosphere would reduce incoming shortwave radiation to the Earth's surface for several subsequent years. Brown (2001) emphasizes that teleconnections with ENSO (Section 2.8) were absent.

The increased storminess resulted in periods of heavy sea flooding in northern and western Europe. The flooding was enhanced by higher sea levels that existed during the MWP. Enhanced erosion, exacerbated by clearing land and draining marshy areas for agriculture, occurred. Flooding was especially prevalent up until 1340, but then drier conditions under greater anticyclonic activity occurred until about 1367.

Impacts on European society

Brown's (2001) detailed review establishes that worsening climate and weather conditions created major agricultural production problems in Europe. Immediate impacts occurred on the more marginal farming areas, at higher elevations for example, that were at least partially viable during the MWP. Because peasants tended to grow crops in strips, using open-field rotation, the risk of harvest failure during wet or cold periods was enhanced. The variability of climate meant harvests were erratic. For example, 1319 to 1322 were particularly bad years. In 1322, grain production was only about 40% of the best years earlier. 1325 and 1326 also produced very poor harvests. Michaelowa (2001) lists the relationship between prices of major agricultural products and weather variations. Wheat correlates negatively with summer and winter temperatures, but positively with summer rainfall. Hay correlates negatively with precipitation, and cattle production is reduced in lower autumn and winter temperatures. Rye, barley, and oats seem to survive better the climate variations without significant change to production. During the fourteenth century, most grain production in Europe was wheat.

Because the feudal system did not encourage innovation and investment, and life for the peasants meant severe poverty and day-to-day survival, a famine year created an immediate impact. There were no food reserves, and diets were poor anyway. Meat and fish were reserved for the rich. While the benefit of growing legumes to enhance soil quality was known, this was not common practice.

Starvation and malnutrition encouraged disease, not only in humans but also in animals. Farms and landholdings were abandoned. Reports of cannibalism abounded. In wet years, ergot fungus destroyed grain crops. Heat spells further encouraged mortality. In 1348–50 the Black Plague struck, killing 25–33% of the already weakened and decimated population. Major peasant uprisings and riots occurred at different times and locations across Europe. While many of these were brutally suppressed, they helped bring the eventual demise of the feudal system, beginning around 1350.

Brown (2001) states that, at least in England, there was some buffering for agriculture against poor climate conditions. A consolidation of land holdings created a more efficient production situation. British farmers constructed ditches for irrigation and water control, and drained marshes to increase arable land. The balance of production was shifted away from grains to sheep (imported from Spain), which survived the variations in climate better. However, the population of Britain peaked about 1310 and decreased thereafter, demonstrating that the country was not immune to disaster. Between 1319 and 1322, wheat prices in Winchester rose from about 3 shillings per quarter to about 17 shillings. Approximately 10–15% of the population died during these critical decades of the fourteenth century.

Wars and the attitude of the church did not help. With the demise of feudalism came the rise of nationalism. England tried to conquer Scotland beginning around 1280, and by 1350 was semi-successful. England also had continuing troubles with France, and the Hundred Years War began in 1337. As Brown (2001) states, it is always easier and less costly to focus on warfare than on the social and economic ills of society. The church blamed the human race in general, and God's wrath in particular, but its influence was on the wane.

The governments attempted to maintain trade and tried to establish currency exchange to replace barter. This worked reasonably well in some places, especially Italy. Wine, wool, and salt were important trade commodities. But increased flooding closed many salt mines and wine production was severely affected by the cold, wet weather. In England, the wine trade collapsed by 1327, partially due to poor quality and partially due to anti-government reactions.

On the more positive side, there were some developments and adaptations that allowed better survival in parts of Europe. Some of these are listed in Table 8.2. However, it seems clear that a worsening climate exacerbated the social tensions, both within and between countries. Therefore, overall adaptation was poor, and by the end of the fourteenth century, the population of Europe was considerably smaller, and economic conditions considerably worse, than those of a century earlier.

8.3.2 The seventeenth century

Between 1600 and 1700, the LIA culminated in the lowest temperatures in Europe since the end the Ice Age, some 11 000 years before. This was a period

Table 8.2   *Examples of adaptations and inventions, fourteenth century*

- Earliest known weather diary in Europe, England, 1337 (Walter Merle, Merton College)
- Introduction of the heavy plow
- Increase in industrial mechanization
- Ratio of grain yield to seed improves
- Recognition of the soil enhancement benefits of legumes and manure
- Development and use of the windmill
- Replacement of oxen by horses for haulage
- Mill power replaces hand threshing
- Blast furnace developed
- Artillery barrel developed
- Spindle wheel developed
- Mechanical clocks in operation by 1350
- Slow expansion of philosophy independent of the church
- Slow increase in academic freedom
- University of Prague founded, 1348
- University of Krakow founded, 1364
- Oxford University founded, 1379

Information from Brown (2001).

of severe hardship for the majority of the population in Europe, but was also a century of major advancements in knowledge and technology. Nationalism dominated the political structure of Europe, global exploration and colonization was a significant feature, and conflict was a major component of life. Copernicus, Newton, Galileo, Kepler, and others made great advancements in science and mathematics, as academia became an important component of society. But the general population suffered from lack of education and an ever-growing gap between the richer royalty and gentry, and the poverty-stricken peasant on the land. Adaptations to climate variability and change were better than in the fourteenth century in some parts of Europe, such as England, but in countries such as France were significantly delayed by government disinterest and lack of technological development.

Climate change and variability

Figures 8.1 and 8.5 demonstrate that temperatures during the seventeenth century were highly variable, and reached their lowest levels during the 1690s. Table 8.3 summarizes some of the major climatic, environmental, and societal events that occurred. Analysis by Lamb (1977), Grove (1988), Wanner *et al.* (1995), Bokwa *et al.* (2001) and others established that this very cold period was dominated by extreme winter and spring temperatures. Luterbacher (2001) states that springs in

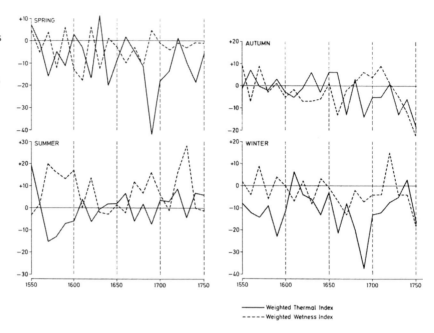

Figure 8.5 Weighted thermal and wetness indices for Switzerland for the seventeenth century. (Adapted from Grove 1988)

the last quarter of the century were the coldest in 500 years. These seasons were usually very dry. Summers often had cool, wet conditions, although occasionally extremely hot and dry weather (and drought) occurred. In England, severe winters occurred in 1600–9, the 1680s, and the 1690s. Fagan (2000) reports that the winter of 1683–4 was so severe that sea ice occurred along the coasts of the English Channel, and the ground froze to depths exceeding one meter. In Switzerland, the 1620s and 1640s were very cold (Figure 8.5), with glacial advances in the Alps. Particularly in the last quarter of the century, temperatures in most of Europe averaged 1.5 to 2.0 °C lower than the mid twentieth century.

Significant periods of severe convection creating stormy weather were also a major feature, leading to periods of high flood frequency in central Europe and along the coast. Precipitation frequency was greater in summer and lower in winter compared to present conditions. Comparisons of precipitation indices between 1675 and 1704 for England and Switzerland by Wanner *et al.* (1995) are representative of most of Europe, and establish that regional variations occurred:

- In winter, precipitation in England was average or slightly lower than normal up to 1692, and higher than average thereafter. According to Grove (1988), in the 1690s there were more snow days in London than ever recorded before or since. In Switzerland precipitation was considerably lower than average for the entire period except for 1682–4.
- Spring rain or snowfall in England was much higher than average except between 1696 and 1700. Spring was delayed in Scotland, and higher than normal snowfall, supported by enhanced lapse rates and instability, lasted into the early summer. Switzerland in spring was drier up until 1691, and wetter in most years until 1700.

Table 8.3   *Major climate and societal events in Western Europe in the seventeenth century*

1600	Eruption of Huanyaputina volcano in Peru, global impact
1600 to early 1630s	High variability in winter temperatures, excess of cool wet summers, cool dry springs
1600–16	Expanding glaciers, French Alps and Scandinavia
1607	Very cold winter, split tree trunks in England due to heavy frost
1608	Very cold winter, poor summer, very poor harvest
1618–48	Thirty Years War, France and England
1621	Late wine harvest
1625	Plague year for England, large number of deaths
1625–9	Cod harvest failure, Faroe Islands, ocean water too cold
1627–8	Late wine harvests
1630	Warm, dry summer, good wine harvest
1636–8	Very hot summers
1641–3	Global volcanic eruptions; glacial advances French Alps and Scandinavia
1643–1715	Louis XIV, The "Sun King" rules France; severe peasant hardship
1645–1715	Maunder Minimum; sunspots, aurora borealis rare
1645–50	Poor harvests
Early 1650s	Abundant wine harvests
Late 1650s	Poor harvests
1659–62	No snow in winter in England (Samuel Pepys diary)
1661	Sharp rise in wine prices
1665	Bubonic Plague kills 57 000 in London
1666	Extremely hot dry summer; Great Fire of London starts September 2
1666–9	Global volcanic eruptions
1670–1700	Warfare between France and other European nations
1673	Poor harvest
1675–1704	Very cold period but some warm years; cod harvest failure, Faroe Islands
1676	Severe winter, but good summer growing season
1678	Poor harvest
1680s	Generally good harvests
1683–4	Severe winter, but good summer growing season; River Thames in England freezes over; famine in France
1686	Very mild winter
1687–1703	Cold, wet summers; wine harvests late
1690s	Coldest period of the Little Ice Age, cold wet summers, bitterly cold winters
1692–3	Very poor harvests, severe storms, famine in France (10% loss of population)
1693–4	Global volcanic eruptions
1693–1700	Harvests fail in Scotland for 7 years out of 8
1697	Famine in Finland kills one-third of the population
1698–9	Global volcanic eruptions

Based on information from Lamb (1977), Grove (1988), Pfister *et al.* (1996), Burroughs (1997), Fagan (2000).

Figure 8.6 Schematic of differences in precipitation between the extremes of positive and negative NAO. H and L indicate areas where high and low pressure dominate. The arrow represents cold dry NE airflow in Europe during strong negative NAO. (Modified from Hurrell *et al.* 2003: permission from American Geophysical Union)

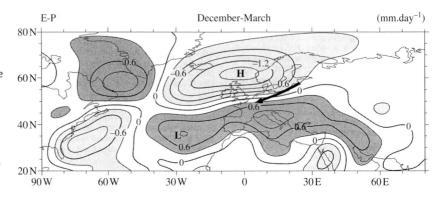

Between 1640 and 1659, drought occurred regularly in Eastern Europe, followed by floods in the 1660s. The winters in England between 1659 and 1662 were mild and dry, without snow.

The circulation variations reflected the dominance of the negative phase of the NAO, especially in the second half of the century. The schematic in Figure 8.6 suggests that high pressure dominated the North Atlantic and extended over northern Europe. Enhanced meridional circulation and blocking encouraged dry, cold north to northeast winds in winter to extend over all of Europe (Wanner *et al.* 1995). Depression tracks and a much weaker westerly flow were shifted well south, to the southern Mediterranean and North Africa. Luterbacher (2001) summarizes the major pressure and circulation features for other seasons during the 1675–1700 period. In spring, high pressure dominated Iceland and the area north of 55° N latitude. To the south, the equivalent of the Icelandic Low shifted toward Europe, creating unsettled, cold and snowy weather. In summer, wet and cool westerlies in west and central Europe prevailed, originating from lows which extended into the continent from further south than normal. In autumn, high pressure occurred north of 60° N latitude, with southern Europe and the Mediterranean under the influence of low pressure systems.

The changes in circulation strength leading to the extreme climate variations progressed from west to east, associated with an expanded CPV, and changes in length of the Rossby and standing waves in the mid-tropospheric westerlies (Lamb 1977; Bokwa *et al.* 2001). Thus, the coldest period of the LIA began in Britain about 1670, but did not reach Eastern Europe until about 1680.

As in the fourteenth century, the major reasons for the climate variations were natural, despite human advances in industry and technology (Wanner *et al.* 1995; Luterbacher 2001). The decrease in ocean temperatures in the North Atlantic reduced themohaline circulation, and assisted the stability of the high pressure systems, enhancing the negative NAO pattern. Sea ice was also extensive, particularly between 1620 and 1640, and 1670 and 1700. This interrupted the normal process of heat transfer from equator to poles. The Maunder Minimum between 1645 and 1715, when sunspots and the aurora borealis were almost absent, indicated a very quiet solar period, correlating with much colder temperatures. Volcanism occurred in several decades during the seventeenth century (see Table 8.3), increasing particulate concentrations in the upper troposphere and lower stratosphere, and reducing incoming shortwave radiation. Despite the cold, the advance of European glaciers was somewhat limited by the lack of winter precipitation. Diaz *et al.* (2001) suggest that ENSO and its teleconnections (see Section 2.8) were active from about 1680 onward.

Impacts on European society

The seventeenth century was one of great agricultural stress. According to Fagan (2000), at the beginning of the century, over 80% of the population of Europe was still operating at subsistence level. Despite a number of inventions and improvements in agriculture since the 1300s, the only places where major innovations were applied were Denmark, Holland, and Britain. The wide range of reports, publications, and economic summaries, which were then newly available in printed form, had only minimal use. As a result, similar to the fourteenth century, the poor tenants were at the mercy of short-term climate fluctuations. Most of Europe and the Mediterranean went into economic decline.

Aside from the cold, agricultural stresses occurred for a number of reasons (Diaz *et al.* 2001). Growing seasons were shorter by at least a month, growing altitudes were reduced by at least 200 meters, snowlines were lower, and summer wetness prevented crop harvests. In southeast Scotland, marginal lands were rapidly abandoned, as average temperatures fell by 2 to 2.5 °C in the Highlands. In France, average September temperatures in bad years fell by 2 to 3 °C, delaying harvests by up to 26 days (Lamb 1977). Grapes did not ripen well, and the wine quality was poor. Wheat prices soared to four times the level during the worst period in the thirteenth century. In the first half of the 1600s, purchasing power for builders in England fell to the lowest level in three centuries (Burroughs 1997), then began rising slowly. Not all of these income variations can be blamed on climate change. Inflation, economics, and political decision making (or lack of it) were also major factors. In the severest years of the 1690s, famine struck in France (10% of the population was lost in 1693) and in Finland (one-third of the population died in 1697).

England, however, was much better adapted to the climate variations than were most of the European countries, especially France. Here, major changes were made that removed the dependences of subsistence agriculture. Fagan (2000) and Michaelowa (2001) describe several of these innovations, which began in Denmark and Holland several decades earlier. There was a deliberate attempt to grow grass and other forage for animal feed. Crops were rotated, to include the planting of legumes to enrich the soil. New crops, turnips to enhance the winter survival of cattle and sheep, and flax and hops for industrial use, were introduced. Animals were used to provide manure on fields when crops were out of rotation, further enhancing the quality of the soil. Farmers diversified away from grain crops, which were too strongly affected by poor weather, into meat, leather and wool. Whole new areas of rich farmland were created by the draining of swampy areas in eastern England. Perhaps most importantly, the enclosure of land was introduced, allowing agricultural specialization and crop cycling, protecting the soil, and eliminating the centuries-old common-law access by the public. Farmers were now tenants to landlords, although relief from severe hardship did not occur immediately.

The tradition of subsistence farming meant that application of the new methods was slow, but by the end of the seventeenth century, most of English agriculture was reborn. In France, it took until a century later, mainly because of disinterest from the monarchy, graft and corruption in the government, the social chasm between the rich and the poor, and no support for innovation. Most income went to support the wars of Louis XIV. The French ignored the benefits of the potato, introduced from South America in 1570, and stubbornly maintained dependence on wheat and wine. As a result, famine and starvation occurred regularly. In Norway the population diversified into forestry, to overcome the problems of loss of crops and land from excess precipitation and glacier advances. By 1700, Norway had developed a major merchant fleet, exporting wood to all of Europe. But across most of Europe, despite major advances in trade, exploration, science, and cultivation, the education needed to survive periods of climate stress had not yet begun for the majority of the population.

8.4 ESSAY: Economic impacts of climate conditions in the United States

Stanley A. Changnon, *Illinois State Water Survey*

8.4.1 Introduction

This essay examines how the climate has affected the United States economy over the past 50 years and how these impacts have been changing in recent years and may change in the future. Comparison of current economic impacts of the nation's climate with the nation's economic status provides a basis for assessing the significance of climate impacts. Climate conditions that produced major economic impacts in the United States have fluctuated over time scales ranging from a decade to centuries. The 1930s and 1950s experienced the worst droughts of the past 200 years, and the 1960s and 1970s had the best Midwestern crop-weather conditions of the twentieth century (Thompson 1986). The United States experienced record high losses from numerous extremes during the 1990s, a situation that severely impacted the insurance

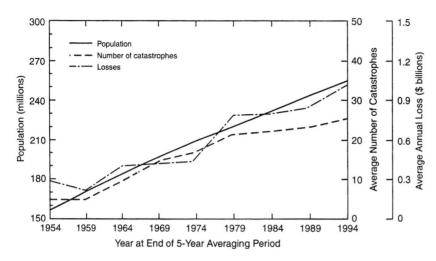

Figure 8.7 Time distribution of catastrophes that have caused 1997 equivalent losses of between $10 million and $100 million, in 5-year periods, and the US population. (Changnon *et al.* 1997: permission from American Meteorological Society)

industry and national and state governments (Changnon 1999a). These large losses led some to believe that the predicted global climate change with more extremes had begun and that the financial impacts would be severe.

Climate-sensitive activities in the United States experience both economic gains and losses, depending on the type and magnitude of the climate conditions. Good growing-season weather conditions bring high crop yields, whereas cold winters bring high heating costs. Major storms produce damage that translates into economic losses (Figure 8.7), but some individuals and institutions also experience financial gain in the aftermath of damaging events (Changnon 1996). Winners often are found later in the storm area and in unaffected regions because most extreme conditions do not cover much of the nation. For example, rebuilding homes after damaging hurricanes benefits the construction industry, and farmers producing crops in unaffected areas during a serious drought benefit from increased incomes as prices are driven upwards by crop losses. Seldom are the economic gains resulting from extremes assessed; and often when they are assessed, they are incomplete (National Academy of Sciences 1999). A few comprehensive impact-oriented studies of recent major events, such as the 1988 drought (Riebsame *et al.* 1991), Hurricane Andrew (Pielke 1995), and the Midwest flood of 1993 (Changnon 1996), found the value of the gains ranged from 30 to 55 percent of the losses.

The complex issue of economic impacts due to the climate has never been studied extensively. The NAS (1999) assessed the nation's economic impacts from natural hazards, noting the paucity of national efforts to collect loss and gain data systematically. However, increasing attention given to the economic impacts of climate during recent years by scientists and economists has led to the collection of data on several events and to studies defining many climate impacts. Nevertheless, many economic impacts are measured poorly, and much existing loss information is based on estimates rather than actual measurements.

This essay focuses on results that have carefully assessed the economic impacts from climate conditions during the past 50 years. Fortunately, recent studies of past impacts have made careful adjustments to the raw historical data. Studies of past economic impacts require careful attention and adjustments for shifting changes in the "target" such as varying crop varieties/hybrids over time, for inflation, and for other changing societal and technological conditions that affect the measurement of impacts. For example, comparison of loss data from a drought in 1950 and losses from a similar drought in 1988 requires adjustment for the changes in the area's population density, in the types of water resource systems, and the levels of insurance coverage.

This essay has five sections. First, the temporal behavior of various extreme climate conditions during 1950–2000 is assessed. Next, national annual losses from damaging extremes, including floods and hurricanes, during 1949–97 are reviewed. In the third section, financial losses and gains are identified for recent major climate extremes, including the 1988 drought, the 1993 flood, and the extremely warm-dry winter from El Niño 1997–8. Factors causing the increased losses during recent decades appear in part four. The final section summarizes national losses and gains, how the impacts rate in the nation's economy, and potential future impacts resulting from possible future climate changes.

8.4.2 Temporal fluctuations in climate extremes

Individual climate conditions have been assessed in a variety of studies. This information provides a basis for comparing the temporal behavior of certain conditions and the impacts created. If losses related to a given climate condition are increasing but the condition itself is decreasing, this difference helps point to other non-atmospheric factors affecting the amount of loss.

Heavy precipitation and flooding

Recent studies have defined trends to more frequent heavy precipitation events. Karl *et al.* (1995) found that 1-day heavy precipitation events exceeding 5.1 cm have made an increasingly large contribution to annual precipitation over the United States since 1910. Heavy precipitation events of 7-day duration are closely related to hydrologic flooding occurrences on small to medium-sized rivers, and trends in 7-day heavy precipitation events for the entire country have increased in recent decades (Kunkel *et al.* 1999).

Lins and Slack (1997) reported upward streamflow trends, consistent with the observed upward trends in heavy precipitation. Extensive human modification of river basins makes it very difficult to assess long-term trends in flood flows on a national basis and relate this to flood damage (Downton and Pielke 2001). The complexities of shifting climate conditions, altered drainage systems and river courses, plus changes in societal vulnerability make identification of the role of changing climate conditions difficult to quantify as the primary cause of changes in flood impacts. Available data indicate that flood-related damage has increased in recent decades.

Hurricanes

The frequency of land-falling intense hurricanes has decreased since 1950 (Kunkel *et al.* 1999). Figure 8.8 shows that an increase in hurricane damage (unadjusted for inflation and land-use changes) over recent decades has occurred during this period of decreasing hurricane frequencies and intensities. This means that fewer storms are responsible for the increased damage, and these storms are no stronger than those of past years. Adjustment of losses for temporal changes in population growth and development in vulnerable coastal locations revealed that losses have not increased over time, and these factors are the key to a steady series of losses, rather than increased storm numbers and strength (Pielke and Landsea 1998).

Thunderstorms, hail, and tornadoes

The national average number of thunderstorm days during 1910–2000 has a downward trend, decreasing from 40 days in 1910 to 38 days in 2000. The thunderstorm decline agrees with the downward trend in national thunderstorm losses over time (Changnon and Hewings 2001). Changnon (2001) found that thunderstorm-created losses had increased in Florida and the West Coast, but storm frequencies had not increased in these areas. This suggested that the increased losses were a result of the large regional growth of population and wealth in these areas.

The average number of hail days per year during 1910–2000 also has a downward trend. This agrees with a downward trend in the crop-hail losses (Changnon *et al.* 2001). Property losses from hail had a major increase during the past 15–20 years, mainly in rapidly expanding cities. The number of tornado days nationally during 1953–97 had a slight increase over time until 1970, and a flat trend thereafter. Tornado losses showed a temporal increase, but the national number of killer tornadoes, those tornadic storms leading to one or more deaths, during 1953–97 show a marked temporal decrease (Changnon and Hewings 2001). The increases noted for all tornadoes and their losses are considered a result of growing population and more attention to tornado occurrences, collectively leading to more tornadoes seen over time (Kunkel *et al.* 1999).

Figure 8.8 (a) Number of intense hurricanes (>2 on Saffir Simpson scale) and (b) normalized losses due to hurricanes per year, in 1997-equivalent billions of dollars. (Adapted from Changnon *et al.* 2001)

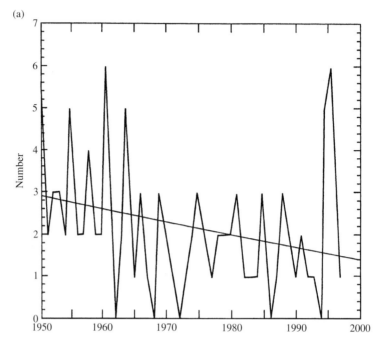

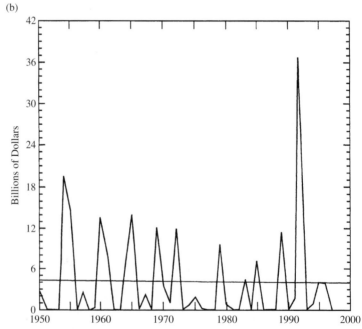

Winter storms

Winter storm losses during 1950–97 underwent a marked increase over time. Very strong winter storms develop along the nation's East Coast where the high population density and extensive coastal development make the region particularly vulnerable to damage from high winds, coastal flooding, heavy snow, and icing. The frequency of strong, damaging East Coast winter storms increased from 1965 into the 1980s (Davis *et al.* 1993). The increase in winter storm losses is associated with both societal and climatological factors.

Droughts

The Palmer drought severity index for each year from 1930 to 2000 defines the percentage of the United States experiencing severe drought. The drought indices for 1950–2000 reveal a steady decline over time after the prolonged severe droughts of the 1930s and the 1950s (Kunkel *et al.* 1999). These two major drought events of the mid century dominate the drought distribution during the twentieth century.

Extremes of heat and cold

Extremes of temperature cause losses to crops and property. Rogers and Rohli (1991) found that six severe freezes occurred in Florida during 1977–89, the most frequent occurrence of freezes since the nineteenth century. These freezes resulted in a significant decrease in the production of citrus and major economic losses (Miller 1988). A heat-wave index, based on each year's frequency of heat waves, revealed that the dominant feature from 1900 to 2000 was a high frequency during the 1930s. A study of trends in extreme temperatures for the northeast United States for 1951–93 found a decrease in the number of days with temperatures exceeding 35 °C (95 °F) (DeGaetano 1996). The study also found a general increase in the number of days with temperatures below freezing. An index of cold-wave frequency for the United States shows no evidence of a major shift since 1980, and heating and cooling degree-days, on a national scale, decreased from 1950 to 1997 (Changnon *et al.* 2001).

8.4.3 National impacts of damaging storms

Quality data have become available allowing definition of losses for major storm conditions during the 1950–97 period adjusted to 1997 dollar values (Figure 8.9; Changnon and Hewings 2001). The leading cause of loss was hurricanes with an annual average of $4.24 billion. Data on losses due to flood damage in the United States ranked as second largest, averaging $3.18 billion annually for 1950–97. Losses from severe thunderstorms (heavy rainfall, lightning, and high winds) averaged $1.632 billion per year, third

Figure 8.9 Annual
average national loss values
in 1997-equivalent billions
of dollars. (After Changnon
and Hewings 2001)

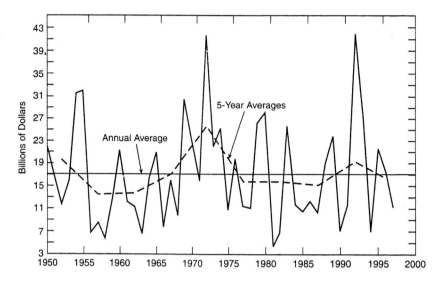

highest. Tornado losses produced an average annual loss of $448 million. Hail and associated wind losses to crops averaged $270 million annually and property losses from hail averaged $174 million per year. Winter storms produced losses averaging $282 million, and windstorm losses averaged $168 million per year. Collectively, all these forms of storms produced $494 billion in losses during 1950–97.

Temporal distributions of these storm variables have been assessed to define their long-term fluctuations and trends (Changnon *et al.* 2001). Normalized (inflation and property at risk) hurricane losses exhibited no change over time, and after comparable normalization, flood losses displayed an upward trend. The trend for property losses due to thunderstorms was slightly downward over time. Winter storm losses continued to increase over time, and the trend was statistically significant at the 1 percent level. Crop losses due to hail showed a significant decline over time. Normalized tornado losses had an upward trend over time for 1950–97. Losses from windstorm catastrophes had a statistically significant downward trend during the 1950–97 period.

8.4.4 Impacts of major recent climate anomalies

Drought of 1987–9

In mid-1987, a drought began in the High Plains and Midwest. By mid-1988, 40 percent of the United States was experiencing severe drought, and drought conditions persisted through most of 1989. Measures of drought intensity and areal extent showed that the 1987–9 drought was one of the ten worst droughts of the century (Riebsame *et al*. 1991). This pervasive drought created major economic, social, and environmental impacts. There were $15 billion in losses in the agricultural sector, but agricultural producers in non-drought areas gained $3.6 billion from their normal crop yields and the high prices caused by the drought. The drought totaled $40 billion in losses and costs, one of the most costly climate anomalies of the twentieth century, but the losses were not a major factor in the nation's economy. The GNP had a 0.4 percent downturn attributed to the drought; the Consumer Price Index in 1988 rose 5 percent, but economists assigned only 0.3 percent to drought effects.

Hurricane Andrew in 1992

Hurricane Andrew struck the Florida coast just south of Miami in August 1992. Costly impacts in Florida were assessed in detail to provide information to help illustrate the importance of better preparedness to reduce vulnerability to extreme weather events (Pielke 1995). Various losses and costs associated with the storm revealed a storm total of $30 billion with $17 billion to private property. The storm assessment pointed to various actions for better preparedness and to mitigate future hurricane losses. A study of past US hurricanes of the twentieth century, based on normalized damages of each storm (Pielke and Landsea 1998), revealed that Hurricane Andrew's losses and costs were the second highest hurricane losses, trailing only those of a 1926 hurricane with normalized losses of $72 billion.

Midwestern floods of 1993

The Midwestern floods of 1993 inundated 10 000 square miles (26 kha) in nine states, creating sizable economic impacts (Changnon 1996). The flood was the nation's record-setting flood of all time with total losses of $20.8 billion. However, certain agricultural interests, businesses, and forms of transportation benefited from the flood. For example, farmers in unflooded areas with good crop yields got higher prices for their harvested crops because the losses in the Midwest drove the prices up. Losses of $2 billion appeared more than a year after the flood and included those resulting from pollutants released by floodwaters, soil losses, and groundwater aquifer damage. Although the most costly flood ever, the 1993 flood had little impact on the nation's economy. The flood did not change the nation's gross domestic product (GDP) in 1993. The flood did not change the 3.3 percent rate of inflation in 1993, but the rate increased to 3.6 percent in 1994 as corn and soybean losses in 1993 caused wholesale farm prices to rise 6 percent.

El Niño 1997–8

Unusual cold season climate conditions of 1997–8 related to a strong El Niño event created major economic impacts (Changnon 1999b). El Niño-influenced climate conditions included damaging weather in the southern and western United States. The West Coast was assaulted by a series of coastal storms, causing floods and agricultural damage, with California losses totaling $1.1 billion. Losses in excess of $0.5 billion came from an unusual snow-storm across the High Plains. Flooding devastated fruit and vegetable crops, and national prices for fresh produce rose 7.9 percent in January, and then rose 5 percent in March 1998. The nation's tourist business suffered from a 30 percent drop in income. National losses from the winter events totaled $4.2 billion.

In contrast, the mild, almost snow-free winter in the northern United States resulted in several major economic gains. Abnormal winter warmth led to major reductions in heating costs, creating national savings of $6.7 billion. Little precipitation and high winter temperatures had a positive influence on construction, retail shopping, and home sales, with total nationwide gain of $5.6 billion.

The effect on the nation's economy included a 4 percent drop in produc-tion at the nation's electricity and gas utilities. Inflation was zero during January–March, the first time in 10 years, and the Consumer Price Index went unchanged due to the falling energy prices. The GNP rose at a rate of 4.2 percent during the first quarter of 1998, as compared to the 3.4 percent expected. Assessment of the national impacts, both losses and benefits, revealed economic benefits of $19.8 billion outweighed losses of $4.2 billion.

Record warm winter of 2001–2

The winter of November 2001–February 2002 was the nation's warmest on record since 1895 and led to reduced energy demand and kept natural gas prices much below normal. Large parts of the nation had below average precipitation, and snowfall was below normal throughout most of the northern United States. Economic impacts included lower heating costs, reduced transportation delays, lower road/highway maintenance costs, added con-struction activities, reduced insurance losses, and increased retail sales, home sales, and tourism. Expenditures for homes and retail products during the November–February period were $5 billion above normal, and the winter created an additional $2 billion income for the construction industry. The low heating bills were a bonanza for consumers, but big utilities lost large sums. One East Coast utility reported a revenue loss of $92 million, an 8.3 percent decrease. However, consumer savings were $7.5 billion due to lower energy costs (Changnon 2002). The total nationwide benefits were $20.6 billion, whereas the winter losses were only $400 million. Some economists claimed the mild winter and its impacts were a key factor in getting the nation's economy out of an on-going recession.

Record high insured climate losses during the 1990s

The United States experienced record-setting high insured property losses from numerous climate extremes during 1990–6. The 7-year total insured loss, after adjusting for inflation and other factors, was $39.65 billion of which $15 billion was due to Hurricane Andrew (Changnon 1999a). Insured losses in the United States represent 70 percent of the total national losses from climate extremes (Changnon *et al*. 1997). Assessment of events causing losses of $100 million or more revealed 72 had occurred during 1990–6, an average of 10.3 per year, as compared to an average of only 3.6 such events annually during 1950–89 (Changnon and Changnon 1991). Annual insured losses during the 1990–6 period averaged $5.665 billion, whereas the 1950–89 loss value (1997 dollars) from events causing $100 million or more damage was $1.7 billion per year. Thus, the 1990s experienced a major increase in the number of costly events and in the magnitude of their annual losses, creating major economic impacts on the insurance industry.

8.4.5 Causes for increasing climate losses

Various factors including shifts in climate conditions have been identified as responsible for the increases in economic losses related to climate extremes in recent decades.

Insurance industry

The record high insured property losses during the 1990s created immense concern amongst the crop insurance, property insurance, and reinsurance industries. They sought explanations for the causes. One that surfaced quickly was that the peaking of losses represented the start of a climate change due to global warming (Swiss Re 1996), whereas others believed the shift was due to natural fluctuations in climate. Studies of storm frequencies did not reveal a major increase in many storm frequencies or intensities. Extensive analysis of when and where the insured losses had increased pointed to shifts in insured risks that the insurance industry had not adjusted for in their rates. The upward trend in insured property losses due to catastrophes (events causing >$1 million in losses) during 1949–94 had a close relationship with the trend in the nation's population. The recent increase in insured losses partly resulted from a lack of adjustment over time by the industry for population growth and shifting risks.

Climate shifts to more extremes

The nation experienced a "climatologically quiet" period from the late 1950s through the early 1970s. It was largely devoid of climatic extremes, such as severe droughts or wet periods that had preceded it during the 1920s, 1930s, 1940s, and early 1950s. This low incidence era was sufficiently long for many climate-sensitive operations, including the insurance industry, to be designed, financed, and operated based on conditions with few climate extremes. Many climate-sensitive operations and managers became attuned to functioning in a period with few major extremes.

Conditions began to change during the late 1970s, as climate aberrations again became commonplace nationwide, comparable to conditions in the 1900–50 period. Many managers of climate-sensitive activities faced problems they did not understand. The list below describes the run of major extremes from the mid 1970s to the early 2000s.

- The late 1970s had a series of four winters that were abnormally severe, setting records in the central United States.
- The early 1980s included the wettest five years on record in the nation, producing record high lake levels on the Great Lakes and Great Salt Lake, with attendant major shoreline damage around the lakes.
- Droughts developed in the Southeast in 1986 and covered half the nation in 1988–9. California had its six consecutive driest years on record before the drought broke in 1992.
- The summers of 1992 and 1993 became the two worst years for hail loss to both crops and property in the High Plains. Prolonged storminess throughout both growing seasons created billions of dollars in crop losses, and major hail damage occurred to property in Denver (1990), Wichita (1992), Dallas-Fort Worth (1994–5), and St. Louis (2001).
- Record Midwestern flooding, in duration and areal extent, occurred in 1993 at the same time that the Southeast had an extreme warm season drought. Severe flooding occurred again in 1996 and 1997 in the Chicago area, California, along the Ohio River, and in the Dakotas.
- Major winter storms and prolonged record cold made the winter 1993–4 the worst on record in the East Coast and parts of the Midwest. Damages reached billions of dollars as the parade of bad storms continued into 1996–7.
- Major droughts became established in the western United States during 1998–9 and persisted into 2004. These created major water shortages and massive wildfires.

The impacts of these events on the government were sizable. Relief programs were employed to help with the trauma of losses, but the multi-billion dollar relief bills to pay for climate-induced losses since 1987 were seen as a threat to the ever-growing national debt. Furthermore, many federal policies relevant to handling these climate anomalies using more sensible fiscal approaches were flawed (Hooke 2000). The floodplain management program was recognized as inadequate as for each of the major floods of 1993, 1996, and 1997, less than 10 percent of those damaged had flood insurance. The crop-weather insurance program was modified, but it required multiple agricultural disasters to bring about more effective legislation and a more stable crop insurance program. After being benign for about 20 years, the climate had become more variable again with conditions more typical of the nation's long-term climate. The enormity of the losses caused by this array of climate anomalies since 1975 is an important part of the story of society's vulnerability to climate.

Societal factors

The 1990s experienced a record number of damaging storms, including 72 storms each with damage exceeding $100 million during 1990–6, whereas only 142 such storms occurred in the preceding 40 years. Trends in insured loss statistics since 1960 display sharp regional differences. On the West Coast, the Arizona–New Mexico–Colorado–Texas area, and the southeastern coastal states, the number of property catastrophes exceeding $100 million in losses during 1990–7 was double the number in the preceding 40 years (Changnon 1999a). Elsewhere in the nation, these costly storms had increased by only 10 to 20 percent. Events with losses exceeding $100 million averaged $551 million in loss per event since 1990, just $12 million more than the average of the 142 events of the prior 40 years. This reveals no increase in storm intensity.

Many climate extreme losses, after careful adjustment for societal and insurance factors, did not display upward trends over time. Comparison of this information with the upward trends in actual dollar losses, and inspection of areas where losses had grown most (Southeast, South, and West) revealed that a major cause of the upward loss trends was societal factors (Changnon 2003). Over time, the nation's society and infrastructure had become more susceptible to climate anomalies. The ever-growing population, with its concomitant demands for food, water, energy, and other climate-influenced resources, became more vulnerable to extremes that reduced these resources. In addition to the increased population, or target at risk, there are other reasons for this increased vulnerability to climate extremes.

- Increased wealth with more valuable property at risk.
- Increased density of property.
- Demographic shifts to coastal areas and to storm-prone large expanding urban areas.
- Aging infrastructure, structures built below standards, and inadequate building codes.
- Interdependency of businesses and product development.

Thus, the results from extensive recent assessment studies show a marked increase in the nation's vulnerability to weather and climate extremes.

8.4.6 An economic perspective on climate impacts

Losses and gains from climate conditions

Annual loss values from all climate extremes that produce major losses in the United States have been assembled. Critical to this endeavor was the use of quality loss data with long historical records. Important variables with good data included losses due to floods, hurricanes, and tornadoes, plus climate-induced losses to the nation's major crops and increased costs of energy usage. These five variables and four others (losses from severe thunderstorms, hail, windstorms, and winter storms) created a total of nine variables that defined most of the national climate-related losses during 1950–2000. Energy use costs ranked highest ($4.65 billion), and three conditions – hurricanes, energy costs, and floods – accounted for 69 percent of total average loss.

Losses not measured by the variables for which quality data existed have been estimated. Secondary and tertiary losses and costs that develop over time (6 months to 5 years after an anomalous event) were estimated based on findings from recent studies. Unmeasured insured property losses were found to average $230 million annually and uninsured property losses averaged $115 million per year. Agricultural losses to livestock and speciality crops were estimated at $450 million annually (Changnon *et al.* 2001).

The losses and gains defined from recent economic assessments were used to develop a list of average annual losses and gains for each sector of the nation's economy. Table 8.4 presents the resulting values, with annual losses of $35.94 billion (2000 dollars) resulting from these conditions (Winstanley and Changnon 2004). The annual loss average from extremes in Canada is

Table 8.4 *Estimated annual national economic losses/costs and gains (billions of 2000 dollars) resulting from climate anomalies during 1950–2000*

Sector	Annual losses & costs	Annual gains
Transportation	1.60	0.30
Retail sales	1.25	3.80
Agribusiness	1.90	1.65
Farmers, crops/livestock	3.32	1.90
Energy use	4.65	3.92
Property damage	10.46	0
Government	7.00	6.60
Tourism	0.20	0.15
Property insurance	4.36	6.50
Construction	1.20	1.50
Total	35.94	26.32

$11.6 billion (Bruce *et al*. 1999). Maunder (1986) carefully analyzed the nation's economic losses due to adverse weather in the 1970s, and adjustment of those total losses for inflation produces a value of $37.59 billion (2000 dollars). This is close to the losses calculated for all recent extremes, totaling $35.94 billion.

Analysis of climate conditions causing the maximum loss each year during 1950–2000 revealed six conditions that rated highest in one or more years. Energy costs were the highest value in 14 of the years; hurricane losses were highest in 12 years; and flood losses were highest in 11 years. Crop losses ranked first in 9 years, severe thunderstorm losses in 4 years, and windstorm losses were highest in 1 year. Temporal distribution of the loss and gain values revealed no significant upward or downward trend during 1950–2000.

Assessments of the economic gains due to many climate conditions have been made (Winstanley and Changnon 2004). For example, the studies of the 1987–9 drought, Hurricane Andrew in 1992, and the 1993 flood revealed economic gains ranging from 30 to 50 percent of the losses, or between $27 and $45 billion. Crop gains in years with good weather averaged $1.9 billion annually, and energy usage reductions in mild years produced gains of $3.9 billion per year. National gains from favorable climate conditions (Table 8.4) average $26.32 billion per year. Annual gains exceeded total losses in 14 years during 1950–2000.

Climate impacts and the national economy

The financial losses and gains from climate conditions were assessed against various measures of the nation's economy. However, it must be remembered that the losses and gains determined are mainly direct impacts measured at the time and do not include all the delayed financial impacts that may develop months and years after an anomalous event. Hence, the total financial impacts of climate are somewhat greater than those shown in Table 8.4.

The $35.9 billion annual average loss for the 1950–2000 period was evaluated against two national economic measures. It was found to be 2 percent of the total federal expenditure in 2000 ($1.6 trillion), and 0.4 percent of the nation's GDP for 2000 ($8.1 trillion). The highest one-year loss of $54.4 billion during 1950–2000 rated as 3.3 percent of the 2000 federal expenditure. Data on losses from storms and climate extremes, as reflected in the federal disaster relief payments made since 1953, revealed loss payments totaling $54 billion for 1953–97 (Sylves 1998). The peak one-year federal assistance payment for climate disasters was $7.1 billion in 1994, 0.5 percent of the 1994 federal expenditure. The Subcommittee on Natural Disaster Reduction (SNDR 1999) estimated that natural disasters (climate extremes, earthquakes, and other non-climate events) averaged $1 billion in losses per week in the United States, only 0.7 percent of the GDP.

Assessment of the recent major climate extremes revealed several short-term national-scale economic impacts. The 1987–9 drought led to a GDP downturn in 1988 of 0.4 percent, and the Consumer Price Index (CPI) rose 0.3 percent. The 1993 flood led the GDP to be increased by 0.01 percent in 1994, and corporate profits were down 0.01 percent in 1993, but up by 0.8 percent in 1994. The odd winter weather of El Niño 1997–8 caused the price of food to increase 0.4 percent in February 1998; inflation held at zero for the first time in 10 years; and the first quarter GDP in 1998 was up 4.2 percent, rather than the 3.4 percent predicted. All of these values are notably small. Agricultural weather losses during 1950–2000 varied between 9 and 14 percent of the annual net cash income for US agriculture in the 1990s.

Climate-related economic impacts are more significant regionally where many losses are often concentrated. The $6 billion in flood-related losses in Illinois in 1993 was 2 percent of the Gross State Product (GSP) for 1993. Losses of $21.9 billion in Florida due to Hurricane Andrew in 1992 rated as 10.2 percent of Florida's GSP. The SNDR (1999) indicated that the largest state losses from natural hazards were typically 5 percent or less of the states' domestic products. Although these state-scale impacts are relatively larger than the national impacts, they are not exceptionally large.

Estimates of future economic impacts from climate change

During recent years, a few economists have generated estimates of financial impacts resulting from future climate change. Economic modeling has been used to derive estimates of future economic impacts. In the early 1990s, economists assessed the national economic impacts of global warming. Annual losses generated, in 1988 dollars, were on the order of $55 billion (Cline 1992). The assumed global warming in each case was from 2.5 °C to 3.0 °C, and all three predictions assumed 1988 levels of outputs and composition of goods and services produced. The major finding of these studies was that the estimated impact values were small in comparison with the total US economy. A critical issue for estimation of future financial impacts is economic models. Burroughs (1997) evaluated economic models and their use in assessing climate change impacts, pointing to modeling weaknesses and the complexities of integrating the outputs of global climate models with those of macroeconomic models. He further pointed to many other unpredictable factors, such as technology developments over the next 50 to 100 years and their unknown influence on economic impacts of future weather.

A recent economic assessment using three climate scenarios and their estimated impacts on the US economy in 2060 revealed a range of outcomes (Mendelsohn and Smith, 2002). The net national annual economic impact was $36 billion (1998 dollars) in benefits with a climate having a 1.5 °C temperature increase and a 15 percent precipitation increase. A scenario with a 5.0 °C increase and no precipitation change was estimated to be (for 2060) a national annual loss of $19.9 billion (1998 dollars). Mendelsohn and Smith noted that these various predicted economic impacts would be about 0.1 percent of the GDP expected by 2060, and they further note their values are about an order of magnitude less than those of the IPCC (1996).

A recent national assessment of climate change involved an in-depth investigation of the consequences of climate change in the United States with impacts assessed based on conditions predicted by two global climate models: the Hadley model and the Canadian climate model (NAST 2001). Thousands of potential effects were identified (for example, less water, more heat waves, and altered crop seasons, etc.), but few of these impacts were translated into financial outcomes. The agricultural sector assessment found that the projected conditions increased yields of many crops, including corn, wheat, and soybeans. The agricultural impact assessment did quantify the economic outcomes showing that the Canadian model conditions, given various adoption strategies, ranged from losses of $0.5 billion (2000 dollars) annually to gains of $3.5 billion, whereas the Hadley climate model conditions resulted in agricultural benefits ranging between $6 billion and $12 billion (2000 dollars) annually. In comparison, an agricultural assessment of the 1950–2000 period found an average annual benefit of $1.9 billion (1997 dollars) with a one-year peak of $4.8 billion (Changnon and Hewings 2001). When losses occurred, the average annual loss was $2.6 billion (1997 dollars). The concluding chapter of the national assessment states, "For the nation as a whole, direct economic impacts are likely to be modest" (NAST 2001).

The above listed economic estimates of losses and gains from a future changed climate must be considered quite speculative. Recent papers by economists, all familiar with the climate change issue, revealed the difficulty of estimating, even crudely, the future economic impacts of global warming and the costs of various approaches for mitigating climate change. As Yohe (2003) stated, "How could we estimate the distribution of costs and benefits (of mitigating climate change) across the wide range of unknown and unpredictable economic and climate futures?"

Furthermore, estimating national impacts based on shifts in the nation's climate does not take into consideration external impacts. Climate changes in other parts of the world, particularly in developing nations, may create financial impacts that greatly influence the United States' economy. All of this adds to the uncertainty in estimated future economic impacts in the United States.

8.5 Conclusions

The discussion in the chapter establishes that adaptation to climate is an essential component in human life, whether the society is based on subsistence

Table 8.5   *Some relationships between climate variability and society*

Climate influences	Direct effects	Flow-on consequences	Indirect effects on population
Climate change	Agricultural production	Food availability and demand	Mortality rates
Climate variability	Disease	Population nutrition	Birth rates
Seasonal weather variability	Environmental damage	Abandonment of farms	Population growth
Extreme weather events	Flood	Starvation	
	Drought	Economic trends	
		Market development	
		Trade	
		Innovation	
		Work availability	
		Labor supply	
		Migration	
		Income	
		Education	
		Technological development	
		Development of buffers	

Based in part on Michaelowa (2001).

agriculture, or is highly technologically advanced. The level of adaptation, and the importance of climate variations to the society, depend on the economic and social structure of the society and how well equipped it is to cope. Table 8.5 provides a summary of how climate change links to society and its activities.

McGovern (1991) and Brown (2001) state that it is in marginal living areas, historically such as Greenland and subsistence Europe, where adaptive strategies of the people are most critical. Here, the combination of population stresses, political and economic inflexibility, and adverse weather can create a major crisis. At this basic level of development, food and water are essential, there are few buffering opportunities, and one bad season can cause famine, disease, and major loss of life. Vail (1998) emphasizes that the deterioration of the environment can be more important than hostile interactions with other cultures. The discussions about the Viking settlements in Greenland and the fourteenth century in Europe highlighted these problems. Some societies in the seventeenth century in Europe survived the depths of the LIA better, as they became more flexible, and the application of new knowledge enhanced adaptability.

Changnon's essay highlights that climates stress an advanced society differently. It is the economic cost to society, rather than loss of life, that becomes the critical component. Buffers, such as insurance, food storage, education, and government support, help overcome difficulties caused by hazardous weather. However, the society must be flexible enough to adjust to changes in food costs, scarcity, economic inflation, and rising insurance premiums. These interactions are very complex, and actions by governments that are seen as unsupportive by the population can have significant political consequences.

8.6 Examples of climate and history websites

The US National Oceanographic and Atmospheric Administration provides links between past Holocene climate periods and variability of climate and drought, www.ngdc.noaa.gov/paleo/ctl/resources1000.html.

The report of the UNEP and WMO Climate Change working group in 2001 includes some analysis of the Little Ice Age and the Medieval Warm Period, www.grida.no/climate/ipcc_tar/wg1/070.htm.

A global summary of the Little Ice Age and the Medieval Warm Period plus some very useful references are on the website of the Center for the Study of Carbon Dioxide and Global Change, www.co2science.org/subject/l/summaries/lianglobal.htm, europemwp.htm.

A report on a 2003 conference on climate of the last millennium, including papers by various authors is on Professor Stephen Schneider's (Stanford University) website, for example http://stephenschneider.stanford.edu/publications/pdf_papers/bradley.pdf.

A useful and colorful summary of the Viking migrations during the Medieval Warm Period, and the Little Ice Age, is available at www2.sunysuffolk.edu/mandias/lia/index.html. This website was created by Professor Scott Mandia at State University of New York in Suffolk.

Warning: There are a large number of websites with links to the LIA and MWP periods. Unfortunately, many of these are used politically to argue about greenhouse warming. Many have incorrect information.

8.7 References

Bokwa, A., Linanówka, D. and Wibig, J., 2001. Preinstrumental weather observations in Poland in the 16th and 17th centuries. In P. Jones, A. Oglivie, T. Davies and K. Briffa, eds., *History and Climate – Memories of the Future*. New York: Kluwer, pp. 9–27.

Brown, N., 2001. *History and Climate Change: A Eurocentric Perspective*. London: Routledge.

Bruce, J. F., Burton, I. and Egener, T., 1999. *Disaster Mitigation and Preparedness in a Changing Climate*. Research Paper 3, Institute for Catastrophic Loss Reduction, Toronto.

Burroughs, W. J., 1997. *Does the Weather Really Matter?* Cambridge: Cambridge University Press.

Changnon, S. A., 1996. Losers and winners: A summary of the flood's impacts. In S. A. Changnon, ed., *The Great Flood of 1993*. Boulder, Colo.: Westview Press, pp. 276–299.

1999a. Factors affecting temporal fluctuations in damaging storm activity in the U.S. based on insurance data. *Meteorological Applications*, **6**, 1–10.

1999b. Impacts of 1997–98 El Nino-generated weather in the United States. *Bulletin American Meteorological Society*, **80**, 1819–1827.

2001. Damaging thunderstorm activity in the U.S. *Bulletin American Meteorological Society*, **82**, 597–608.

2002. Weather on our side. *Weatherwise*, **55**, 11–12.

2003. Shifting economic impacts from weather extremes in the United States: A result of societal changes, not global warming. *Natural Hazards*, **29**, 273–290.

Changnon, S. A. and Changnon, J. M., 1991. Storm catastrophes in the U.S. *Natural Hazards*, **2**, 612–616.

Changnon, S. A. and Hewings, G. D., 2001. Losses from weather extremes in U.S. *Natural Hazard Review*, **2**, 113–123.

Changnon, S. A., Changnon, D., Fosse, E. R., *et al.*, 1997. Effects of recent extremes on the insurance industry: Major implications for the atmospheric sciences. *Bulletin American Meteorological Society*, **78**, 425–435.

Changnon, S. A., Changnon, J. M. and Hewings, G. D., 2001. Losses caused by weather and climate extremes: A national index for the U.S. *Physical Geography*, **22**, 1–27.

Cline, W., 1992. *The Economics of Global Warming*. Washington, DC: Institute of International Economics.

Davis, R. E., Dolan, R. and Demme, G., 1993. Synoptic climatology of Atlantic coast north-easters. *International Journal of Climatology*, **13**, 171–189.

DeGaetano, A., 1996. Recent trends in maximum and minimum temperature thresholds exceedences in the Northeast U.S. *Journal of Climate*, **9**, 1648–1660.

Diaz, H., Kovats, R., McMichael, A. and Nicholls, N., 2001. Climate and human health linkages on multiple time scales. In P. Jones, A. Oglivie, T. Davies and K. Briffa, eds., *History and Climate: Memories of the Future*. New York: Kluwer, pp. 267–289.

Downton, M. and Pielke, R. A. Jr., 2001. Discretion without accountability: Politics, flood damage, and climate. *Natural Hazards Review*, **2**, 157–166.

Fagan, B., 2000. *The Little Ice Age: How Climate Made History 1300–1850*. New York: Basic Books.

Grove, J., 1988. *The Little Ice Age*. Methuen: London.

Hooke, W. H., 2000. U.S. participation in the International Decade for Natural Disaster Reduction. *Natural Hazards Review*, **1**, 2–9.

Hurrell, J., Kushmir, Y., Ottersee, G. and Visbeck, M., 2003. An overview of the North Atlantic Oscillation. In J. Hurrell, Y. Kushmir, G. Ottersee and M. Visbeck, eds., *The North Atlantic Oscillation: Climatic Significance and Environmental Impact*. Geophysical Monograph 14, Washington, DC: American Geophysical Union, pp. 1–36.

Intergovernmental Panel on Climate Change (IPCC), 1996. *Climate Change 1995: The Science of Climate Change*. Cambridge: Cambridge University Press.

Jensen, K., Kuijpers, A., Koc, N. and Heinemeier, J., 2004. Diatom evidence of hydrographic changes and ice conditions in Igaliku Fjord, South Greenland during the past 1500 years. *The Holocene*, **14**(12), 152–164.

Jones, P., Ogilvie, A., Davies, T. and Briffa, K., 2001. Unlocking the doors to the past: recent developments in climate and climate impact research. In P. Jones, A. Oglivie, T. Davies and K. Briffa, eds., *History and Climate: Memories of the Future*. New York: Kluwer, pp. 1–8.

Karl, T. R., Knight, R. W., Easterling, D. W. and Quayle, R. G., 1995. Trends in U.S. climate during the twentieth century. *Consequences*, **1**, 3–12.

Kunkel, K. E., Pielke, R. A. Jr. and Changnon, S. A., 1999. Temporal fluctuations in weather and climate extremes that cause economic and human health impacts: A review. *Bulletin American Meteorological Society*, **80**, 1077–1098.

Lamb, H., 1977. Climate in historical times. Chapter 17 in *Climate History and the Future*. London: Methuen, pp. 423–473.

Lins, H. F. and Slack, J., 1997. *Flood Trends in the U.S. during the 20th Century*. Reston, VA: U.S. Geological Survey.

Luterbacher, J., 2001. The Late Maunder Minimum (1675–1715): Climate of the Little Ice Age in Europe. In P. Jones, A. Oglivie, T. Davies and K. Briffa, eds., *History and Climate: Memories of the Future*. New York: Kluwer, pp. 29–54.

Maunder, W. J., 1986. *The Uncertainty Business: Risks and Opportunities in Weather and Climate*. London: Methune & Co.

McGovern, Y., 1991. Climate, correlation, and causation in Norse Greenland. *Arctic Anthropology*, **28**(2), 77–100.

McGovern, T. and Pendikaris, S., 2000. The Viking's silent saga. *Natural History*, **109**(8), 50–58.

Mendelsohn, R. and Smith, J. B., 2002. Synthesis. In *Global Warming and the American Economy*. Northampton, Mass.: E. Elgar Publishers, pp. 187–201.

Michaelowa, A., 2001. The impact of short-term climate change on British and French agriculture and population in the first half of the 18th Century. In P. Jones, A. Oglivie, T. Davies and K. Briffa, eds., *History and Climate: Memories of the Future*. New York: Kluwer, pp. 201–216.

Miller, K. A., 1988. Public and private sector responses to Florida citrus freezes. In *Societal Responses to Regional Climate Change: Forecasting by Analogy*. Boulder, Colo.: Westview Press, 375–405.

NAST, 2001. *Climate Change Impacts in the U.S.* National Assessment Synthesis Team, U.S. Global Change Research Program, New York: Cambridge University Press.

National Academy of Sciences, 1999. *The Costs of Natural Disasters: A Framework for Assessment*. National Research Council, Washington, DC: National Academy Press.

Ogilvie, A., Barlow, L. and Jennings, A., 2000. North Atlantic Climate c. AD 1000: Millennial reflections on the Viking discoveries of Iceland, Greenland, and North America. *Weather*, **55**, 34–45.

Pfister, C., Schwartz-Zanetti, G. and Wegmann, M., 1996. Winter severity in Europe: The fourteenth century. *Climate Change*, **34**, 91–108.

Pielke, R. A. Jr., 1995. *Hurricane Andrew in South Florida: Mesoscale Weather and Societal Responses*. Boulder, Colo.: National Center for Atmospheric Research.

Pielke, R. A. Jr. and Landsea, C. W., 1998. Normalized hurricane damages in the U.S.: 1925–1995. *Weather & Forecasting*, **13**, 621–632.

Riebsame, W. E., Changnon, S. A. and Karl, T. R., 1991. *Drought and Natural Resources Management in the U.S.: Impacts and Implications of the 1987–1989 Drought*. Boulder, Colo.: Westview Press.

Rogers, J. C. and Rohli, R., 1991. Florida citrus freezes and polar anticyclones in the Great Plains. *Journal of Climate*, **4**, 1103–1113.

SNDR, 1999. Progress and challenges in reducing losses from natural disasters. *Natural Disaster Managment*. Washington, DC: Subcommittee on Natural Disaster Reduction.

Sylves, R. T., 1998. *Disasters and Coastal States: A Policy Analysis of Presidential Declarations of Disasters 1953–1997*. Del-SG-17-98, University of Delaware, Newark, Del.

Swiss Re, 1996. Natural catastrophes and major losses in 1995: Decrease compared to previous year, but continually high level of losses since 1989. *Sigma*, **2**, 1–50.

Thompson, L. T., 1986. Climatic change, weather variability, and corn production. *Agronomy Journal*, **78**, 649–653.

Vail, B., 1998. Human ecological perspectives on Norse settlement in the North Atlantic. *Scandinavian Studies*, **70**, 293–313.

Wanner, H., Pfister, C., Brázdil, R. *et al.*, 1995. Wintertime European circulation patterns during the Late Maunder Minimum cooling period (1675–1704). *Theoretical and Applied Climatology*, **51**, 167–175.

Winstanley, D. and Changnon, S. A., 2004. *Insights to Key Questions about Climate Change*. Champaign, Ill.: Illinois State Water Survey.

Yohe, G. W., 2003. More trouble for cost-benefit analysis. *Climatic Change*, **56**, 235–244.

Chapter 9

ESSAY: Model interpretation of climate signals: an application to the Asian monsoon climate

William Lau, *NASA Goddard Space Flight Center*

9.1 Introduction

Numerical modeling is a powerful tool to provide better understanding of the modus operandi, and the prediction of the Earth's climate system. However, a climate model's usefulness is limited by its crude representations of physical processes, most of which we do not understand very well. Since models are only crude approximations of the real system, model results must be validated against observations to ensure reliability. The scarcity of detailed observations for climate processes with the high spatial and temporal resolutions needed for model validation and improvement has been a major impediment for advancement in climate model simulation capability and model predictions.

Climate modeling is an attempt to mimic the evolution of the real climate states, which are described by a vast set of long-term global and regional observations in the atmosphere, ocean and land, from both *in situ* and satellite observations. Given that there are large uncertainties both in observations and in models, and that even the best model is simply a crude approximation of the real world, models and observations should be used in a synergistic manner for better understanding and for improved prediction. The relationship between observations, climate models, data assimilation, process studies, and climate predictions is shown schematically in Figure 9.1. A climate model consists of a dynamical core represented by governing equations of climate state variables, and physics modules of varying complexity (see next section for further discussion). The physics modules, which appeared in the form of numerical sub-models or parameterizations, are the drivers of a climate model. The modules are developed and continuously improved from knowledge gained from field measurements and related process studies. Long-term monitoring refers to observations that are made repeatedly for sustained periods to track the evolution of key parameters of the Earth system. Because models are imperfect, and observations have inherent errors and inadequate coverage, neither model nor observations alone will provide a

Figure 9.1 Synergistic application of observations and models for climate diagnostic and prediction studies.

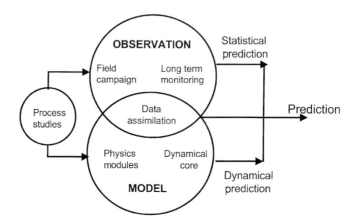

full, comprehensive description of the Earth's climate system. For such a description, data assimilation plays a critical role. Data assimilation is the numerical process by which observations are assimilated into models to produce a complete set of dynamically consistent data sets for the entire climate system (Kalnay *et al.* 1996). Climate predictions can be derived either from observations through statistical techniques or from climate models, or from a combination, i.e. statistical-dynamical predictions. Data assimilation can provide models with appropriate initial states to produce more skilful predictions.

In this chapter, we address the various issues arising from using models to detect, understand, and predict climate signals. This chapter consists of two main parts. The first part is devoted to discussions of climate models as a tool for climate studies, including a brief history of the development of climate models, model basics, and modeling methodologies used in modeling studies. The second part is an illustration of the use of climate models to study the anomalous climate of the Asian monsoon.

9.2 A climate model primer

9.2.1 A brief history

Climate models originate from atmospheric general circulation models (AGCM) used in numerical weather forecasting. AGCMs for numerical weather forecasting were developed during the 1950s and 1960s. (Charney *et al.* 1950; Smagorinsky *et al.* 1965; Bengtsson and Simmons 1983). By the early 1970s most weather services around the world had adopted numerical weather prediction models for short-term (days) to medium-range (weeks) weather forecasting. During that period, climate modelers first began to explore the use of AGCM to study climate anomalies through numerical experimentation with various prescribed forcing functions in the atmosphere, land, and oceans (Manabe and Wetherald 1975; Manabe *et al.* 1979; Gilchrist 1977, 1981). A climate model differs from a weather prediction model in that the former has to be integrated for an extended period of time (multi-years), whereas the latter is generally integrated for a few days at the most. Some of the current climate models to study global change have carried out integration up to thousands of simulated years in order to determine the reliability of long-term climate signals. Because of the requirement for long-term integration, climate models are most sensitive to the conservation of mass, energy, and moisture. Small imbalance in any of the conserved properties can introduce substantial errors that may amplify during the course of the integration to produce severe model systematic bias – a problem known as "climate drift." In contrast, for numerical weather prediction, the accuracy of the initial conditions is more critical, and simulations generally cover a period too short for the climate drift to be an issue.

One of the problems facing the climate modeling community in the 1970s and 1980s was the enormous demand on computation resources required to carry out long-term climate simulations. As a result, for most early applications, climate models with coarse resolutions of the order of 250–500 km, with 2–10 vertical layers, were used and the simulation periods limited to a few years. At such coarse resolution, many physical processes are grossly under-represented. For this reason, many of the early climate model results can only be regarded at best as exploratory. With the advent of computer technology, and more efficient computation codes, climate models can now be run at increasing spatial and temporal resolutions, and with ever more complex physics modules. At present, climate models are currently being run with resolution higher than 50 km at operational centers such as ECMWF. Integrations have been carried out for hundreds of years, such as those used for the IPCC climate assessments, and other global change scenarios by many climate modeling groups around the world. Currently, the Earth Simulator Project at the Frontier Global Change Research Program of Japan is running global climate models at approximately 10 km resolution, with over 200 layers in the vertical and for hundreds of simulated years.

However, even with very high-resolution climate models, large uncertainties remain with regard to prediction of future climate change, especially in the projection of statistics of increased hazards on regional and subregional scales due to extreme weather events. This shortcoming stems from our very limited understanding of the physics of the real climate system, which makes it impossible to include all the details required at the higher model resolution. Hence, merely increasing resolution is not a panacea to the problems of

Figure 9.2 An illustration of the model grid and basic physical processes in a climate model consisting of the atmosphere, ocean, and land. (Reproduced from McGuffie and Henderson-Sellers 2001: Copyright John Wiley & Sons Ltd., with permission)

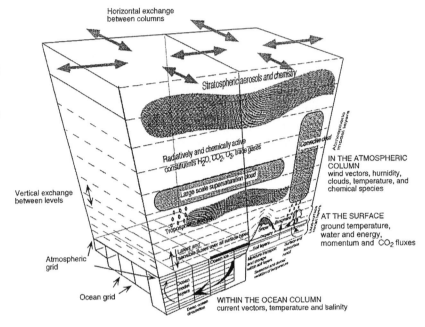

climate modeling. Our knowledge of the physics of the Earth system still needs to catch up with the advances in computer technology. At a more fundamental level, climate signals at the regional and subregional scale may be inherently chaotic and therefore unpredictable. Even if the local signals are there, climate information downscaled from global climate models may be masked by large random local fluctuations. To unmask the regional and subregional scale signals from noise, a number of modeling strategies have been adopted.

9.2.2 Elements of a climate model

A climate model is derived from an atmospheric general circulation model, with a dynamical core consisting of a set of primitive equations for the atmospheric state variables such as temperature, pressure, wind, and water vapor, which govern the fluid motions, thermodynamics, and conservation properties for fluid motion of air parcels in the troposphere and stratosphere on the rotating Earth. The equations are extremely complex and can only be solved numerically on a horizontal grid system with different vertical levels (see Figure 9.2).

The AGCM is coupled to component models of the oceans, the land, and the biosphere. Each component model has its own governing equations for its state variables and physical processes. The grid spacing and vertical interval in the atmosphere, ocean, and the land are often different because of the different fundamental spatial and temporal scales of dynamical and physical processes in each component, as shown schematically in Figure 9.2. The dynamical equations are driven by physical processes that constitute the forcing functions of the climate system. These processes, which are represented as physics modules or "parameterization", include absorption and reflection of solar energy, emission of terrestrial radiation, aerosols, atmospheric composition and chemistry, latent heat release, moisture transport, processes underlying the formation of clouds, rain and water vapor, boundary layer processes, surface fluxes of heat and water, ocean salinity, temperature and currents, and sea ice as well as land surface processes including soil moisture, river run-off, land vegetation and biomass photosynthetic processes, and many others. Given the proper initial and boundary conditions, and external forcing functions, e.g. solar radiation, and time history of its atmospheric composition, a climate model can be integrated forward in time, starting at some time in the past up to the present to simulate past history of the Earth's climate. These climate history simulations are important to ensure that the climate models have the capability to predict future climates. Currently, climate models are routinely being run in major research institutions to provide guidance for seasonal-to-interannual, e.g. El Niño and related regional climate, predictions. Under the Intergovernmental Panel on Climate Change (IPCC) initiative, models have also been run for hundreds of simulated years into the future, subject to different scenarios of climate change regarding the different rate of increase of carbon dioxide in the atmosphere, to provide projections for future climates associated with global warming.

9.2.3 Experimental design

In a fully coupled model, all component models are interactive. In principle, once the initial and boundary conditions are specified, a climate model can be integrated indefinitely into the future, to produce the so-called "nature" or control run. Very often to test out a given hypothesis, a climate model has to be re-run with one or more components held fixed, and the results compared with the control run. Table 9.1 shows possible configurations in which a climate model with three major components, i.e. atmosphere, ocean and land, can be run to test climate sensitivity to anomalous SST forcings due to El Niño, and land surface processes.

In all model sensitivity or climate simulation studies, a control experiment has to be defined first. In Table 9.1, the control experiment (Exp-I) is one in which the SST field is prescribed as the climatology, i.e. the mean over many years. The climatological SST describes the part of the variation that is due to

Table 9.1 *Possible experimental designs for climate sensitivity experiments.*
The letters C and O denote climatology and observation respectively.
A check mark indicates an interactive component.

	Atmosphere	Ocean	Land
Exp-I (control)	✓	C	✓
Exp-II	✓	O	✓
Exp-III	✓	O	C
Exp-IV	✓	C	C
Exp-V	✓	✓	✓

forcings external to the climate system, e.g. annual cycle of solar radiation. Here, deviations from the climatology are due either to internal dynamics within the atmosphere or its interaction with the land surface. In the control, the model can be integrated for typically 50 simulated years to obtain a stable climatology. In Exp-II, the model is run under identical conditions as in the control, except that the SST is obtained from actual observations that cover a 50-year period, which include several major El Niños. The anomalies in Exp-II will then be computed with respect to the control. The impacts of El Niño SST forcings on global climate can be estimated from the anomaly fields of rainfall, temperature, and wind, and from comparison with the actual observations. It is possible that some of the regional impacts over land may be due to land processes feedback induced by the SST. To estimate the effect of land–atmosphere interactions, results from Exp-III and Exp-I need to be analyzed. If the interest is in isolating the natural variability generated by land–atmosphere interactions alone, without SST anomaly forcings, Exp-IV should be compared to Exp-I. Finally, comparing the fully interactive run Exp-V with Exp-II will provide insight on the role of coupled ocean–atmosphere processes in producing the model climate anomalies.

An example of a set of experiments to show the effect of ocean forcings vs. land–atmosphere interactions on the generation of the Southern Oscillation (SO) is shown in Figure 9.3 (Lau and Bua 1998). The SO (see Section 2.8) is known to have strong impacts on the Asian monsoon climate anomalies. Comparing Exp-II (ALO, in Figure 9.3a) with Exp-I (AL in Figure 9.3c), it can be seen that the eastern portion of the see-saw is missing in the latter when the anomalous SST forcing is withheld. This suggests that the SO arises primarily from anomalous SST forcing. The similarity in the SO in Exp-III (AO in Figure 9.3b) with Figure 9.3a implies that land–atmosphere interaction is not important in generating the SO, but may have some impact on the signal at higher latitudes. Finally, the negative anomalies over the extratropical North Pacific and the North Atlantic in Exp-III (Figure 9.3c) and

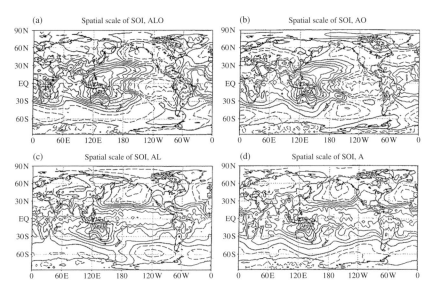

Figure 9.3 Sea-level pressure patterns showing the spatial structure of the Southern Oscillation for (a) ALO, (b) AO, (c) AL and (d) A. See text for definition of symbols. Contour interval is 1 mb. (Adapted from Lau and Bua 1998)

Exp-IV (A in Figure 9.3d) suggest that there is an intrinsic inverse variation of the tropical and extra-tropical atmospheres even in the absence of any anomalous SST forcings.

If the objective is to examine the impact of a particular land surface anomaly, such as snow cover, as a climate forcing, then it is possible to run a new set of experiments in which the land and ocean forcing conditions can be reversed. The design of the numerical experiments using climate models will depend on the objectives of the experiments, and on the hypotheses being tested.

9.2.4 Ensemble simulations

Given that atmospheric variations have a large chaotic component, it is possible that even when forced by a specified lower boundary anomaly such as SST, the atmosphere may respond differently depending on the initial conditions. Often, the real climatic signals are obscured by the large variability due to internal dynamics of the atmosphere. Ensemble forecasts have commonly been used in numerical long-range weather forecasts since the late 1980s to extend the lead time for useful forecasts and to evaluate the skill of the forecasts using some measure of the spread among the ensemble members (Hoffman and Kalnay 1983; Palmer 1993; Tracton *et al.* 1993). To increase the signal-to-noise ratio and to extend predictability, ensemble approaches are increasingly being used in long-term climate simulations and projections (Shukla *et al.* 2000; Kawamura *et al.* 1998). Typically, an ensemble climate simulation calls for a set of control experiments and a set of anomalous experiments. The control consists of at least 5 to 10 members subject to identical climate forcings, e.g. SST, sea ice, or present-day CO_2 composition, etc., but different initial conditions, to ensure that the model results span the range of possible realizations of the model climate.

In the anomaly experiments, the ensemble integrations are repeated as in the control, but the forcing function is varied in some specified but identical manner, but with different initial conditions. The sensitivity of the forcing function on the climate system can then be evaluated based on the variance of the ensemble mean and the spread of the ensemble members about the mean. For a climate variable X_{ij}, where the index i ($= 1, 2 \ldots N$) is the time index, say at yearly intervals, and the ensemble number is j ($= 1, 2 \ldots n$), the ensemble mean $\overline{X}_i$ and the climatological mean $\overline{\overline{X}}$ are defined by:

$$\overline{X}_i = \frac{1}{n}\sum_{j=1}^{n} X_{ij}; \qquad \overline{\overline{X}} = \frac{1}{nN}\sum_{i=1}^{N}\sum_{j=1}^{n} X_{ij}$$

An unbiased estimator of the variance of the noise and of the ensemble is given respectively by:

$$\sigma_{\text{noise}}^2 = \frac{1}{N(n-1)}\sum_{i=1}^{N}\sum_{j=1}^{n}(X_{ij} - \overline{X}_j)^2$$

$$\sigma_{\text{E}}^2 = \frac{1}{N-1}\sum_{i=1}^{N}(\overline{X}_i - \overline{\overline{X}})^2$$

The climate forced variance and the total variance are obtained respectively as:

$$\sigma_{\text{signal}}^2 = \sigma_{\text{E}}^2 - \frac{1}{n}\sigma_{\text{noise}}^2$$

$$\sigma_{\text{total}}^2 = \sigma_{\text{noise}}^2 + \sigma_{\text{signal}}^2$$

The climate signal-to-noise ratio is then defined as $S = \sigma_{\text{signal}}^2/\sigma_{\text{total}}^2$. The statistical significance of the signal for a given ensemble climate experiment can then be tested using the F-test (Von Storch and Zwiers 1999). The larger the ratio, the more likely it is that the signal is detectable in the real world.

In most applications, the ensemble mean is computed with equal weights for each ensemble member. In more recent applications, when the ensemble comprises not only outputs from the same models with different initial conditions, but also different models, it may be necessary to assign a different weight to each model ensemble member. In the so-called multi-ensemble super-ensemble approach, weights for each model variable and for each model grid are assigned based on past model performance (Krishnamurti et al. 2000; Stefanova et al. 2002). In this way, models with strong bias tend to be weighted less than those with weak bias. The super-ensemble approach has

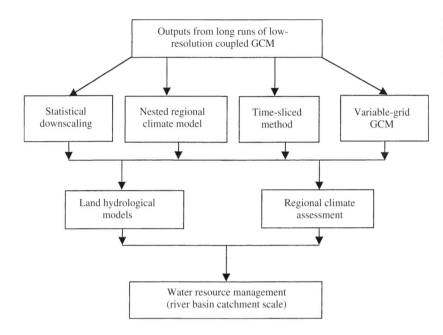

Figure 9.4 Downscaling global climate signals for regional and subregional scale applications.

produced remarkable improvement in short-term weather forecasting and is promising as a tool in multi-model climate projections.

9.2.5 Climate downscaling

To produce multiple realizations of climate variability and to obtain robust statistics, climate models have to be run in the fully coupled ocean–atmosphere–land mode for many simulated years. For seasonal-to-interannual time scales, typically 40–50 simulated years are needed. For decadal scales and climate change scenarios, several hundred or even thousands of simulated years have to be carried out (cf. IPCC reports). Because of the enormous computation resources required for such long-term integrations, coupled models are typically run in the low-resolution mode to capture only the slow physics of the system, which is deemed to be important for long-term climate change. As a result, regional and subregional scale processes are unresolved. To obtain regional and local climate information from the long-term integrations, climate downscaling is necessary.

Climate downscaling is the procedure by which climate signals at the scale of GCM grid size are translated into regional and subregional scales, which are unresolved by GCMs (Giorgi and Mearns 1991; Hewitson and Crane 1996). Downscaling is usually applied to a pre-selected region, in which climate and/or water resource assessment need to be estimated. As shown in Figure 9.4, downscaling may take place in single or multiple stages. At present, there are four basic approaches: statistical, nested models, time-sliced method, and variable-grid GCM. In the statistical approach, cross-scale relationships, known as "transfer functions" are first derived from large-scale observational and local-scale data, and checked for consistency with the synoptic-scale forcings of the GCM. For a given climate scenario provided by the low-resolution GCM (typically with horizontal resolution of 250–500 km), the transfer functions are used to generate the statistics from global GCM outputs to regional scales. This approach is limited by the amount of available global and regional data required for robust statistics, and the possible inconsistency between model and observational data. It has the clear advantage of computational ease. In the nested regional model approach, the large-scale forcing functions derived from the GCM are used as lateral boundary conditions to drive a regional climate model (with typical resolution of 20–50 km) applied to a specific geographic region usually of continental to sub-continental scale, and for a chosen time period of interest. This time period may be related to the occurrence of a devasting drought or flood in a certain region, and one wants to see what are the causes, and if they are related to the underlying large-scale climate forcings or to local feedback processes. Multiple nesting grids, with increasing resolution are sometimes used to zoom in on a subregion to resolve even smaller scale features. The nested regional models may have numerical instabilities at the lateral boundaries, so appropriate buffer zones have to be designed (Giorgi and Mearns 1991).

Alternatively, to avoid the lateral boundary problems, a time-sliced approach is used, by re-running the atmospheric component of the coupled GCM, at a higher spatial resolution and for a shorter time period, using the large-scale lower boundary forcings, such as sea surface temperature, from the coupled model. This approach has the disadvantage of "wasting" valuable computational resources outside the region of interest. More recently, a new strategy has been developed to use GCMs with variable resolution, or so-called stretched or "telescoping grids," in which the GCM can zoom in on a specific region, with high resolution to resolve local features, while keeping the computations elsewhere at the coarse resolution (Fox-Rabinovitz et al. 2001). This approach can achieve considerable savings in computation resources, while achieving the desired higher spatial resolution in the region of interest. The variable-grid GCM requires the redesign of the model numerics, as well as the physical parameterizations to maintain dynamical consistency between the regions with high and low resolutions.

Depending on the space-time resolution, outputs from the regional models can be used for climate assessments, and for water resource management. For applications to river-basin catchment scales, further downscaling may be needed (Lattenmaier et al. 1999). For that purpose, outputs from the regional climate models are used to drive macro-scale hydrology models (<1 km resolution) to provide information, such as stream-flow, surface runoff, or subsurface water storage, needed for management of irrigation, flood control, hydropower production, municipal and industrial supply, navigation, and recreation.

9.2.6 Model intercomparison and validation

Given the large uncertainties in GCMs, it is clear that results from a single model cannot be interpreted too literally and that an estimate of the model reliability has to be included. This consideration has led to the Atmospheric Model Intercomparison Project (AMIP), which was initiated in 1989 under the auspices of the WCRP with the aim to systematically validate, diagnose, and intercompare the performance of AGCMs in various simulated aspects of the climate system (Gates *et al.* 1992, 1999). During AMIP-I, over 30 AGCMs around the world were organized to carry out simulation of the evolution of the Earth climate from 1979 to 1988, subject to identical prescribed observed monthly sea surface temperature, sea ice, CO_2 concentration and solar constant. A large number of model output variables are archived and standardized and made available to the scientific community. Thanks to the AMIP climate model, users have gained a better appreciation of the strengths and weaknesses of climate models. More importantly, AMIP allows modelers to learn more about their own model from having independent examination of their own model outputs in comparison with other models. It is the driving force behind many efforts in model improvement at research institutions. Moreover, AMIP results have shown that even if an individual model does not perform well, the ensemble means of all models can do a better job than individual models in simulating the evolution of climate. This is because model errors tend to cancel out in large model ensembles, so that the signal-to-noise ratio can be increased. The use of super-ensemble techniques (Krishnamurti *et al.* 2000), whereby statistical weights are assigned to each model variable, at each grid point, hold promise for more reliable simulations, and climate projections on regional scales. Following on the success of AMIP-I, an expanded AMIP-II is now underway to include a wider range of variability, to accelerate model physics improvement and to improve the infrastructure for model diagnostics, validation, and experimentation. Various model intercomparison projects (MIPs), tailored to various modeling communities, have emerged in recent years. These include the Coupled Model Intercomparison Project (CMIP), the Seasonal Model Intercomparison Project (SMIP), the Project of Intercomparison of Land Parameterization Schemes (PILPS), the Paleoclimate Model Intercomparison Project (PMIP), and many others.

9.3 Modeling the Asian monsoon climate

The Asian monsoon (AM) encompasses the vast region spanning the Indian subcontinent, Southeast Asia and East Asia, surrounded by the Indian Ocean and the western Pacific Ocean. It is home to more than half of the world's population. The socioeconomic infrastructure of the mostly agrarian societies in the AM region has been built, in large part, on the basis of a highly reproducible annual cycle of rainfall. Agriculture, drinking water, health, energy generation, and more generally the livelihood and well-being of this vast human population all depend on monsoon rains. Imbedded in the large-scale monsoon circulation are powerful rain-producing weather systems, known as monsoon depressions. An anomaly of the AM in the form of a slight shifting of the monsoon rain system will cause major flooding in one place and drought in another. Droughts and floods are the major causes of extensive destruction of the ecosystem, property damage, collapse of regional economies, and loss of human life in the AM region.

While drought can cause crippling long-term effects on a country or a region, a single season of flooding can be devastating. For example, the widespread monsoon flood over central and eastern Asia during the summer of 1998 was responsible for the loss of over 3000 human lives, damaged more than 30 million acres (12 Mha) of farmland, and ruined over 11 million acres (4.5 Mha) of crops. In all, the flooding inflicted an economic loss totaling over 12 billion US dollars to China, and brought the country's economy to its knees. Since much of the world's productivity relies on the natural, economic and human resources residing within the thriving economies of the AM region, our ability to predict the interannual changes in monsoon circulation and rainfall is a critical requirement for the sustainable development not only of the AM region, but of the world.

Understanding, modeling, and predicting monsoons is also of great importance to the projections of future climate change due to the increase in the concentration of the greenhouse gases. The AM region is one of the few places in the world for which nearly all climate models predict increased rainfall in association with global warming. Furthermore, it has been observed that the SST warming trend during the past 20 years has been largest in the tropical oceans, especially in the Indian Ocean. Increased convection in the Indo-Pacific region associated with the warming of the Indian Ocean may be linked to long-term climate change in the North Atlantic (Hoerling *et al.* 2001a, b). Therefore an understanding of the possible effects of a warmer Indian Ocean on the AM is essential to understand the regional and global effects that may stem from global warming.

There is a large body of observational and modeling research that suggest strong interaction between the ENSO and the AM. The ENSO and AM cycles mutually affect each other. While neither ENSO nor the monsoon owe their origin and existence to the other, there is clear evidence that their variations are affected by interactions between them (Kirtman and Shukla 2000; Lau and Wu 2001). Most intriguing is the recent observation of a dramatic drop in the relationship between Indian monsoon rainfall and ENSO in the past two decades. It has been hypothesized that the drop in the correlation between Indian monsoon rainfall and ENSO may be related to a shift of the Walker circulation, or temperature changes over Eurasia due to global warming, or the interaction of the monsoon with North Atlantic oscillations (Kumar *et al.* 1999; Chang *et al.* 2001). Appropriately designed modeling experiments will shed light on the possible dynamical mechanism underlying these observations and hypotheses.

9.3.1 Modeling the mean climate

The prediction of monsoon rainfall using dynamical models has been a major challenge for the climate research community in general, and the climate modeling community in particular. While it has been generally recognized that the mean monsoon system is highly stable and predictable (in the sense of a highly reproducible annual cycle), state-of-the-art climate models have been singularly unsuccessful in predicting the fluctuations about the mean annual cycle.

Because of the complex physical processes involved, interannual variability of the AM is generally only poorly simulated by climate models (Gadgil and Sajani 1998; Sperber and Palmer 1996). In the following, we discuss results of the WCRP/CLIVAR Monsoon Model Intercomparison Project (MMIP). In MMIP, 10 international modeling groups collaborated in carrying out ensemble integrations for a two-year period to assess the ability of climate models to simulate the impacts of the 1997–8 El Niño on AM anomalies (Kang *et al.* 2001a). Each model ensemble consists of 10 members, and each member is subject to the same lower boundary forcings from SST and sea ice, but with different initial conditions. While the ensemble mean rainfall distribution is broadly similar to the observed (Figure 9.5a and b), the difference between the model mean and the observed is quite large over the AM region, as well as the eastern equatorial Pacific and Central America. Compared to the observations, the model mean tends to overestimate the rainfall over the land and underestimate that over the adjacent oceanic regions. In addition, the model random noise, as measured by the deviations from the model mean, is also largest in the AM regions. In some AM regions, the model mean bias (Figure 9.5c) and the random errors (Figure 9.5d) are as large as, or larger than the mean rainfall (Figure 9.5b). The MMIP results show that modeling the mean climate and its annual variation correctly is a prerequisite for improving simulations and predictions of climate variability and global change.

Figure 9.5 Spatial distribution of climatological rainfall for (a) observations for 1997–8, (b) ensemble mean of all models (c) model minus observation differences, and (d) rms deviations from all-model mean.

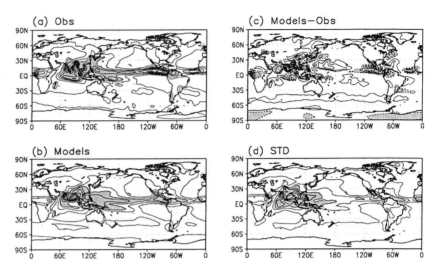

9.3.2 Pattern correlations

The performance of models can be evaluated against the similarity of the model rainfall patterns to the observed by the pattern correlation (P_{cor}) and the root-mean-square ratio (R_{rms}) defined by:

$$P_{cor} = \frac{\sum_{r} (X_r - \overline{X_r})(O_r - \overline{O_r})}{\sigma_X \sigma_O}, \qquad R_{rms} = \frac{\sigma_X}{\sigma_O}$$

where X is a model variable, and O is the corresponding observation. The summation is over the spatial coordinate **r**, over the chosen domain, the overbar represents the spatial average, and σ is the spatial standard deviation. P_{cor} and R_{rms} have been computed for each ensemble member, for each model, and for different seasons. The closer P_{cor} and R_{rms} are to unity, the better is the model performance.

Figure 9.6a and b show the model ensemble mean values of P_{cor} and R_{rms} for each model as bar charts, for December-January-February (DJF) and June-July-August (JJA) for two years, over the Indo-Pacific region $30°$ S–$30°$ N, $60°$ E–$90°$ W). The standard deviation of P_{cor} and R_{rms} for each model and for each season is indicated by the vertical lines inside the bars. The all-model ensemble mean is show in the far right column. From Figure 9.6, the performance of individual models can be compared with the others and to the all-model ensemble mean. It can be seen that the mean P_{cor} for individual models ranges from 0.2 to 0.8 (Figure 9.6a). The correlations seem to be higher during the boreal winter compared to the boreal summer, indicating that the models tend to capture the physics of the wintertime rainfall and circulation regimes better than that for the summer. All models seem to have a higher correlation for DJF 1998,

(a)

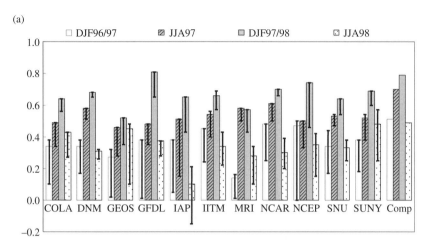

(b)

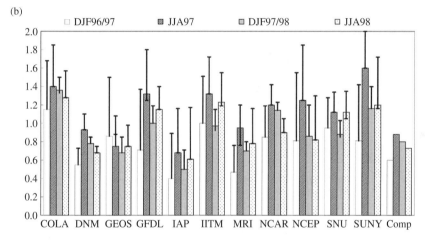

Figure 9.6 (a) Pattern correlation coefficient between the simulated and observed precipitation anomalies for each model and each season over the monsoon–ENSO region (30° S–30° N, 60° E–90° W). (b) Root-mean-square (rms) of the simulated precipitation anomalies over the monsoon–ENSO region, normalized by the observed rms. The vertical line in the bar indicates the range of the correlation and rms values of individual runs. (Adapted from Kang *et al.* 2001b)

when the El Niño signal is at a peak, suggesting that all models are responsive to the warm phase of the El Niño. If the eastern portion of the domain (east of the dateline) is excluded in the pattern correlation calculation, P_{cor} reduces dramatically (not shown), indicating that most of the good correlation is contributed by the rainfall directly responding to the El Niño SST over the central and eastern Pacific. The ensemble mean P_{cor} for each model tends to be higher than for most individual ensemble members. Similarly, P_{cor} for the model mean (columns to the extreme right) are generally among the top tier of the better performing models. The picture is quite similar for R_{rms} (Figure 9.6b). Here, about half of the models have R_{rms} greater than 1.0, and about half less than 1.0. The rms ratios for the individual members (whose range is indicated by the vertical line) tend to be larger than the model mean. This is because the model mean tends to smooth out the spatial features. The all-model R_{rms} is less than 1.0, suggesting that the models collectively underestimate the observed variability of the rainfall

Figure 9.7 Pattern correlation of ensemble anomalies of each model over the ENSO–monsoon domain (*y*-axis) vs. that of the corresponding climatology (*x*-axis) for the 1997 summer (open circle) and 1997–8 winter (dark circle) seasons. (b) As in (a), except for the normalized rms.

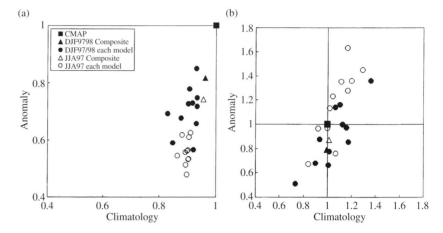

anomalies in the Indo-Pacific region. This is expected, because the all-model mean is derived from a large number (in this case, for 10 models and 10 members per ensemble, $10 \times 10 = 100$) of model realizations, while the observation is just a one-time realization of the real system. In all, the multi-model ensemble mean provides a simulation with more reliability and skill comparable to the top performing models. Other more sophisticated ensemble means, such as the super-ensemble procedure, can produce simulations with skills that exceed all the individual models (Krishnamurti *et al.* 2000).

Another important finding of MMIP is that models which can simulate a realistic climatology generally have better skill in simulating interannual variability. Figure 9.7a shows scatter plots of the climatological pattern correlation vs. the anomaly pattern correlation of rainfall of the models with respect to observations over the Indo-Pacific region. Figure 9.7b shows the same, but for R_{rms}. The climatological quantity is a measure of how good the models are in simulating the annual cycle, and the anomaly, of how well the models simulate interannual variability. Prediction skills are based on the ability of the models to simulate the interannual anomalies above and beyond those provided by the climatology. The positive slope of the regression lines in Figure 9.7a and b suggests that a good simulation of climatology generally implies a good simulation of the interannual anomaly. A model that has a good climatology is an indication that the physical and dynamical processes are well represented, and therefore provides some assurance that the model may be used for climate anomaly predictions on interannual or longer time scales.

9.3.3 Response to 1997–8 ENSO

Intercomparison of the simulations of the impact of the 1997–8 El Niño indicates that AGCMs generally simulate reasonably well the eastward shift

VELP(x10⁶ m²sec⁻¹). Diff (JJA1998−JJA1997)

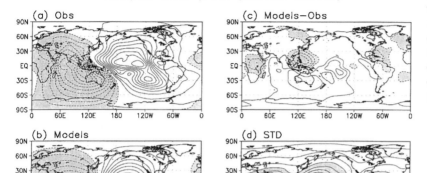

Figure 9.8 Anomalous velocity potential (JJA 1998 minus JJA 1997) for (a) observed, (b) all-model mean, (c) model minus observed and (d) standard deviation from all model-mean.

of the Walker circulation associated with the anomalous warm SST of the central and eastern Pacific (Ju and Slingo 1995; Lau and Nath 2000). Figure 9.8 shows a comparison of the observed and the simulated model mean velocity potential difference (JJA 1997 minus JJA 1998) at 200 mb. Here, the positive contours indicate anomalous large-scale ascent over the central and eastern Pacific, and the negative contours anomalous descent over the Indian Ocean. However, the difference map between the model mean and observation (Figure 9.8c) indicates that large errors are found in the Asian monsoon region, with the model overestimating the anomalous sinking motion over the maritime continent and the rising motion over the South China and East China Seas. This may be interpreted as the AGCM's inability to model the regional anomaly induced by the Walker circulation (Soman and Slingo 1997; Lau and Wu 2001). The models tend to disagree most among themselves over the Indian Ocean and the tropical western Pacific (Figure 9.8d).

Many modeling studies have been conducted to unravel the causes of the record summer monsoon flooding over central East Asia (Shen *et al.* 2001; Wang *et al.* 2000). Figure 9.9 (Plate 7) shows the ensemble mean of the rainfall and 850 mb wind anomalies (1998 minus 1997) for JJA and for each model participating in the CLIVAR MMIP. For comparison, the observation is shown in the bottom right panel labeled "obs". The observation shows a zonally oriented rainfall anomaly pattern with reduced rainfall over the tropical western Pacific, and Indo-China along 10° N, increased rainfall over the maritime continent/equatorial eastern Indian Ocean, and the sub-tropics between 30 and 40° N. The regions with rainfall increases are located on the northern and southern flanks of a subtropical anticyclone, which is

Figure 9.9   Rainfall (mm/ day) and 850 mb wind (m/s) differences (JJA 1998 minus JJA 1997) for individual models participating in CLIVAR MMIP. The observation is shown in the right-hand bottom panel. For color version see Plate 7.

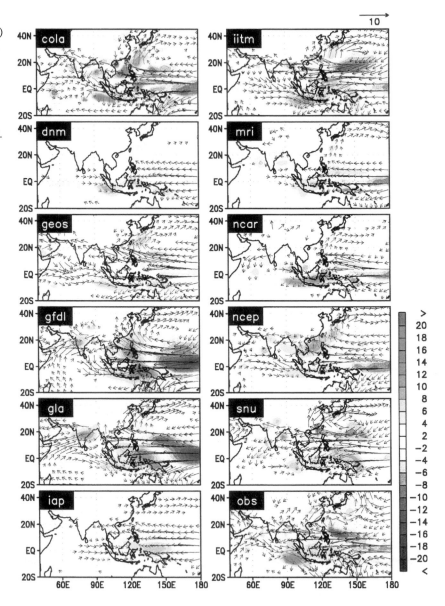

very pronounced during JJA 1998. The former is related to the *Mei-yu* rain-belt of East Asia and the latter to the development of the ITCZ over the eastern Indian Ocean and the Maritime Continent (MC). Both features were enhanced during JJA 1998. Anomalous low-level westerlies were found along 10° N from the central/western Pacific across Indo-China to the Bay of Bengal and India. Near Japan and the northwestern Pacific, the large-scale circulation shows wave-like features, associated with fluctuation of the subtropical jet stream (Lau and Weng 2002). During JJA 1997, the Walker circulation shifts eastward in response to the El Niño SST forcing, inducing strong downward motion over the maritime continent and suppressing the ITCZ. The anomalous anticyclone is the cause of the major flooding over the Yangtze River Valley during JJA 1998 (Lau and Wu 2001; Lau and Weng 2001; Shen *et al.* 2001). The establishment of the anticyclone is related to descending motion associated with the eastward shift of the Walker circulation, and also to amplification by local air–sea interactions.

The models' ability to simulate the aforementioned features is generally not very impressive. While most models show the correct sign of the large-scale response in the AM region, most models, except perhaps SNU, fail to simulate the observed zonally oriented rainfall structure. It is clear from the results shown that the simulation of the East Asian monsoon rainfall anomalies is critically dependent on the anomalous anticyclone, which governs the moisture available for precipitation. Because the anticyclone is generated by large-scale dynamics, its broad feature is represented in most models. However, it is the simulation of the exact location and magnitude of the anticyclone that is required in order for climate models to simulate the regional AM rainfall anomalies, and hence the severe floods and droughts in the AM region. The use of higher resolution climate models, or the use of downscaling methodologies is required.

The performance of AGCMs to simulate the 1997–8 rainfall anomalies can also be evaluated from examination of the distribution of model climate states represented as two-dimensional scatter plots along axes representing key climate variables. Shown in Figure 9.10 (Plate 8) are model rainfall anomalies averaged over selected domains (labeled by latitude–longitude boundaries in Figure 9.10), plotted against the SOI. The domains are for the AM region as a whole, and for its component parts over the MC, the South Asian Monsoon (SAM) and the South East Asian Monsoon (SEAM). Each data point represents one ensemble member of each model. The heavy shaded symbols represent the observed states. It can be seen that the models are quite responsive to the El Niño signal, in that there is a clear separation of model states between 1997 and 1998 along the SOI axis in all panels. Indeed, the models tend to overestimate the east–west see-saw, as evident in the larger spread along the SOI axis compared to the observations. For the AM region as a whole (Figure 9.10a (Plate 8a)), a reduction in rainfall during 1997 compared to 1998 can be discerned. This reduction is mainly due to the rainfall anomalies over the MC, which is situated at the descending branch of the Walker circulation. For the SAM and SEAM region, the large clusters of model states for 1997 and 1998, and the lack of obvious shift of the center of gravity of the clouds on the y-axis during these two years suggest that there is large rainfall variability, but there is no significant climate impact from the

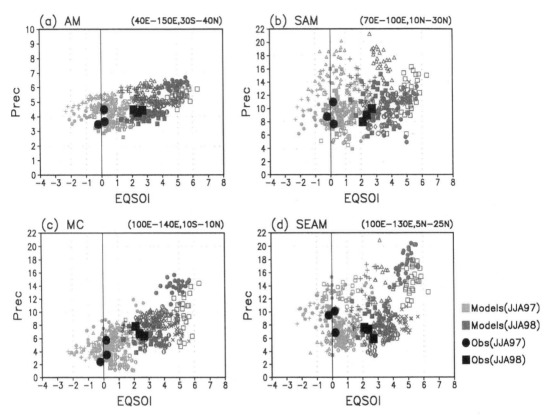

Figure 9.10   Scatter plots showing the distribution of anomalous precipitation vs. the Southern Oscillation Index for (a) the entire AM monsoon region, (b) the South Asian Monsoon region, (c) the Maritime Continent (MC), and (d) the South East Asian Monsoon region. The latitude–longitude boundary of each domain is indicated. Unit for precipitation is mm/day and for SOI is in mb. For color version see Plate 8.

El Niño in these two monsoon subregions, in agreement with observations. The observed AM anomalies represent only a single realization drawn from an intrinsic distribution effected by SST forcings identical to the 1997–8 El Niño. To the extent that the model can mimic the real climate, the cluster of model climate states around the observation provides a measure of that intrinsic distribution. AMIP results suggest that models that simulate well the seasonal cycle are a prerequisite for better simulation of the interannual variability of AM rainfall.

9.3.4 Intraseasonal variability

One of the key characteristics of the AM is the presence of a rich spectrum of subseasonal-scale variability, generally referred to as intraseasonal variability. These include quasi-periodic oscillations of 30–60 days, or 10–20 days and transient waves of 3–5 days. The intraseasonal variability is generated by

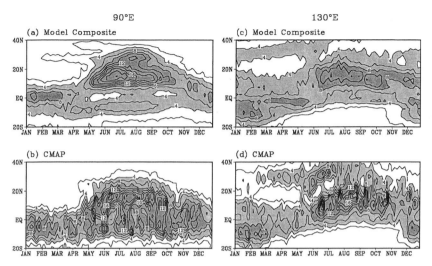

Figure 9.11 Time–latitude cross section of climatological pentad-mean precipitation. (a) and (b) are for the model composite and the CMAP observations along the longitude of 90° E. (c) and (d) are for longitude 130° E. (Adapted from Kang *et al.* 2001)

internal atmospheric dynamics but strongly modified by sea surface temperature and land surface processes. It is responsible for the modulation of monsoon onsets, breaks, and evolution regionally. Intraseasonal variability, especially at the lower frequency end of the spectrum, can have strong impacts on the seasonal mean monsoon climate. Over different regions, it can either strengthen or weaken the direct influence by ENSO on the AM. It has been suggested that the near normal monsoon rainfall over India during the strong El Niño of 1997–8 may be due to the effects of pronounced intraseasonal variability, which brought copious rainfall to many parts of India, in spite of the tendency of ENSO to weaken the AM (see discussion in previous section).

Modeling intraseasonal variability of the AM is a very challenging problem. Models generally fail to capture the phase locking between intraseasonal variability and the seasonal cycle. For example, at the longitude of the Bay of Bengal (Figure 9.11), the ensemble mean of MMIP models depicts an over-simplified picture of the monsoon evolution with a sudden onset of the South Asia monsoon near the middle of May and beginning of June. However, the models fail to reproduce the complex intraseasonal structure associated with the evolution of the monsoon rainbelt as observed. During JJA, the models show a much more quiescent atmosphere over the oceanic regions near the equator compared to the observed. Similarly, at 130° E, the models seem to capture the broad seasonal evolution, but they fail to capture the climatological intraseasonal variability during JJA, which is very prominent in the observation. In particular, the models fail to reproduce the development of the monsoon rainbelt associated with the *Mei-yu* front from 20° N to 40° N. The *Mei-yu* front is the major climate feature that dominates the climate of continental East Asia, Korea and Japan. The absence of such features in the coarse resolution climate models is an endemic problem in almost all AGCMs (Lau *et al.*

Figure 9.12 Schematic showing possible land–atmosphere feedback mechanisms associated with fluctuations of the water and energy cycles, in leading to extreme floods or prolonged drought events in monsoon regions. (Adapted from Lau and Bua 1998)

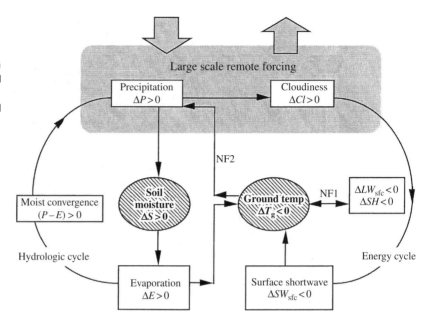

1996; Liang *et al.* 2001) indicating that downscaling approaches may be needed to capture this unique feature of the East Asian monsoon.

9.3.5 Land–atmosphere feedback

Recent modeling studies have shown that land–atmosphere processes can affect monsoon and monsoon–ENSO relationships by altering the energy and water cycles within the AM regions, through surface heat fluxes and hydrologic feedback mechanisms. Lau and Bua (1998) carried out a series of numerical experiments using a NASA global climate model, similar to those described in Section 9.3.3. Their results suggest an atmosphere–ocean–land feedback scenario as illustrated in Figure 9.12. If the soil moisture content of the Asiatic land mass is abnormally high during the start of a monsoon season, land surface evaporation will be increased. This will lead to increased moistening of the atmospheric boundary layer, more unstable air masses and hence more convection and rainfall, resulting in a positive feedback leading to further moistening of the land region. However, the cloudy sky condition stemming from enhanced convection will shield off and reduce solar radiation reaching the land surface, causing the land to cool. As the land mass cools off, the resulting decreased land–sea thermal contrast can only support a weaker large-scale monsoon circulation, with reduced monsoon rainfall, thus producing a negative feedback, which halts further increase in soil moisture. These feedback mechanisms are dependent not only on local processes but also on remote forcing such as forced large-scale descent or ascent over the AM region by ENSO. The large-scale vertical motions provide a strong control on atmospheric stability and initiation of convection. Even though the ENSO remote forcing has relatively slow time scales, its impact may be sufficient to tip the delicate balance of the aforementioned local feedback processes causing either the amplification of a given climate state or transition from one state to the other. The hypothesis needs to be verified with additional data, analyses, and further experimentations with other climate models with detailed land surface processes.

In summary, studies up to now have shown that a large portion of the predictable part of interannual variability of the monsoon rainfall is forced by the slowly varying boundary conditions at the Earth's surface. However, no climate model has been able to replicate even the simplest empirical relationship between the SST anomalies and monsoon rainfall anomalies. It is unclear at this stage whether the inability of current models to simulate and predict monsoon rainfall is due to model deficiencies or due to intrinsic lack of predictability of the monsoon. It is likely that both play important roles.

9.4 Future challenges

In this chapter, we have discussed the importance of modeling in providing better understanding of causes of regional climate anomalies, and in predicting future climate evolution, using the AM climate as a specific example. Given that models will be increasingly used for climate predictions on all time scales, it is important to keep in mind that large uncertainties exist and that not all aspects of model predictions have the same degree of reliability. The challenge is how to reduce these model uncertainties, and to make climate forecasting more reliable and useful. The following are suggested steps that should be taken to move in that direction. Because each of the steps involves complex procedures and organized efforts, successfully implementing these steps will take years of sustained efforts by the science community.

- *Use finer spatial and temporal resolution.* As stated previously, one of the major uncertainties in climate models stems from the lack of spatial and temporal resolution, and as a result regional to subregional scale features are not well represented. Yet it is these subregional features and short-term events that cause the most socioeconomic damage. With the advances in computational power, it is now possible to run climate models with high resolution globally for extended periods. For example the European Centre for Medium-Range Weather Forecasts is running its operational model for medium and long-range weather predictions of the order of 50 km resolution, and the Frontier Climate Change Research Program of Japan is running the Earth Simulator at 10–20 km resolution. One of the most obvious improvements in going to higher resolution is the better simulation of orographic rain, especially in regions of complex topography. Even with the increased resolution, for some applications such as catchment-scale water resource management, resolution of less than a few kilometers may be required. In this respect, the downscaling techniques using regional climate models and macroscale hydrology models will be important.

- *Improve model physics.* While increasing resolution will reduce model uncertainty and improve the geophysical fluid dynamic aspects of climate models, the major culprit of model uncertainty is still the inadequacy of representation of the physical processes that determine the forcing functions of the model. If a model is driven by erroneous forcing functions, no matter how well the flow fields can be simulated, it will not give the right answers. Improvement of physical representation in models is therefore paramount and should be focused on processes that are key drivers of the Earth's hydrologic cycles. These include cumulus heating in the tropics, aerosol–cloud–radiative processes, fluxes at the air–sea and air–land interfaces, land surface and vegetation processes. Improving model physics is an extremely difficult and tedious endeavor, because the physics of climate is very complex and interwoven. Improving a physical process in a stand-alone model does not necessarily mean that it will give better performance in a coupled model. Likewise, improving one part of the system does not always lead to improvement of other parts. Hence the process of improving model physics can be very arduous, calling for multiple tests and validation with observations under a variety of conditions. Substantial improvements are not likely to come in the short term, but sustained organized efforts by the scientific community are required. In some sense, we have exhausted much of the reliable information that can be derived from current climate models. Unless model physics improvement is taken seriously, model uncertainties will remain unacceptably large.
- *Improve data for model validation.* One major stumbling block for model improvement is the lack of detailed data suitable for model validation and improvement. Given the vast amount of data obtained from ground-based and satellite atmospheric and oceanic observations, field campaigns, and special measurement platforms, it may seem a bit puzzling that there is still a shortage of data for model validation. The reason is that for model physics improvement very specialized data with high spatial and temporal resolutions, directly relevant to the model parameters, are required. These data are often not direct observables in the climate system, but are quantities derived from the observables and therefore have large uncertainties themselves. Often, they require special intensive observation platforms, which for practical purposes can only be carried out over a short period in field campaigns. To be sure, besides new data from future field campaigns, there are data that can be extracted from the vast satellite and operational historical database, as well as various enhanced observational sites to be used for model validation and physics improvement. There is a need for coordinating and extracting global, regional, and site data from various sources, and making them available for model improvement and prediction.

- *Model prediction applications.* Climate forecasts both short and long term have tremendous potential benefits for society. Accurate seasonal forecasts of winter storms, summer droughts and floods, and hurricane frequencies will result in substantial savings and in reduction of damaged properties and loss of human life. Skilful prediction of El Niño using coupled ocean–atmosphere models has resulted in limiting its adverse impacts on food production and fisheries by advanced planning, and implementation of mitigating measures in many regions around the world. To realize the full benefit of climate forecasts, climate models should be coupled with cost and risk models for agriculture, food production, water resource management, and other societal applications, to reduce vulnerabilities to natural hazards and climate change.

9.5 Acknowledgment

This work is supported by the Global Modeling and Analysis Program of the NASA Earth Science Office.

9.6 Examples of climate modeling websites

The Hadley Centre, part of the Met office in the United Kingdom, has many good examples of climate models and their results, www.metoffice.com/research/hadleycentre/models/modeltypes.html.

The Division of Marine and Atmospheric Research in Australia's Commonwealth Scientific and Industrial Research Organisation (CSIRO) models global and Southern Hemisphere climate, www.dar.csiro.au/earthsystems/.

The Climate Modeling and Diagnostics Group at Columbia University, USA, models past and present climate distributions and variations, http://rainbow.ldeo.columbia.edu/.

The Climate Modeling Section of the National Center for Atmospheric Research (USA) simulates future climates and potential social impacts of climate variability and change, www.cgd.ucar.edu/cms/.

Regional climate variations are modeled through the Canadian Regional Climate Modelling Network, based at the University of Quebec in Montreal, www.mrcc.uqam.ca/E_v/index_e.html.

There is a wide variety of climate modeling websites from universities and government research and forecasting agencies, all over the world.

9.7 References

Bengtsson, L. and Simmons, A. J., 1983. Medium range weather prediction – operational experience at ECMWF. In B. J. Hoskins and R. P. Pearce, eds., *Large-Scale Dynamical Processes in the Atmosphere*. Academic Press, pp. 337–63.

Chang, C. P., Harr, P. and Ju, J., 2001. Possible role of Atlantic circulations on the weakening Indian monsoon rainfall–ENSO relationship. *Journal of Climate*, **14**, 2376–2380.

Charney, J. G., Fjortoft, R. and von Neumann, J., 1950. Numerical integration of the barotropic vorticity equation. *Tellus*, **2**, 237–254.

Fox-Rabinovitz, M. S. *et al.*, 2001. A variable-resolution stretched-grid general circulation model: Regional climate simulation. *Monthly Weather Review*, **129**, 453–469.

Gadgil, S. and Sajani, S., 1998. Monsoon precipitation in the AMIP runs. *Climate Dynamics*, **14**, 659–689.

Gates, W. L., 1992. AMIP: The atmospheric model intercomparison project. *Bulletin of the American Meteorological Society*, **73**, 1962–1970.

Gates, W. L. *et al.*, 1999. An overview of the results of the Atmospheric Model Intercomparison Project (AMIP-I). *Bulletin of the American Meteorological Society*, **80**, 29–55.

Giorgi, P. and Mearns, L. O., 1991. Approaches to simulations of regional climate change: a review. *Reviews of Geophysics*, **29**, 191–216.

Gilchrist, A., 1977. An experiment in extended range prediction using a general circulation model and including the influence of sea surface temperature anomalies. *Beiträge zur Physik der Atmosphare*, **50**, 25–40.

Gilchrist, A., 1981. Simulation of the Asian summer monsoon by an 11-layer general circulation model. In M. J. Lighthill and R. P. Pearce, eds., *Monsoon Dynamics*. Cambridge: Cambridge University Press, pp. 131–145.

Hewitson, B. C. and Crane, R. G., 1996. Climate downscaling: techniques and application. *Climate Research*, **7**, 85–96.

Hoerling, M. P. *et al.*, 2001a. The midlatitude warming during 1998–2000. *Geophysical Research Letters*, **28**, 755–758.

Hoerling, M. P., Hurrell, J. W., and Xu, T., 2001b. Tropical origin for recent North Atlantic climate change. *Science*, **292**, 90–92.

Hoffman, R. N. and Kalnay, E., 1983. Lagged average forecasting, an alternative to Monte Carlo forecasting. *Tellus*, **35a**, 100–118.

Ju, J. and Slingo, J., 1995. The Asian summer monsoon and ENSO. *Quarterly Journal of the Royal Meteorological Society*, **121**, 1133–1168.

Kalnay, E. *et al.*, 1996. The NCEP/NCAR 40-year reanalysis project. *Bulletin of the American Meteorological Society*, **77**, 437–471.

Kang, I. S., Jin, K., Wang, B. and Lau, K., 2001a. Intercomparison of the climatological variations of the Asian summer monsoon rainfall simulated by 10 GCMs. *Climate Dynamics*, **19**, 383–395.

Kang, I.-S. *et al.*, 2001b. Intercomparison of GCM simulated anomalies associated with the 1997–98 El Niño. *Journal of Climate*, **15**, 2791–2805.

Kawamura, R., Sugi, M., Kayahara, T. and Sato, N., 1998. Recent extraordinary cool and hot summers in East Asia simulated by an ensemble climate experiment. *Journal of the Meteorological Society of Japan*, **76**, 597–617.

Kirtman, B. P. and Shukla, J., 2000. Influence of the Asian summer monsoon on ENSO. *Quarterly Journal of the Royal Meteorological Society*, **126**, 213–239.

Krishnamurti, T. N., Kishtawal, C. M., Shin, D. W. and Williford, C. E., 2000. Improving tropical precipitation forecasts from a multi-analysis superensemble. *Journal of Climate*, **13**, 4217–4227.

Kumar, K. K., Rajagopalan, B. and Cane, M. A., 1999. On the weakening relationship between the Indian Monsoon and ENSO. *Science*, **284**, 2156–2159.

Lattenmaier, D. P., Wood, A. W., Palmer R. N., Wood, E. F. and Stakhiv, E. Z., 1999. Water resources implications of global warming: A US regional perspective. *Climate Change*, **43**, 537–579.

Lau, K.-M. and Bua, W., 1998. Mechanism of monsoon–Southern Oscillation coupling: insights from GCM experiments. *Climate Dynamics*, **14**, 759–779.

Lau, K. M. and Weng, H., 2001. Coherent modes of global SST and summer rainfall over China: an assessment of the regional impacts of the 1997–98 El Nino. *Journal of Climate*, **14**, 1294–1308.

 2002. Recurrent teleconnnection patterns linking summertime precipitation variability over East Asia and North America. *Journal of the Meteorological Society of Japan*, **80**, 1309–1324.

Lau, K. M. and Wu, H. T., 2001. Intrinsic modes of coupled rainfall/SST variability for the Asian summer monsoon: a re-assessment of monsoon–ENSO relationship. *Journal of Climate*, **14**, 2880–2895.

Lau, K.-M., Kim, J. H. and Sud, Y., 1996. Intercomparison of hydrologic processes in AMIP GCMs. *Bulletin of the American Meteorological Society*, **77**, 2209–2227.

Lau, N. C. and Nath, M. J., 2000. Impact of ENSO on the variability of the Asian-Australian monsoons as simulated in GCM experiments. *Journal of Climate*, **13**, 4287–4309.

Liang, X., Wang, W. C. and Samel, A. N., 2001. Biases in AMIP model simulations of the east China monsoon system. *Climate Dynamics*, **17**, 291–304.

Manabe, S. and Wetherald, R. T., 1975. The effects of doubling the CO_2 concentration on the climate of a general circulation model. *Journal of Atmospheric Science*, **32**, 3–15.

Manabe, S., Bryan, K. and Spelman, M. J., 1979. A global ocean-atmosphere climate model with seasonal variation for future studies of climate sensitivity. *Dynamic Atmospheres and Oceans*, **3**, 393–426.

Palmer, T. N., 1993. Extended range atmospheric prediction and the Lorenz model. *Bulletin of the American Meteorological Society*, **74**, 49–66.

Shen, X., Kimot, M., Sumi, A., Numagauti, A. and Matsumoto, J., 2001. Simulation of the 1998 East Asian summer monsoon by the CCSR/NIEW AGCM. *Journal of the Meteorological Society of Japan*, **79**, 741–757.

Shukla, J. *et al.*, 2000. Dynamical seasonal prediction. *Bulletin of the American Meteorological Society*, **81**, 1593–2606.

Smagorinsky, J., Manabe, S. and Holloway, J. L., 1965. Results from a nine-level general circulation model of the atmosphere. *Monthly Weather Review*, **93**, 727–768.

Soman, J. K. and Slingo, J., 1997. Sensitivity of Asian summer monsoon to aspects of sea surface temperature anomalies in the tropical Pacific Ocean. *Quarterly Journal of the Royal Meteorological Society*, **123**, 309–336.

Sperber, K. R. and Palmer, T. N., 1996. Interannual tropical rainfall variability in general circulation model simulations associated with the atmospheric model intercomparison project. *Journal of Climate*, **9**, 2727–2750.

Stefanova, L. and Krishnamurti, T. N., 2002. Interpretation of seasonal climate forecast using Brier skill score, the Florida State University superensemble and the AMIP-I data set. *Journal of Climate*, **15**, 537–544.

Tracton, M., Kalnay, S. and Kalnay, E., 1993. Operational ensemble prediction at the National Meteorological Center: Practical aspects. *Weather Forecasting*, **8**, 379–398.

Von Storch, H. and Zwiers, F. W., 1999. *Statistical Analysis in Climate Research*. Cambridge: Cambridge University Press.

Wang, H.-J., Matsuno, T. and Kurihara, Y., 2000. Ensemble hindcast experiments for the flood period over China in 1998 by use of the CCSR/NIES AGCM. *Journal of the Meteorological Society of Japan*, **78**, 357–365.

Chapter 10
Conclusions and the future of climate research

10.1 Introduction

We have emphasized in this book the importance of considering climate, climate change, and climate variability in terms of interacting components of a system, with many different, often simultaneous, impacts. This chapter describes some of the important research areas for the future which should lead to better understanding of the climate system.

The climate system can be considered complex, highly non-linear, and chaotic, with variations consisting broadly of two types. The first is longer-term transition changes, such as some of those described in Chapter 6, relating to forces such as polar wandering, changes in the Earth's tilt, changes in the Earth's orbit around the Sun, and the ice ages. The second is short-term abrupt changes, relating to rapid variations in forcing functions, such as the radiative output from the Sun and particulate matter from volcanic eruptions. The most recent short-term forcing function is human activity and its impact on the atmosphere.

10.1.1 Programs and organizations

Over the past three decades, there have been major advances in understanding how the climate system operates. These include improvements in forecasting, understanding of the role of sea ice, interactions between the atmosphere and ocean, the role of atmospheric chemistry, and so forth. The more we learn, the more we seem to need to know. Over the past 20 years, the study of the climate system has moved strongly onto the international stage. Table 10.1 presents a summary of major organizations and projects that cover a wide range of climate and related research. The scientific activities can be divided into government and non-government organizational components. The non-government components operate under the broad goals and coordination of ICSU, the International Council for Science. The atmosphere and climate make up only one component of ICSU, which represents the scientific community in global affairs, supports a

Table 10.1   *Structure of major organizations and programs within the international research community, with focus on the atmosphere (courtesy Henning Rodhe, Department of Meteorology, University of Stockholm, and adapted from IGAC Conference discussions in Crete, 2002).*

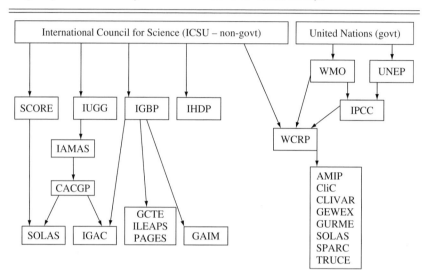

Organizations and programs

SCORE	Scientific Committee on Ocean Research
IUGG	International Union of Geodesy and Geophysics
IGBP	International Geosphere/Biosphere Program
IHDP	International Hydrological Development Program
WCRP	World Climate Research Programme
WMO	World Meteorological Organization
UNEP	United Nations Environment Programme
IPCC	International Panel on Climate Change
IAMAS	International Association of Meteorology and Atmospheric Science
CACGP	Commission on Atmospheric Chemistry and Global Pollution
IGAC	International Global Atmospheric Chemistry Program
GAIM	Global Analysis, Integration, and Modelling Program

International projects (WCRP and IGBP)

AMIP	Atmospheric Model Intercomparison Project
CliC	Climate and Cryosphere
CLIVAR	Climate Variability and Predictability
GEWEX	Global Energy and Water Cycle Experiment
GCTE	Global Chemistry Tropospheric Experiment (completed Dec 2003)

GURME	Global Urban Research Meteorology and Environmental Project
ILEAPS	Integrated Land Ecosystem–Atmospheric Processes Study
PAGES	Past Global Changes
SOLAS	Surface Ocean–Lower Atmosphere Study
SPARC	Stratospheric Processes and their Role in Climate
TRUCE	Tropical Urban Climate Experiment

very wide range of scientific research, encourages the exchange of ideas, and promotes many scientific conferences (www.icsu.org). Table 10.1 lists, under ICSU, a number of major programs and projects that represent climate research at different scales.

Government support for climate and other scientific research is coordinated internationally through the United Nations (UNEP) and the WMO. At this level, the IPCC and its work defining the impacts of global warming and human activities is supported (IPCC 2001). Both organizations support the WCRP, which provides the organizational framework for projects in areas such as energy, chemistry, the cryosphere, past climates, urban climate, and interactions with other global systems. The climate research "umbrella" goals of the WCRP are listed in Table 10.2, along with those of the United States Climate Change Science Program. These goals define in the broadest sense an international framework for more specific research, through international projects with more specific goals, to improve understanding of the climate system and its components. Patterns, processes, and teleconnections are crucial aspects of this research.

Overall, major aims include developing a complete understanding of "... the dynamical, radiative, and chemical processes in the atmosphere–land–system ..." (IPCC 2001) and "... to understand and predict to the extent possible climate variability and change, including human influences" (Grassl 2001). Steffen (2001) states that critical aspects of these goals and aims include understanding the stability and foundation of the entire atmospheric system, the triggers causing change, how change and transition occur, the human influence, how changes in climate influence other systems, and how the mix in forcing functions works.

10.2 Understanding the global climate system

10.2.1 Radiative forcing and climate change over time

Radiative forcing is a key factor in the overall climate system. It is important to the variations associated with oscillations and teleconnections (Chapter 2), helps determine the strength and variations in circulation between the equator and the poles (Chapters 4 and 5), and is a critical determinant of past climate variations

Table 10.2   *Broad research goals related to climate and climate variability from the World Climate Research Programme (WCRP) and US Climate Change Science Program (CCSP), as examples of "umbrella" goals*

World Climate Research Programme (WCRP)

Goal 1   Develop the fundamental scientific understanding of the physical climate system and climate processes to determine to what extent climate can be predicted and the extent of human influences on climate

Goal 2   Find quantitative answers to climate and the range of natural climate variability

Goal 3   Establish the basis for predictions of global and regional climate variations and of changes in the frequency and severity of extreme events

Goal 4   Address forcefully outstanding issues of scientific uncertainty in the Earth's climate system (i.e. transport and storage of heat by the ocean; global energy and hydrological cycle; formation of clouds and their effect on radiative transfer; role of the cryosphere in climate change)

US Climate Change Science Program (CCSP)

Goal 1   Improve knowledge of the Earth's past and present climate and environment, including its natural variability, and improve understanding of the causes of observed variability and change

Goal 2   Improve quantification of the forces bringing about changes in the Earth's climate and related systems

Goal 3   Reduce uncertainty in projections of how the Earth's climate and environmental systems may change in the future

Goal 4   Understand the sensitivity and adaptability of different natural and managed ecosystems and human systems to climate and related global changes

Goal 5   Explore the uses and identify the limits of evolving knowledge to manage risks and opportunities related to climate variability and change

---

www.climatescience.gov/Library/sap/sap-summary.htm;
www.wmo.ch/web/wcrp/about.htm

(Chapter 6). Figure 10.1 (from IPCC 2001) is a well-known summary of the major components causing forcing in the global climate system, and indicates the level of scientific understanding associated with each. Only forcing from the greenhouse gases, where major understanding has improved strongly over the past 20 years, is ranked high. The role of ozone, both in the stratosphere and troposphere, is listed as medium. For all other radiative factors, the level of understanding is listed as very low. Radiative forcing is also critical in understanding the link between climate, climate variability, and other components of the Earth's environment. These include biogeochemical cycles, ocean–atmosphere and land–atmosphere interactions, and sources and sinks of atmospheric components. Full understanding requires a multi-disciplinary approach,

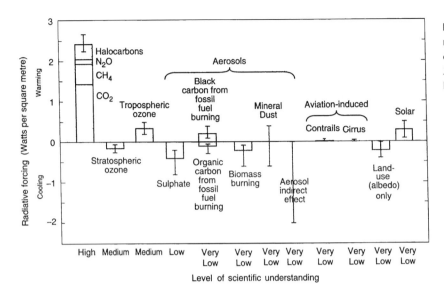

Figure 10.1 The global mean radiative forcing of the climate system for the year 2000, relative to 1750. (See IPCC 2001 for details)

investigating multiple scenarios to establish a range of responses. A good example is SPARC, which concentrates on interactions between radiation, chemistry, and dynamics in the stratosphere.

10.2.2 Patterns of climate change over time

Chapter 6 provides an indication of climate variations over geological time, and more recently during the Holocene, demonstrating the value of good quality proxy data. A complete understanding of the patterns of climate change requires a full investigation of climate transition periods over various time scales, and definitions of the impacts of long-term and more abrupt variations. The research associated with PAGES is dedicated toward this goal. The periods of extreme climate, drought and flood, cold and heat, and what causes them, can then be better explained and predicted. On a short-term scale, understanding teleconnections associated with oscillations, such as ENSO, NAO, PDO, and QBO (see Chapters 2 and 3), and atmospheric impacts in different part of the globe, are critical. While the impacts that occur are apparent, establishing why and how they are created requires considerably more research.

10.2.3 Feedbacks

Feedbacks are changes that occur in the climate system when some component of that system varies. Examples of such components include water vapor, the carbon cycle, atmospheric chemistry, and ocean circulation. Feedbacks can involve all components of the atmospheric system. They affect the sensitivity of climate processes and are not well understood. They may be positive (enhancing) or negative (reducing). A controversial current example is the role of infrared radiation absorption by greenhouse gases in global temperature and moisture change. Incoming shortwave radiation is absorbed by the Earth's surface, and re-released as longwave energy (see Chapter 1). Increased greenhouse gas levels in the atmosphere lead to greater infrared absorption, creating higher atmospheric temperatures. A warmer atmosphere can hold more water vapor, and thus evaporation from surface water sources is enhanced. The higher moisture levels in the atmosphere, enhanced by increased tropospheric instability due to the warming, enhance cloud formation. The increased clouds (along with increased particle matter from human sources) reduce incoming shortwave radiation through backscattering and reflection. Theoretically, then, less longwave radiation becomes available from a cooler Earth's surface. But the increased clouds are more efficient at absorbing longwave radiation than the atmospheric gases, potentially increasing atmospheric warming.

Within this very general feedback framework are a whole series of further complexities and questions. What impacts do variations in atmospheric chemistry have on the feedback process? What is the role of dominant processes, such as the ocean and the biosphere? How do the cloud processes work and why? These kinds of questions are being addressed through projects such as SOLAS and ILEAPS, which are studying biochemical feedbacks between surfaces and the atmosphere and how transformation processes work. The IPCC (2001) acknowledges that of all the feedback components, it is the role of clouds that is the least understood.

10.2.4 Climate variability and change at the extremes of the Earth

Despite the advent of satellites, and major increases in research and knowledge over the past 20 years, understanding of climate and its variability in the polar regions is in many ways the last great climate frontier. Modeling results reported by IPCC (2001), and also in Chapter 9, plus a wide range of measurements, have identified the potential for major climate change associated with warming in the polar regions, especially in the Arctic. Chapter 5 and the essays by Wendler highlight some important areas where information is lacking.

The polar ice caps are physically the most demanding, and logistically the most difficult, environments on the globe in which to measure and assess climate. CliC provides the major international research focus on the cryosphere and climate. There are three main goals. The first is to enhance the observations and monitoring to support accurate detection of climate change. Long-term data sets are lacking, and early data are basically meteorological, and very patchy. More recent data from automatic weather stations, supplemented by satellite information, and models (NCEP/NCAR reanalysis for example, Chapter 4), have created a more geographically representative set of data in both the Arctic and the Antarctic. However, more long-term data are needed to obtain a true picture of polar climates and how they change over time.

The second is to increase understanding of the ice/ocean/land/atmosphere relationships and the feedback processes and how they change over time. Examples of questions include: if the Arctic ice pack melts, how will the radiation and energy budgets be affected? What will happen to the AO and the general circulation? How will teleconnections between the polar regions and lower latitudes be changed? What effects do surface contrasts and changes to them have on the regional and local climate? If in the Antarctic, warming brings increases in precipitation and an expanded volume of ice, what will happen to the circumpolar trough? How will the mid-latitude circulation structure be changed? What will be the impact on cyclogenesis and the strength and geographic distribution of extra-tropical lows?

The third CliC goal is to elevate the accuracy of climate simulations and predictions by providing a more accurate or representative picture of cryospheric processes in models. Currently, various models show a wide range of polar sensitivities to the climate. Better depiction of the current influences by the equatorial–polar temperature gradient, the AO and AAO, and the surface–atmospheric interactions are essential, if we are to be able to properly estimate climate change.

Projects such as the Arctic Climate System Study (ACSYS) are focused on providing progress toward the CliC goals. Under ACSYS, a series of research activities is planned that will evaluate the sensitivity of the Arctic ice cover to climate change, through improved measurement and modeling. More detailed assessment of the components of the hydrological cycle, especially river runoff and precipitation will be completed. The potential impacts of climate change on the great global ocean conveyor belt circulation, currently subject to much theory and speculation, will be evaluated.

10.2.5 Air pollutants and climate change

Basic atmospheric chemistry, and the forcing impacts of air pollution from human activities, are listed in Figure 10.1 as having a medium to very low level of scientific understanding. Chapters 5 and 7 establish the importance of pollutants in atmospheric chemical processes, associated with various scales of atmospheric transport. On a global scale, greenhouse gases and fine particulates play complex and potentially opposite roles on global warming. On a regional scale, pollutants such as ozone, carbon monoxide, and particulate matter strongly influence the quality and chemistry of urban and downwind atmospheres. Better estimates of future emissions will support more representative depictions of the impacts on climate.

A key objective of the IGAC program is to provide more accurate determinations of how the chemical process works in the atmosphere, and the resultant impacts on climate patterns. Improved knowledge of the global distribution of both short- and long-lived chemical species, and how these change over time will provide valuable support. On a more regional scale, GURME and TRUCE addresss interactions between the urban environment and the atmosphere, and incorporate resultant impacts on the community. Integration of atmospheric chemistry with other atmospheric processes, and the complexities associated with feedbacks and surface coupling will lead to an increased holistic understanding of the climate system.

10.2.6 Quantifying climate change uncertainties

Accurate determinations of climate change and variability, and projections for the future, require appropriate quantification of uncertainties. Figure 10.1 establishes that much work is needed before this task can be accomplished. Quantification can improve through better measurements over longer periods of time, and through more improved modeling of the current and future climate system (see Section 10.2.7). One of the major goals of the WCRP and the WMO is to improve the temporal and spatial coverage of the measurement network around the globe. While advances in satellite technology and data have increased exponentially over the past three decades, the near-surface measurement network has been in decline. The network must produce spatially compatible data, with climate and its variability a major purpose, and be supported by adequate computational resources. Problems such as political differences, inaccurate use of instruments, inaccurate data collection and analysis, and clashing demands for financial support must be overcome.

A network on its own has limited value if there is no integrated strategy to share and assess that data. Consistent statistical methods which accurately define variations that are acceptable to the international scientific community are needed. Uncertainties in the data must be defined and reduced. Network information can provide the basis for sensitivity studies, to define the impacts of a range of forcing variables, and to establish the levels of climate system perturbations and teleconnections. More accurate and more complete data are also needed to support modeling.

10.2.7 Improving climate modeling

Chapter 9 establishes the importance of global general circulation models (GCMs) and nested regional models to enhancing knowledge about the climate system. Over the past 20 years, models have become a major tool to assess climate variability and change, and their causes. The improvement in hardware and software technology has allowed rapid model development, and increased the skill of model results. Those models that can provide an assessment of dynamic coupling between the atmosphere–ocean, atmosphere–land systems are especially useful. IPCC (2001) emphasizes that incorporation of the consequences of human activities is critical.

Lau in Chapter 9 states that large uncertainties still exist, and while major improvements in model reliability have been made, there are important limitations to resolve. It is essential to remember that climate models are a tool, and will not produce final answers. Errors in one part of the model can propagate, and magnify inaccuracies in the overall results, which can then lead to incorrect conclusions and the loss of credibility. The approach taken by IPCC (2001) is to recommend that models be used to develop multiple ensembles of probability scenarios for future climate change, to assess the potential impacts created by changes to forcing functions.

Challenges to future improvements in modeling range across a number of areas. These are being addressed through projects such as AMIP:

- improvements in reliability of simulating both short-term abrupt and longer-term transitional climate change;
- better use of fine spatial and temporal resolution, which will allow more accurate details of change on regional and smaller scales;
- improved simulation of extreme events and their causes, which would help define vulnerability and plan for the economic and social costs of climate and weather disasters, such as those described in Changnon's essay in Chapter 8;
- better incorporation of the physics of how the atmosphere works, to more accurately determine the impacts of forcing;
- more accurate incorporation of the complexities of atmospheric chemistry, especially in critical areas such as the stratosphere (Chapter 5) and the urban atmosphere (Chapter 7);
- increasingly accurate sensitivity studies, not only for the overall model, but for its interactive components;
- better determination of the role of coupling and feedbacks (i.e. convection, advection, clouds, sea ice, and other processes) between the atmosphere and the earth and ocean, and how each influences the other.

Further detailed evaluations of climate research needs in the future can be found in IPCC (2001), Grassl (2001) and other climate review publications, plus position statements on the websites of climate science organizations such as the World Meteorological Organization (www.wmo.ch), the American Meteorological Society (www.ametsoc.org) and the American Geophysical Union (www.agu.org).

10.3 The importance of communication

There is now overwhelming evidence that human beings and their activities are creating important influences on climate and its variability. The major arguments are about the scale of the influence, how it affects the patterns and processes within the climate system, the scale of the either positive or negative effects on societies, and the level of mitigation necessary (see Glantz essay, Chapter 1). IPCC (2001) and others recognize the need to incorporate social, behavioral and economic aspects within the physical models to obtain a holistic understanding. Progress here has been slow, limited by difficulties in matching time scales, methodological approaches, and incorporation of human behavior within a physical system. However, a beginning has been made, through incorporating aspects such as land-use changes and air pollution emissions, and through a better understanding of the beliefs, attitudes, level of education, and decision-making processes of different cultures (Chapter 8).

While the scientific understanding of climate, its variability, and the reasons for change is improving, communication of the importance of these changes to political and economic decision makers can be very difficult. The risks related to climate change, despite their potential global and regional impacts, are often a very low priority compared to economic development, social tensions, and other environmental stresses.

Climate scientists need to develop ways of communicating more effectively on the outside. Kinzig and Starrett (2003) state that differences in handling uncertainty and establishing proof are major barriers between science and policy (see Table 10.3). For example, scientists are often unwilling to provide definite conclusions about climate change because they recognize that limitations and errors in the methodology will affect their results. Policy makers want more black and white conclusions, upon which they can make economic and political decisions. They expect these results to be correct, since often political, and possibly economic, survival is at stake. If decisions supporting mitigation are not possible or considered not important, then a better understanding of how different societies will adapt to climate change is essential.

The communication process is improving, but much more needs to be done. Organizations such as the IPCC and their reports are a major step in the right direction. Further enhancement of the cooperation and coordination process depends on such things as free data exchange (a major goal of ICSU and the WMO), more globally accurate data sets, and better understanding between countries. Acceptance that climate change is as important a problem as warfare, global poverty, and economic development may never occur. However, incorporating the consequences of policy decisions on climate and its variability must be a major goal for the decision-making process in the twenty-first century.

Table 10.3 *Some problems and difficulties in communication between scientists and policy decision makers (based on concepts from Kinzig and Starrett 2003)*

Science	Policy
Priorities	*Priorities*
• advance knowledge	• address challenges to society
• investigate what is not known	• ensure human welfare
• build on earlier knowledge	• timely responses
• accuracy (the cost of incorrect knowledge is high)	• actions may need to precede knowledge
• protect against inaccurate knowledge	• avoid political and social costs
• identify error	• ensure security of government
• ensure small probability of error	• economics and development

Communications problems occur because of:
- differences in ranking priorities
- differences in language
- differences in standards
- lack of scientific consensus
- scientific quantification of uncertainties difficult
- scientific information not useable for policy
- scientific evidence not important or acceptable to policy
- special interest lobby groups have priority to policy
- lack of understanding and knowledge
- political and economic differences

Ways of overcoming communications problems:
- combined science-policy forums at highest levels
- incorporate science into all decision making analysis
- develop national and international consensus on appropriate responses
- translate science-speak into policy-speak
- consider the consequences
- improve methods of presenting uncertainties
- accept compromise where necessary
- improve willingness to make decisions within uncertainty
- improve public education and understanding
- develop policy/planning approaches that are flexible and can change direction
- encourage holistic solutions
- community involvement

10.4 References

Grassl, H., 2001. Research objectives of the World Climate Research Program. In L. Bengtsson and C. Hammer, eds., *Geosphere-Biosphere Interactions and Climate*. Cambridge: Cambridge University Press, pp. 280–284.

IPCC, 2001. *Climate Change 2001: The Scientific Basis. Contribution of Working Group I to the Third Assessment Report of the Intergovernmental Panel on Climate Change.* J. Houghton, Y. Ding, D. Griggs, *et al.* eds., Cambridge: Cambridge University Press.

Kinzig, A. and Starrett, D., 2003. Coping with uncertainty: a call for a new science-policy forum. *Ambio*, **32** (5), 330–335.

Steffen, W., 2001. Toward a new approach to climate impact studies. In L. Bengtsson and C. Hammer, eds., *Geosphere–Biosphere Interactions and Climate*. Cambridge: Cambridge University Press, pp. 273–279.

Other books on climatology and the climate system

Akin, W. E., 1991. *Global Patterns: Climate, Vegetation, and Soils*. Norman, Oklahoma: University of Oklahoma Press. Chapter 1.

Ayoade, J. O., 1983. *Introduction to Climatology for the Tropics*. New York: John Wiley. Chapter 3.

Barry, R. G. and Carleton, A. M., 2001. *Synoptic and Dynamic Climatology*. London; New York: Routledge. Chapter 1.

Barry, R. G. and Chorley, R. J., 1998. *Atmosphere, Weather, and Climate*, 7th edn. London: Routlege. Chapter 1.

Beniston, M. and Verstraete, M. M., 2001. *Remote Sensing and Climate Modeling: Synergies and Limitations*. Boston: Kluwer Academic Publishers. Chapter 2.

Bigg, G. R., 2003. *The Oceans and Climate*, 2nd edn. Cambridge: Cambridge University Press. Chapter 1.

Bolle, H.-J., 2003. *Mediterranean Climate: Variability and Trends*. New York: Springer. Chapter 3.

Bonan, G. B., 2002. *Ecological Climatology: Concepts and Applications*. New York: Cambridge University Press. Chapter 3.

Boucher, K., 1975. *Global Climates*. New York: Halstead Press. Chapter 1.

Bradley, R. and Jones, P., eds., 1995. *Climate Since AD 1500*, London: Routledge. Chapter 8.

Brown, N., 2001. *History and Climate Change: a Eurocentric Perspective*. London: Routledge. Chapter 4.

Bryant, E., 1997. *Climate Process and Change*. Cambridge: Cambridge University Press.

Budyko, M. I., 1982. *The Earth's Climate: Past and Future*. New York: Academic Press. Chapter 6.

Burroughs, W. J., 2003. *Weather Cycles: Real or Imaginary*, 2nd edn. Cambridge: Cambridge University Press. Chapter 2.

2004. *Climate into the 21st Century*. World Meteorological Organization, Cambridge: Cambridge University Press. Chapter 10.

Chunzai, W., Shang-Ping Xie and Carton, J. A., eds., 2004. *Earth's Climate: The Ocean-Atmosphere Interaction*. Washington, DC: American Geophysical Union. Chapter 1.

Collier, M. and Webb, R. H., 2002. *Floods, droughts, and climate change*. Tucson: University of Arizona Press. Chapter 8.

Curry, J. A. and Webster, P. J., 1999. *Thermodynamics of Atmospheres and Oceans*. San Diego, California: Academic Press. Chapter 1.

Díaz, J. I., 1997. *The Mathematics of Models for Climatology and Environment*. Berlin; New York: Springer-Verlag. Chapter 2.

Diaz, H. F. and Morehouse, B. J., eds., 2003. *Climate and Water: Transboundary Challenges in the Americas*. Dordrecht: Kluwer Academic. Chapter 4.

Dietz, A. J., Ruben, R. and Verhagen, A., 2004. *The Impact of Climate Change on Drylands with a Focus on West Africa*. Dordrecht: Kluwer Academic. Chapter 3.

Fagan, B. M., 2004. *The Long Summer: How Climate Changed Civilization*. New York: Basic Books.

Geer, I. W., 1996. *Glossary of Weather and Climate*. Boston: American Meteorological Society. Chapter 1.

Geiger, R., Aron, R. H. and Todhunter, P., 2003. *The Climate Near the Ground*, 6th edn. Lanham, Md.: Rowman & Littlefield. Chapter 1.

Glantz, M. H., 2003. *Climate Affairs: A Primer*. Washington, DC: Island Press. Chapter 1.

Glickman, T., 2000. *Glossary of Meteorology*, 2nd edn. Boston: American Meteorological Society. Chapter 1.

Goody, R., 1995. *Principles of Atmospheric Physics and Chemistry*. New York: Oxford University Press.

Graedel, T. E. and Crutzen, P. J., 1993. *Atmospheric Change: An Earth System Perspective*. New York: Freeman and Co. Chapter 1.

Griffiths, J. F. and Driscoll, D. M., 1982. *Survey of Climatology*. Columbus: C. E. Merrill Pub. Co. Chapter 1.

Grove, J. M., 1988. *Little Ice Age*. London: Methuen. Chapters 6, 8.

Guyot, G., 1997. *Physics of the Environment and Climate*. New York: Wiley. Chapter 1.

Hansen, J. E. and Takuhashi, T., eds., 1984. *Climate Processes and Climate Sensitivity*. Geophysical Monograph 29. Washington, DC: American Geophysical Union. Chapter 3.

Haragan, D. R. ed., 1990. *Human Intervention in the Climatology of Arid Lands*. Albuquerque: University of New Mexico Press. Chapter 3.

Harman, J. R., 1991. *Synoptic Climatology of the Westerlies: Process and Patterns*. Washington, DC: Association of American Geographers. Chapter 1.

Hartmann, D. L., 1994. *Global Physical Climatology*. San Diego: Academic Press. Chapter 1.

Henderson-Sellers, A., 1995. *Future Climates of the World: A Modelling Perspective*. New York: Elsevier.

Henderson-Sellers, A. and McGuffie, K., 1997. *A Climate Modelling Primer*, 2nd edn. Chichester; New York: Wiley. Chapter 2.

Hermann, Y., ed., 1989. *The Arctic Seas, Climatology, Oceanography, Geology, and Biology*, New York: Van Nostrand Reinhold. Chapter 5.

Hobbs, J. E., Lindsay, J. A. and Bridgman, H. A., eds., 1998. *Climates of the Southern Continents: Past, Present and Future*. London: Wiley. Chapters 3, 4, 5, 10.

Holton, J., 1992. *An Introduction to Dynamic Meteorology*, 3rd edn. New York: Academic Press, Inc. Chapter 1.

Jones, P., Bradley, R. and Jouzel, J., eds., 1996. *Climatic Variations and Forcing Mechanisms of the Last 2000 Years*. NATO ASI Series I: Global Environmental Change, Vol. 41, Berlin: Springer-Verlag. Chapter 6.

Karoly, D. and Vincent, D., eds., 1998. *Meteorology of the Southern Hemisphere*. Boston: American Meteorological Society. Chapters 3, 4, 5.

Lamb, H. H., 1967. *Climate: Present, Past and Future*. Vol. 1, *Fundamentals and Climate Now*. London: Methuen. Chapter 1.

   1977. *Climate: Present, Past and Future*. Vol. 2, *Climatic History and the Future*. London: Methuen. Chapters 6, 8.

   1995. *Climate History and the Modern World*, 2nd edn. London: Routledge. Chapter 8.

Landsberg, H. E., ed., 1969–1984. *World Survey of Climatology*. Amsterdam: Elsevier. Chapter 1.

Lewis, W. M., Jr., ed., 2003. *Water and Climate in the Western United States*. Boulder, Colo.: University Press of Colorado. Chapter 4.

Lovejoy, T. E. and Hannah, L., 2004. *Climate Change and Biodiversity*. New Haven, CT: Yale University Press. Chapter 3.

Macdougall, J. D., 2004. *Frozen Earth: The Once and Future Story of Ice Ages*. Berkeley: University of California Press. Chapter 6.

McGregor, G. R. and Nieuwolt, S., 1998. *Tropical Climatology: An Introduction to the Climates of the Low Latitudes*, 2nd edn. New York: Wiley. Chapter 3.

McIlveen, R., 1992. *Fundamentals of Weather and Climate*. London: Chapman and Hall. Chapter 1.

McLaren, S. J. and Kniveto, D. R., 2000. *Linking Climate Change to Land Surface Change*. Dordrecht: Kluwer Academic. Chapter 3.

McMichael, A. J. *et al.*, 2003. *Climate Change and Human Health: Risks and Responses*. Geneva: World Health Organization. Chapters 6, 8.

Morgan, M. D. and Moran, J. M., 1997. *Weather and People*. Upper Saddle River, NJ: Prentice Hall Kendall/Hunt Publishing. Chapter 1.

Nagle, G., 2002. *Climate and Society*. London: Hodder & Stoughton.

Oliver, J. E. and Fairbridge, R. W., 1987. *Encyclopedia of Climatology*. New York: Van Nostrand Reinhold. Chapter 1.

Orme, A. R., ed., 2002. *The Physical Geography of North America*. New York: Oxford University Press. Chapter 4.

Orvig, S., ed., 1970. *Climates of the Polar Regions*. Vol. 14, *World Survey of Climatology*. Amsterdam: Elsevier. Chapter 5.

Owen, A. D. and Hanley, N., 2004. *The Economics of Climate Change*. London: Routledge. Chapter 8.

Pap, J. M. and Fox, P., eds., 2004. *Solar Variability and its Effects on Climate*. Washington, DC: American Geophysical Union. Chapter 1.

Pfister, C., Brázdil, R. and Glaser, R., eds., 1999. *Climate Variability in Sixteenth Century Europe and its Social Dimension*. Dordrecht: Kluwer. Chapter 8.

Philander, S. G., 2004. *Our Affair with El Niño: How We Transformed an Enchanting Peruvian Current into a Global Climate Hazard*. Princeton, N. J.: Princeton University Press. Chapter 2.

Ravindranath, N. H. and Sathaye, J. A., 2002. *Climate Change and Developing Countries*. Dordrecht: Kluwer Academic. Chapter 3.

Rayner, J. N., 2001. *Dynamic Climatology: Basis in Mathematics and Physics*. Malden, Mass.: Blackwell Publishers. Chapter 1.

Roberts, N., 1998. *The Holocene: An Environmental History*. Oxford: Blackwell. Chapters 6, 8.

Robinson, P. J. and Henderson-Sellers, A., 1999. *Contemporary Climatology*, 2nd edn. Harlow, Essex, UK: Longman. Chapter 1.

Ruddiman, W. F., 2001. *Earth's Climate Past and Future*. New York: W. H. Freeman and Co. Chapters 1, 6.

Schwerdtfeger, W., 1984. *Weather and Climate of the Antarctic*. New York: Elsevier. Chapter 5.

Sturman, A. and Tapper, N., 2001. *Weather and Climate of Australia and New Zealand*, 2nd edn. Melbourne: Oxford University Press. Chapters 3, 4.

Thompson, R. D., 1998. *Atmospheric Processes and Systems*. London: Routledge. Chapter 1.

Thompson R. D. and Perry A., 1997. *Applied Climatology: Principles and Practice*. London; New York: Routledge. Chapter 1.

Turco, R. P., 1996. *Earth Under Siege: From Air Pollution to Global Change*. New York: Oxford University Press. Chapters 1, 10.

Tyson, P. and Preston-Whyte, R., 2000. *The Weather and Climate of Southern Africa*. Cape Town: Oxford University Press. Chapters 3, 4.

Voituriez, B., 2003. *The Changing Ocean: Its Effects on Climate and Living Resources*. Paris, France: UNESCO. Chapter 2.

Wallace, J. M. and Hobbs, P. V., 1997. *Atmospheric Science: An Introductory Survey*. San Diego, Calif.: Academic Press. Chapter 2.

Wright, H. E., Jr., Kutzbach, J. E., Webb, T., III *et al.*, eds., 1993. *Global Climates Since the Last Glacial Maximum*. Minneapolis: University of Minnesota Press.

Yarnal, B., 1993. *Synoptic Climatology in Enviromental Analysis: A Primer*. Boca Raton, Fla.: Belhaven Press. Chapter 1.

Yoshino, M. M., 1975. *Climate in a Small Area*. Tokyo: University of Tokyo Press. Chapter 1.

Index